Chemical Thermodynamics

A Series of Books in Chemistry
LINUS PAULING, Editor

THIRD EDITION

Chemical Thermodynamics

A COURSE OF STUDY

Frederick T. Wall

RICE UNIVERSITY

W. H. FREEMAN AND COMPANY
San Francisco

Library of Congress Cataloging in Publication Data

Wall, Frederick Theodore, 1912-
   Chemical thermodynamics.

   Bibliography: p.
   1. Thermodynamics.   I. Title.
QD504.W34   1974   541′.369   73-13808
ISBN 0-7167-0173-1

Printed in the United States of America

1 2 3 4 5 6 7 8 9

Contents

Preface

The third edition of *Chemical Thermodynamics* is not markedly different from the earlier editions, since the basic goals remain unchanged. Some new material has been added and changes in notation have been made to conform more closely to recommended international practice.

The book is not intended to be a treatise on chemical thermodynamics, nor is it designed to be a reference book for experts in the field. Its purpose is to present a reasonably complete story of the principles of chemical thermodynamics, including the essentials of statistical mechanics. It is assumed that the reader already knows something about work, heat, the gas laws, and the basic laws of thermodynamics. The treatment in the early part of the book is therefore rather brief, the emphasis being on logical coherence rather than on detailed explanation. Material likely to be entirely new to the student is, of course, dealt with in more detail.

The underlying plan of the book is to establish means of calculating reaction potentials, or Gibbs free energies of reactions—in other words, to formulate procedures for deciding when and to what extent chemical reactions may be expected to occur. To this end it is necessary to review briefly some basic physical-chemical concepts and then to formulate the laws of thermodynamics that constitute the postulatory basis for the development of the subject. To lend color to the treatment and to stimulate the student's full appreciation of the subject, I refer to the molecular picture as soon as possible; for, although one can formulate thermodynamics without even assuming the existence of atoms and molecules, the subject is obviously much more interesting if one can talk about molecular motions and molecular configurations when discussing thermodynamic properties. After we have established the basis of thermodynamics, we apply the concepts developed therefrom to ideal gaseous systems. From ideal gaseous systems we proceed to statistical mechanics in order to develop means of calculating, from molecular-structure data, the thermodynamic properties of ideal systems. After the basic treatment of statistical mechanics is completed, we take up nonideal systems from the phenomenological point of view and, when feasible, from the standpoint of statistical mechanics.

Important derivations are not left to the student, although some problems and exercises involve derivations of propositions that are not a necessary part of the continuous story. The exercises interspersed in the text are, for the most part, easy, often requiring mere substitution into formulas; on the other hand, the problems at the ends of the chapters are usually more challenging. An instructor can readily prepare additional numerical exercises by drawing on data from tables. For thermochemical data, the most useful compilations are those published by the U.S. National Bureau of Standards. Data sufficient for most study purposes appear in Appendix D.

Thermodynamics is a subject that can be studied more than once to good advantage, for many of the abstract concepts become fully appreciated only after repeated consideration. The student should be encouraged to try alternative derivations of the various propositions, for doing so will surely increase his understanding. The use of alternative derivations also shows how closely knit the various parts of thermodynamics really are. The practical vapor-pressure equations, for example, can be derived from either the Clapeyron equation or the Gibbs-Helmholtz equation, provided the same approximations and assumptions are made.

By reducing the space devoted to some of the old topics (aqueous solutions of electrolytes, for example, which are not treated here in great detail), I have been able to make room for other subjects. I have, for example, treated the effect of differences in molecular size on solutions of nonelectrolytes from a theoretical point of view, while cutting down consideration of the effect of charge on the behavior of solutions of electrolytes.

Without cutting out traditional derivations of the statistical mechanical distribution laws, a chapter has been introduced to show how those laws can be derived without recourse to the methods of variational calculus, Lagrangian multipliers and Stirling's approximation. The distribution laws are also treated from the standpoint of microscopic equilibrium. It is hoped that these approaches will contribute to a better understanding of the concepts basic to statistical mechanics.

In keeping with the recommendations of the International Union of Pure and Applied Chemistry, the symbol U is used instead of E to denote internal energy. Moreover, the names "Helmholtz free energy" and "Gibbs free energy" are used in place of the older "work content" and "free energy." Sign conventions concerning electrode potentials are also in accord with the IUPAC recommendations.

Data tables have been largely revised, thanks to generous help from Professor Bruno Zwolinski and Dr. Jing Chao of the Thermodynamics Research Center at Texas A&M University. I am also indebted to Professor Frederick Rossini of Rice University for helpful comments and assistance in assembling data. I am likewise grateful to users of the earlier editions for helpful comments, including the identification of errors.

In the course of my academic work at several universities, I have taught a number of subjects in physical chemistry, but the one that has given me the greatest satisfaction has been chemical thermodynamics. Teaching chemical thermodynamics has been a pleasure for a number of reasons. First, the subject has an excellent balance of logical deduction and widespread application. Second, the material is of such a character and at such a level that it is attractive to a great variety of students—not only physical chemists, but chemical engineers, organic chemists, and others. Third, each time I have taught the course, some new features have arisen, both in subject matter and in manner of presentation. Because of the stimulation of variety, no course of instruction in chemical thermodynamics need ever settle into a rigid pattern.

During 1972–73, I taught thermodynamics using the second edition of my book as a text. This proved most helpful, since classroom discussions disclosed some of the ambiguities and inadequacies of my treatment. Accordingly, certain clarifications have been made in response to student comments.

Of course an attempt has been made to correct errors discovered in the earlier editions. Unhappily, new ones will doubtless appear, so I would appreciate having them called to my attention.

June 1973   *Frederick T. Wall*

Chemical Thermodynamics

1

Introduction

Thermodynamics is a physical science concerned with the transfer of heat and the appearance or disappearance of work attending various conceivable chemical and physical processes. The processes subject to thermodynamic considerations include not only the natural phenomena that occur about us every day, but also controlled chemical reactions, the performance of engines, and even hypothetical processes, such as chemical reactions that do not occur but can be imagined. Thermodynamics is exceedingly general in its applicability, and this makes it a powerful tool for attacking many kinds of important problems; at the same time, however, this generality renders it incapable of answering many of the specific questions that arise in connection with those problems. Thermodynamics can often tell us, for example, that a process will occur but not how fast it will occur, and it can often provide a quantitative description of an over-all change without giving any indication of the character of the process by which the change might take place.

Historically, the science of thermodynamics was developed to provide a better understanding of heat engines, with particular reference to the conversion of heat into useful work. From the basic principles that were found to govern the operation of such engines, conclusions that find their most striking applications in the chemical world have since been deduced. For this reason we speak of "chemical thermodynamics" almost as if it were a completely separate discipline. Actually it stems from the same basic considerations as any other branch of thermodynamics, but with an unequaled profusion of examples and applications.

Although a typical chemical reaction, such as the oxidation of iron or the hydrolysis of an ester, may appear far removed from the working of an engine, the same fundamental principles of heat and work apply to both. To facilitate the development of thermodynamics, however, we find it convenient and sometimes necessary to define numerous other concepts that are derived from, and related to, those of heat and work. Among the concepts to be so defined are the energy and entropy functions, which are suggested by, and rendered meaningful by, certain laws of thermodynamics; these laws also provide a postulatory basis for the logical development of the subject. Other functions, such as enthalpy and the Gibbs free energy, will also be defined—but principally for convenience, and not because they are directly suggested by laws.

At the outset it should be emphasized that there are many ways in which thermodynamics can be developed, for the subject can be effectively treated in a highly sophisticated manner or in an exceedingly pragmatic way. Accordingly, the order to be followed and the methods to be employed in our development are a compromise between pedagogy and logical science. Once a student has completed his initial study of thermodynamics, he will find, upon further reflection, that many of the various propositions can be succinctly reformulated without recourse to the lengthy methods that appear to be best for the first approach to the subject.

The textual material is intended to be sufficiently complete to provide a basic knowledge of chemical thermodynamics, but a full appreciation of the subject can be gained only by working numerous problems. A substantial number of problems and exercises have therefore been incorporated into this book.

The main body of the book is written on the assumption that the reader is well grounded mathematically and, in particular, that he is familiar with partial differentiation and certain theorems dealing with line integrals. Since line integrals may be new to some readers, a purely mathematical discussion is given in Appendix A.

Temperature

One of the concepts to be used with great frequency in our discussion of thermodynamics is that of temperature. We have all experienced certain physiological reactions or sensations that we qualitatively accept as indicative of hot or cold. Since it may be important to know just how hot or cold an object is, one of our first tasks will be to develop a quantitative definition of temperature that will be more reliable than our physiological responses.

It is generally observed that material objects change their shapes or dimensions when their temperatures, as perceived through our senses, are changed. A rod of copper, for example, will have a certain length when it is cold and a greater length when it is hot. Utilizing this fact, we can construct a temperature-measuring device, or thermometer, by relating the length of such a rod to the temperature. If such a thermometer is to be useful for the quantitative measurement of temperature, it is necessary that an appropriate scale be attached to, or in some way connected with, the length (or other physical change) that is being observed. To establish what is known as the Celsius scale, we assign the value zero to the temperature of the object after it has been in intimate contact, for a sufficiently long time, with melting ice exposed to the atmosphere. (By "a sufficiently long time" we shall mean a period of time beyond which further exposure produces no change in the physical property being observed.) We then fix another point on the scale by assigning the value 100 to the temperature of the object in similar contact with the vapor of water boiling at one atmosphere pressure. Now, by assuming a linear relationship* between temperature and the physical dimensions or shape of the thermometric substance, and by using the two afore-mentioned fixed points for reference, we can complete the construction of the thermometer.

If we employ the length of a metal rod for thermometric purposes, the quantitative scale will be expressed by the equation

$$t = \frac{(L - L_0)100}{L_{100} - L_0} \tag{1.1}$$

in which t is the Celsius temperature, L_0 the length of the rod at zero degrees, L_{100} the length at 100 degrees, and L the length at temperature t. It will be observed that this thermometer correctly reproduces the temperatures arbitrarily chosen for calibration purposes—namely, 0 and 100 °C—and will give other numbers for temperatures within and outside the calibration range.

Instead of using the length of a rod as the basis for constructing a thermometer, one might equally well employ the volume of a fluid,

---

* A nonlinear relationship might also be assumed, but there is no advantage in doing so.

gaseous or liquid. If one works with a gas, the pressure could be used or, in some instances, the pressure-volume product. Equally good, both in principle and in practice, is the electrical resistance of some conducting material; indeed, one might employ any physical property that responds to changes in temperature.

From the foregoing considerations it is evident that one can make as many different kinds of thermometers as there are different substances or different physical properties to be measured. The number of possibilities being endless, we need some agreement on an acceptable standard. The necessity for such agreement becomes evident when we realize that different thermometers will, in general, give different readings at the same temperature—except, of course, at the fixed calibration points. This lack of agreement among different kinds of thermometers is distressing from a logical point of view, and without doubt an arbitrary thermometer would have been selected as a standard long ago were it not that a fundamental principle, to be discussed later, provides a basis for establishing an unequivocal, or thermodynamic, scale of temperature. Meanwhile, before establishing such a thermodynamic scale, we shall need some kind of provisional scale to use in our discussions, and to this end we shall utilize the properties of the so-called permanent gases* at sufficiently low pressures. Thermometers made from any of these gases, employing volumes measured at constant pressure, are found to behave identically or very nearly so. This remarkable fact suggests that a gas thermometer would not be an illogical one to use as a standard. An even greater advantage of a gas thermometer is that under appropriate limiting conditions its scale can be shown to be identical with the thermodynamic scale that will be developed later.

Careful measurements of gases disclose still another important general feature. Not only do dilute gases give rise to equivalent temperature scales, but the relative coefficients of thermal expansion of the gases with respect to such a scale of temperature are all identical. This means, for example, over the range from 0 to 100 °C, that

$$\frac{V_{100} - V_0}{V_0} = \frac{1}{A} \tag{1.2}$$

in which expression the V's denote volumes measured at a fixed low pressure, and A is a constant with the *same value for all gases*. (Experimentally, A is found to equal 2.73.†) Now it should be clear that two

---

* These are gases like hydrogen, oxygen, and nitrogen, which early investigators were unable to liquefy.
† More precisely, A approaches 2.7315 when the gas pressure approaches zero.

substances might have different coefficients of expansion and yet give identical thermometric readings; all that is required for this to be true is that the volume of one be linearly related to the temperature as measured by the other. The afore-mentioned generalization relating to dilute gases—that the coefficients of expansion are identical—suggests an even more far-reaching conclusion. By extrapolating the temperature measured by any dilute gas back to the point where the volume would become zero, one finds a common intercept that can be designated as a point of absolute zero. By referring temperatures to this point, instead of to melting ice, we can then assert that the volumes of gases measured at constant pressure are proportional to this new temperature scale, which will be called the absolute scale. Replacing the L of equation 1.1 by V and making use of equation 1.2, we find that

$$t = \frac{100(V - V_0)}{V_{100} - V_0} = 100A\left(\frac{V}{V_0} - 1\right) \tag{1.3}$$

Evidently the value of absolute zero will equal $-100A$ (or $-273\ °C$), a conclusion reached simply by measuring the volumes of dilute gases at the fixed temperatures, 0 and 100 °C, without utilizing any established thermometer.

Ideal Gases

The foregoing arguments with respect to volumes of gases are equally valid when applied to pressures measured at constant volume. This is fully compatible with Boyle's Law, which states that the pressure-volume product of a gas is constant at a given temperature. Combining Boyle's Law with the absolute temperature scale defined above, we can write an equation describing a gas,

$$pV = \text{constant} \cdot T \tag{1.4}$$

in which p is the pressure, V the volume, and T the absolute temperature. This equation involves two distinct concepts, (1) an empirical generalization known as Boyle's Law and (2) a definition of absolute temperature. To establish the size of the degree for this absolute scale, we will turn to melting ice and boiling water for a difference of 100 units, thus rendering $T = t + 273$ in accordance with equation 1.3. The absolute scale so obtained is also known as the *kelvin* scale, denoted by K.

We can now complete the equation describing a dilute gas by introducing a principle that requires, among other things, a recognition of the molecular hypothesis. This principle, generally known as Avogadro's Law, states that equal volumes of gases measured at the same temperature and pressure contain equal numbers of molecules. Since equal numbers of molecules imply equal numbers of moles, we can write a complete gas-law equation as follows:

$$pV = nRT \qquad (1.5)$$

In this equation, n is the number of moles of gas, and R is a universal constant, simply called the gas constant. (Values of the gas constant in different units are given in Table 1-1.) The equation is experimentally

TABLE 1-1

Some Useful Constants and Conversion Factors

1 calorie (cal)	= 4.184 joules (J)
	= 0.041 293 liter atm
0 °C (ice point)	= 273.15 kelvin (K)
R (gas constant)	= 8.314 3 J deg^{-1} mol^{-1}
	= 1.987 2 cal deg^{-1} mol^{-1}
	= 0.082 056 liter atm deg^{-1} mol^{-1}
RT (at 0 °C)	= 2271.1 J mol^{-1}
	= 542.80 cal mol^{-1}
	= 22.414 liter atm mol^{-1}
RT (at 25 °C)	= 2 478.9 J mol^{-1}
	= 592.47 cal mol^{-1}
	= 24.465 liter atm mol^{-1}
1 atm (pressure)	= 1 013 250 dyn cm^{-2}
	= 760 Torr (mm Hg)
1 Torr	= 1 333.2 dyn cm^{-2}

Additional values will be found in Appendix C.

found to describe the behavior of actual gases quite accurately, though departures from it are observed at high pressures and low temperatures, particularly at temperatures near the liquefaction points of the gases. Its general applicability to all gases at sufficiently low pressures suggests that it can be regarded as representing some kind of ideal behavior, and so the equation is said to be the ideal or perfect gas-law equation. In the course of our development of chemical thermodynamics, we shall have frequent occasion to refer to ideal gases almost as if they were actual gases, readily available. This is fully justified on theoretical grounds because the ideal equation describes an actual gas in its asymptotic behavior, and on practical grounds because the departures from ideality are generally small for most real gaseous substances.

Mixtures of ideal gases also obey equation 1.5, provided n measures the total number of moles of gases in the mixture. Each component of the mixture is assumed to exert a partial pressure precisely equal to the pressure the individual gas would exert if it occupied the entire volume by itself. Then, according to Dalton's Law, the total pressure of the mixture will equal the sum of the partial pressures.

If n_1 and n_2 equal the numbers of moles of each of two gases, and p_1 and p_2 denote the corresponding partial pressures, we can assert that

$$p_1 V = n_1 R T \quad \text{and} \quad p_2 V = n_2 R T \tag{1.6}$$

V being the total volume of the mixture. The total pressure p, moreover, will equal $p_1 + p_2$, so that

$$p V = (n_1 + n_2) R T \tag{1.7}$$

Finally, if we introduce mole fractions, $x_1 = n_1/(n_1 + n_2)$ and $x_2 = n_2/(n_1 + n_2)$, we observe that

$$p_1 = x_1 p \quad \text{and} \quad p_2 = x_2 p \tag{1.8}$$

These relationships can be readily generalized for a system consisting of more than two components.

System and Surroundings

Before we discuss processes from a thermodynamic point of view, it becomes necessary to define certain terms that will frequently recur in our vocabulary. Among these are *system* and, complementary thereto, *surroundings*. By "thermodynamic system" we shall mean a certain portion of the universe bounded by real or imaginary boundaries and thus separated from the rest of the universe, which we shall call the surroundings. Across the boundaries of such a system heat will flow, work will appear or disappear, and sometimes even matter will move. The complementary nature of system and surroundings becomes clear when we realize that they exhibit no fundamental differences in character but are mutually reciprocal. For convenience, however, we shall generally use the designation "system" for that portion of the universe which is enclosed by the boundaries and upon which we can more or less focus our attention as external observers.

Thermal Equilibrium, Heat, and Work

When two objects at different temperatures are placed and left in contact with each other, their temperatures eventually become equal. We say that the objects are then in thermal equilibrium with each other.*

The process accompanying the attainment of thermal equilibrium is called "heat flow," and we say that heat flows across the boundary separating the two objects, from the object of higher temperature to that of lower temperature. Apparently, therefore, heat can be regarded as something that is transferred because of temperature differences and that usually produces a rise in temperature when it enters a system. This should not be construed to mean that temperature changes can be brought about only by the transfer of heat; this implication is correct, however, if no work (see below) is involved in the process.

Heat can be measured by the change in temperature it produces upon flowing into a definite amount of a standard substance. The commonest unit of heat is the calorie, which was originally defined as the quantity of heat that would raise the temperature of one gram of water under atmospheric pressure from 14.5 to 15.5 °C. This amount of heat, also known as the 15 degree calorie, is very nearly equal to the mean calorie, which is 1/100 of the heat required to raise the temperature of one gram of water at atmospheric pressure from 0 to 100 °C.

When a thermodynamic process takes place, *work* is said to appear in the surroundings if a weight is lifted in a gravitational field, if the kinetic energy of some macroscopic object is increased, if a spring is wound, if an electrical condenser is charged, or if any changes that can be readily and completely undone to produce one or more of these effects have occurred in the surroundings. All the different forms of work must be completely interchangeable, at least in principle, and anything not so convertible shall not be considered work. By "disappearance of work" we shall simply mean the negative of its appearance. In this connection,

---

* At this point the careful reader may feel that we tacitly assumed the concept of thermal equilibrium when we constructed our thermometers and defined our temperature scales. In a sense this is true, but we must further recognize that we cannot initiate any development of this kind without introducing a certain number of indefinable concepts and taking cognizance of physical observations. Numerous alternative procedures can be employed to reach a given end, but they all involve the same logical suppositions. Thus we might, before we discuss temperature, talk about thermal equilibrium and hope that the reader would know what we mean. The order adopted in this text was chosen because it is readily accepted and because it quickly gets us to the point where we feel that we all mean the same thing when we use the same words.

however, we shall say nothing about how any work that has disappeared from the surroundings manifests itself within the system; nor shall we speak of work being done inside the system, lest we be misguided into ascribing erroneous characteristics to the concept of work. Work appears or disappears in the surroundings because of forces operating across the boundaries of the system, not through the transfer of any material entity.

It is experimentally observed that work can be dissipated through friction to produce certain effects identical with those resulting from the transfer of heat. We can readily convert the kinetic energy of an object into heat by causing it to come to rest by an inelastic impact, or by slowing it down by friction, either event bringing about the disappearance of work as we have defined it. Likewise, we can dissipate electrical work by passing an electric current through a resistor. Such disappearance of work is invariably accompanied by an effect equivalent to the addition of heat to some part of the system under consideration.

The conversion of work into heat by the means just described has led to precise experiments, originally carried out by Joule and subsequently by many others, for the determination of the so-called mechanical equivalent of heat. These experiments have shown that a given quantity of work always produces the effect of a definite quantity of heat; a reliable value for the mechanical equivalent of heat so obtained appears to be 4.1858 joules per 15 degree calorie.

For experimental convenience, however, most heat measurements are referred to work measured by electrical means. Accordingly, a recalculation of a great mass of thermal data (expressed in 15 degree calories) would be required every time a better value for the mechanical equivalent of heat was determined. Current practice avoids this difficulty by *defining* a mechanical equivalent of heat and making the calorie exactly equal to 4.184 absolute joules.* With this understanding the experimental determination of the mechanical equivalent of heat no longer remains a problem. This also means that the calorie, so defined, will not necessarily raise the temperature of one gram of water by exactly one degree Celsius in the neighborhood of fifteen degrees.

Although the conversion of work into heat is readily accomplished, the reverse is subject to certain practical and theoretical limitations. It will be one of our tasks to investigate the reasons for, and the quantitative nature of, such limitations on the conversion of heat into work.

---

* The number 4.184 was chosen to keep most of the older thermochemical data valid. For a long time the situation was confused by the use of several kinds of calories and joules.

State of a System

Another concept we must now define is that of the *state* of a system. We say that the state of a system is determined when we have specified a certain minimum number of independent variables that characterize the system with respect to the features that affect our thermodynamic considerations. To illustrate, let us suppose that our system consists of one mole of hydrogen gas. We can ordinarily describe the state of this system by specifying two characteristic variables, such as temperature and pressure, or temperature and volume, or pressure and volume. It should be noted that only two variables need be specified, the third then being fixed by an equation connecting the three variables. (We shall assume that such an equation exists whether we know it precisely or not. The ideal gas-law equation, in fact, provides an excellent approximation for the description of hydrogen gas at about room temperature and one atmosphere pressure.)

Under some circumstances we might want to know more about the hydrogen gas than just its temperature and pressure. It is conceivable that knowledge of the shape of the container holding it would also be necessary for a precise thermodynamic description. Usually this is not so; but there are systems—for example, emulsions—for which the extent of the surfaces or the shapes of the phases are thermodynamically important. Under those circumstances, if one wished to characterize the system completely, one would be obliged to specify not only temperature and pressure but other quantities as well. For this reason we cannot assert with finality that a predetermined and fixed number of independent variables will always be sufficient to specify the state of a system. Practically speaking, the problem is not so complex. Depending upon the magnitudes of the effects involved, we can usually arrive at a reasonable decision as to what constitutes the correct number of independent variables required for specifying the state of the system in relation to the immediate questions that need to be answered.

Equations of State

We have already mentioned that we can specify the state of a thermodynamic system by stipulating values for a certain number of independent variables. Having specified the independent variables, we can, in principle, then determine all the other variables pertaining to the system. One way of expressing a connection between independent and nonindependent variables involves the use of equations of state, which are equations relating the pressure, volume, temperature, and amounts of the substances involved. The perfect gas-law equation, $pV = nRT$, is a particularly important equation of state. Its approximate applicability to almost all gases makes it one of our most useful equations. Since actual gases depart more or less from this ideal behavior, numerous modifications of the ideal equation have been proposed for the sake of increased accuracy. One of the simple modifications of the ideal gas-law relationship, written for one mole, is the equation

$$pV = RT + \alpha p \tag{1.9}$$

in which α is a function of the temperature. This equation is suggested by the experimental fact that at moderate pressures the pressure-volume product of an actual gas is a linear function of the pressure. If we plot pV against p, for example, we obtain straight lines (Fig. 1-1) with intercepts equal to RT but with slopes not necessarily equal to zero. The

FIGURE 1-1   Pressure-volume product of a nonideal gas plotted against pressure for several temperatures near the Boyle temperature, T_B. The slopes of the lines were calculated from van der Waals' equation. The straight lines are, of course, valid only for moderate pressures.

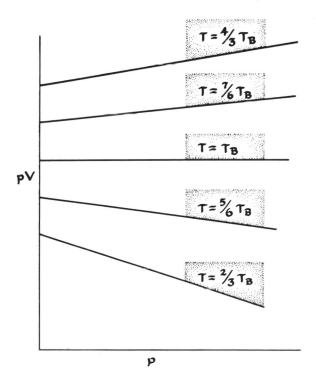

slopes of the lines, which are precisely equal to α, increase with rising temperature, and at a particular temperature, known as the Boyle temperature, α equals zero and the gas appears to obey Boyle's Law.

Another modification of the ideal gas-law equation, which is substantially equivalent to that expressed by equation 1.9, is

$$pV(1 - \beta p) = RT \tag{1.10}$$

in which β is a function of temperature. To a first approximation, α and β are related to each other by the equation $\alpha = \beta RT$.

Perhaps the most widely used modification of the ideal gas-law equation is that of van der Waals, which qualitatively describes many features pertaining to the behavior of both gases and liquids. This equation,

$$(p + a/V^2)(V - b) = RT \tag{1.11}$$

takes into account two reasons for departures from gaseous ideality. The first of these, which gives rise to the term a/V^2 as an additive correction to p, is intended to take cognizance of intermolecular attractive forces. These forces, although relatively feeble, can be considered responsible for the ultimate liquefaction of any gaseous substance when it is sufficiently cooled. By purely statistical considerations these forces, which can be translated into the equivalent of an internal pressure, can be shown to be proportional to the square of the density of the gaseous molecules and hence inversely proportional to the square of the volume of the mass of material.

The second reason for departures from ideality, represented by b in van der Waals' equation, is that actual molecules are not point particles but have finite size. This means that the molecules will collide with one another and will bounce off the walls of the containing vessel more frequently than would be expected from purely kinetic arguments applied to point masses. This is equivalent to saying that the volume available for the motion of the gaseous molecules is somewhat less than the measured volume of the container; hence a negative correction, denoted by b, is introduced into the equation.

Although only two simple considerations were employed for modifying the ideal gas-law equation to arrive at that of van der Waals, the resulting equation represents a remarkable advance in the qualitative description of fluid systems. Among other things, we can use van der Waals' equation of state to predict the existence of a critical temperature, below which liquefaction can occur and above which two phases cannot co-exist. It is also possible, as we shall see later, to predict the vapor pressures of liquids in terms of van der Waals' constants and to deduce certain otherwise empirical rules.

A more general equation of state for gases is the so-called *virial* equation, which for one mole assumes the form

$$pV = RT + Ap + Bp^2 + Cp^3 + \ldots \tag{1.12}$$

in which A, B, C, . . . are functions of temperature.* A is said to be the second virial coefficient, B the third, etc.; these coefficients are determined empirically by precise measurements of gases at several pressures for each temperature of interest. Equation 1.9 is obviously a special case of the virial equation. By including enough terms in equation 1.12 we can presumably describe the behavior of any gas exactly.

For liquids and solids, a crude but often satisfactory equation of state is the simple statement that the volume is constant. A better approximation, involving empirical constants, is generally written as

$$V = V_0(1 + \alpha t - \kappa p) \tag{1.13}$$

in which α and κ are constants. This equation assumes that the volume is linearly dependent upon both the temperature and the pressure, which is equivalent to saying that the coefficients of expansion and compressibility are constant. The equation can be generalized, of course, to include higher-power terms in t and p, and this is often done when enough data of sufficient accuracy are available to warrant the evaluation of the additional coefficients thus introduced.

When we deal with solids, the situation can become complicated by the fact that the applied pressures need not be the same in all directions; moreover, if the material is anisotropic, it will exhibit different coefficients of expansion and compressibility in different directions. Throughout this book, unless we state the contrary, we shall be dealing with isotropic systems that are subject to uniform pressure in all directions.

Description of Various Kinds of Processes

Among the infinite number of processes by which changes in state can be brought about, several, of general importance, are frequently encountered. Among these are *isothermal* processes, which are processes carried out at constant temperature. Similarly the term *isopiestic* is sometimes used to denote constant pressure and *isochoric* for constant volume. An *adiabatic* process is one that occurs without flow of heat across the

---

* Another form of the virial equation is

$$pV = RT + \frac{A'}{V} + \frac{B'}{V^2} + \frac{C'}{V^3} + \ldots$$

in which A', B', C', etc. are functions of temperature.

boundary separating the system from its surroundings. A *cyclical* process is one in which the initial and final states of the system are identical. We shall encounter still other kinds of processes as the subject of thermo-dynamics is developed, but their definitions cannot be made meaningful at this stage.

Problems

1. The linear expansion of solids can be characterized by equations of the type

$$l_t = l_0(1 + \alpha t + \beta t^2)$$

in which l_0 is the length at 0 °C and l_t the length at t °C. If we use the ideal gas thermometer for reference, the coefficients α and β for copper and aluminum are as follows:

	α	β
Cu	0.160×10^{-4}	0.10×10^{-7}
Al	0.222×10^{-4}	0.11×10^{-7}

   (A) How would the volume of an ideal gas depend upon the temperature measured by a copper-rod thermometer?
   (B) How would the length of an aluminum rod depend upon the copper-rod temperature?

2. A mercury-in-glass thermometer consists of a bulb and a uniform stem on which degrees are indicated by uniformly spaced marks. When used and calibrated, all of the thermometer is at the same temperature. Moreover, the thermometer is so calibrated that it indicates the correct temperatures at 0 and 100 °C.

   (A) If both the mercury and the glass have constant coefficients of cubical expansion with respect to the ideal gas temperature scale, what will the thermometer read when the ideal gas temperature is t °C?
   (B) If the coefficient of cubical expansion of the glass is 2.5×10^{-5} per degree, what will be the error in the reading of the thermometer at 50 °C?

3. An aluminum rod is 0.23 percent longer at 100 than at 0 °C, whereas a rod of copper expands by 0.17 percent over the same temperature range. If rods of these metals were used for constructing thermometers, and linear extrapolations were carried back to the points of "zero length," what would "absolute zero" be for each of these substances? Liquid mercury has a volume 1.8% greater at 100 than at 0 °C. What would "absolute zero" be on a mercury volume scale? (Note that the three figures are widely different and hence suggest nothing of general significance. Different gases, on the other hand, extrapolate back to the same point.)

4. Suppose a thermometer were constructed from a metal rod with the temperature, τ, defined in terms of the *square* of the length as follows:

$$\tau = \frac{L^2 - L_0^2}{L_{100}^2 - L_0^2} \times 100$$

(A) What is the relationship between τ and the temperature, t, defined in the usual linear fashion?

(B) At what point between 0 and 100 °C will the difference between the two scales be a maximum?

5. Suppose that, instead of assigning 100 degrees to the difference between the freezing and boiling points of water, we fix the absolute ice point (0 °C) at 273.15 kelvin. With this understanding, the boiling point of water would become an experimentally measurable quantity subject to change as the coefficients of expansion of ideal gases are remeasured. Under this system, what would the boiling point of water become if new measurements of the expansion of ideal gases increased the coefficient of such expansion by x percent over the range now designated as 0–100 °C? (The advantage of fixing the ice point is that temperatures near absolute zero would not be subject to serious fluctuations in value as new values for the coefficients of expansion of gases are obtained. The boiling point of water might be changed, of course, but only by insignificant amounts.)

6. Some methane is burned with twice the minimum amount of air necessary for its complete combustion to carbon dioxide and water. If the total pressure of the products of combustion (including inert materials and excess oxygen) is 1 atm, what are the partial pressures of the gaseous components if (A) the water is completely in vapor form and (B) all the water is removed by condensation? Assume air to consist of 80 mole percent nitrogen and 20 mole percent oxygen.

7. (A) At 1 atm pressure the volume of one mole of nitrogen gas is 22.401 liters at 0 °C and 30.627 liters at 100 °C. Using only these data, what value would one calculate for absolute zero?

(B) At 0.1 atm pressure the molar volume of nitrogen is 224.13 liters at 0 °C and 306.20 liters at 100 °C. From these data what would one calculate for absolute zero?

(c) If the results of parts A and B are extrapolated to zero pressure, what is the calculated value of absolute zero?

8. (A) Expand van der Waals' equation into each of the forms

$$pV = RT + Ap + Bp^2 + \ldots$$

and

$$pV = RT + A'/V + B'/V^2 + \ldots$$

(B) Using the abbreviated form $pV = RT + \alpha p$, show that $\alpha = (b - a/RT)$ and that the Boyle temperature of a van der Waals gas is $T_{Boyle} = a/Rb$.

9. Using the data given in parts A and B of Problem 7, calculate α for nitrogen at 0 °C and at 100 °C. Proceed by calculating the pV product for 0.1 and

1.0 atm at each of the temperatures. From the values so obtained, calculate α for each temperature, without assuming anything about the value of pV extrapolated to $p = 0$. From those results calculate values for van der Waals' a and b, assuming that $\alpha = b - a/RT$ in accordance with the result of Problem 8.

10. Prove the statement in the text to the effect that the quantities α of equation 1.9 and β of equation 1.10 are approximately related to each other by the equation $\alpha = \beta RT$.

11. At 0 °C and 1 atm for coefficient of thermal expansion of acetone is 1.324×10^{-3} per degree, and the coefficient of compressibility is 82×10^{-6} per atm. Calculate the value of $(\partial p/\partial T)_V$ for acetone at 0 °C and 1 atm. Compare this value with the corresponding derivative for an ideal gas at 0 °C and 1 atm. [The coefficient of expansion is defined as $(1/V)(\partial V/\partial T)_p$ and the coefficient of compressibility as $-(1/V)(\partial V/\partial p)_T$.]

12. Prove that, if a gas obeys the equation of state $pV = RT + \alpha p$,

$$\left(\frac{\partial p}{\partial T}\right)_V = \frac{p}{T}\left\{1 + \frac{p}{R}\frac{d\alpha}{dT}\right\}$$

2

The First Law of Thermodynamics

Work

It has already been suggested that the work attending a thermodynamic change in state is a quantity of fundamental significance. We shall now turn our attention to the evaluation of the work involved in various conceivable processes. Throughout our treatment of the subject we shall consider work to be positive if it appears in the surroundings and negative if it disappears from the surroundings.

The simplest representation of work involves the operation of a force through a distance in such a way as to produce an increase in the potential or kinetic energy of some object. In our thermodynamic discussions we shall find it more convenient to think of a pressure operating through a change in volume than of a force acting through a distance, even though the two are substantially equivalent. For this reason we shall frequently refer to pressure-volume work, and we shall evaluate it by recourse to the arguments illustrated in Figure 2-1. This diagram shows a cylinder of uniform cross-sectional area, A, provided with a piston that is, at any instant, at some distance, L, from the closed end of the cylinder. To simplify our calculations, we shall assume that the mechanism is free of friction and that any motion of the piston takes

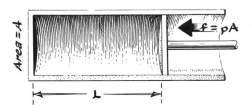

FIGURE 2-1   Cylinder of cross-sectional area
A provided with movable piston at distance
L from end of cylinder.

place infinitely slowly. Under these circumstances the force, f, that must be applied to the piston to keep it in a state of balance will equal the pressure of the fluid contained in the cylinder multiplied by the cross-sectional area, and the element of work, dW, involved in pushing the piston out a distance, dL, will be given by the expression

$$dW = pAdL \qquad (2.1)$$

Since the volume is equal to the cross-sectional area times the length, this element of work can equally well be written as

$$dW = pdV \qquad (2.2)$$

If a change in state that produces a change in volume from value V_1 to value V_2 has been brought about, the total work that can appear in the surroundings from this process will be given by the integral

$$W = \int_{V_1}^{V_2} pdV \qquad (2.3)$$

Since p is not, in general, a function of V alone, the integral cannot be evaluated unless a definite path or manner of bringing about the change in state has first been stipulated; this means that W will depend upon the path.

That the work depends upon the path is demonstrated by graphical considerations shown in Figure 2-2. Obviously, the integral of pdV from the initial to the final state will be given by the area under the curve obtained by plotting p against V. Because an infinite number and variety of curves can be drawn from one point to another, it is clear that the work involved can assume any value whatsoever from minus to plus infinity. Hence it is meaningless to talk about a maximum or minimum amount of work attending a given change in state. Later we shall indicate a maximum for processes that are restricted in certain ways; but now, in the absence of any stipulation regarding the kind of path to be followed, no upper or lower limits can be specified.

Since the quantity of work that appears in the surroundings depends upon the way in which the change in state is brought about, the integral representing work is said to be a line integral whose value depends upon the path. The expression being integrated, pdV, is the kind of expression that cannot be regarded as the differential of any function of the state of the system; it is an "inexact" differential. Since p is, in general, a function of two variables, say T and V, it is evident that the integral of pdV is meaningless unless some functional relationship between T and V is specified, for only under those circumstances will the inexact differential expression become integrable. The preceding argument will now be amplified by recourse to certain theorems relating to line integrals. (See Appendix A.)

FIGURE 2-2   Three different paths in p-V plane for going from point A to point B. For path 1 the work will equal the area under the curve and between the broken lines *plus* the area ACD. For path 2 the work will simply equal the area under the curve. For path 3 the work will equal the area under the curve and between the broken lines *minus* the area BFE. For path 1 the contribution of portion AC is negative and that of CD is positive; for path 3 the contribution of FE is positive and that of EB is negative.

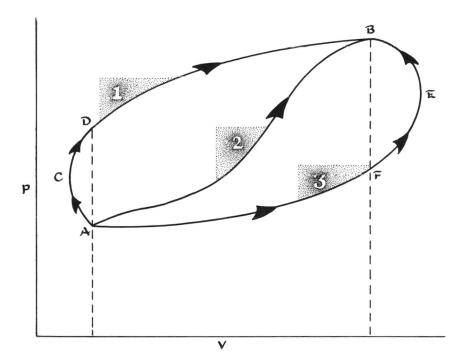

Consider a differential expression such as

$$M(V,T)dT + N(V,T)dV \tag{2.4}$$

The line integral of expression 2.4 along a path C,

$$\int_C M(V,T)dT + N(V,T)dV \tag{2.5}$$

will, in general, depend upon the path of integration. If, however, $M(V,T)$ and $N(V,T)$ are so related that

$$\frac{\partial M}{\partial V} = \frac{\partial N}{\partial T} \tag{2.6}$$

then the line integral will be independent of the path* and will be determined only by the initial and final points. Under these circumstances, there must be some function, $f(V,T)$, such that

$$\frac{\partial f}{\partial T} = M \quad \text{and} \quad \frac{\partial f}{\partial V} = N \tag{2.7}$$

and

$$df = MdT + NdV \tag{2.8}$$

If, however, equation 2.6 does not hold, then there is no function of V and T that will give, upon differentiation, expression 2.4. If equation 2.6 holds, we say that $MdT + NdV$ is an "exact" differential, the differential of f; if equation 2.6 does not hold, the expression is said to be "inexact."

We now observe that, if $M = 0$ and $N = p$, expression 2.4 becomes precisely what we have designated by dW. Since

$$\left(\frac{\partial p}{\partial T}\right)_V \neq 0 \tag{2.9}$$

it follows from the argument pertaining to equation 2.6 that dW is not an exact differential and that its line integral will depend upon the path.

It should be noted that we use the symbol dW (and later dQ) simply as an abbreviation for an infinitesimal to be summed up by line integration. The symbol dW does not represent the differential of any function, W, of the state of the system, for, as we have just seen, there is no such function. The notation dW is nevertheless convenient, for it suggests infinitesimal character of the correct order. †

---

* It is assumed that M, N, and their partial derivatives are continuous throughout the domain of integration and that the region is simply connected.

† Numerous other notations for dW have been employed by different authors; these include DW, $d'W$, δW, and w. All authors are agreed on the idea that must be conveyed, but no notation yet proposed is free from objections or possible misinterpretations.

Let us now consider a few examples to illustrate the calculation of work attending certain processes. Suppose we *isothermally* expand n moles of an ideal gas from volume V_1 to volume V_2. The work involved in such a process is easily determined as follows:

$$W = \int p\,dV = nRT \int_{V_1}^{V_2} \frac{dV}{V} = nRT \ln \frac{V_2}{V_1} \qquad (2.10)$$

If, on the other hand, we carry out the process at *constant pressure*, or *isopiestically*, the work will simply equal the pressure times the change in volume, as seen from the following integration:

$$W = \int_{V_1}^{V_2} p\,dV = p(V_2 - V_1) = p\Delta V \qquad (2.11)$$

This relationship holds for constant-pressure processes whatever the equation of state of the material involved, and, indeed, even if phase or chemical changes accompany the process. Returning, however, to the ideal gas, we see that the constant-pressure expansion of such a gas must be accompanied by an increase in temperature; the work involved can therefore also be written as

$$W = p\Delta V = nR(T_2 - T_1) = nR\Delta T \qquad (2.12)$$

For substances not obeying the ideal gas-law equation, work can still be calculated by direct integration along any specified path provided the equation of state is known. The work attending an *isothermal* expansion of one mole of a van der Waals gas, for example, is given by the expression

$$W = RT \ln \left(\frac{V_2 - b}{V_1 - b} \right) + a \left(\frac{1}{V_2} - \frac{1}{V_1} \right) \qquad (2.13)$$

which takes into account both the effect of the volume occupied by the molecules and the intermolecular attractive forces that must be overcome in the expansion process.

For liquids and solids the same principles and arguments are valid, but one noteworthy difference in degree should be mentioned. Generally speaking, the work accompanying the compression or expansion of a liquid or solid, or the work attending a transition between them, is very small—indeed, often negligible when compared with the corresponding quantities for gases. The reason is simply that, since the total volumes of liquids and solids are small compared with the volumes of gases, their volume changes are necessarily small. Accordingly, one can generally neglect, in a first approximation, the volume changes of liquids and solids in relation to those of gaseous materials.

EXERCISE:

Prove equation 2.13.

EXERCISE:

One mole of an ideal gas is heated at a constant pressure of 1 atm from 0 to 100 °C. (A) Calculate in calories the work involved. (B) If the gas were expanded isothermally at 0 °C from 1 atm to some other pressure, p, what must that final pressure be if the isothermal work is to equal the work of part A?

EXERCISE:

A liquid film (such as might be obtained from soapy water) is held in a wire framework of width w and length L as illustrated. The force of the film on the wire will equal 2γ per unit length of wire if γ is the surface tension of the film. (The factor 2 comes in because the film has two sides.)

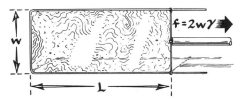

If the wire on the right-hand side of the rectangular framework is now moved (like a piston) through distance dL, show that $dW = -2w\gamma dL = -\gamma dA$ where A is the total film area, counting both sides. (Why the minus sign?) If the surface tension is independent of area, show further that the work attending the formation of new surface is simply $W = -\gamma \Delta A$.

EXERCISE:

When a strip of rubber (or a wire or spring) is stretched, it exerts a restoring force, f, which is a function of the length, L, and the temperature, T. Neglecting any volume changes upon stretching, we see that $dW = -f dL$. (Note the minus sign.) Now an "ideal" rubber strip has the equation of state

$$f = CT\left[\frac{L}{L_0} - \left(\frac{L_0}{L}\right)^2\right]$$

in which C is a constant and L_0 is the length under zero force. Show that the work attending the isothermal stretching of a piece of rubber from L_0 to L will be

$$W = -\frac{CL_0 T}{2}\left\{\left(\frac{L}{L_0}\right)^2 + \frac{2L_0}{L} - 3\right\}$$

Reversible Processes

The foregoing calculations of pressure-volume work have all been made with the tacit assumption that the external force on the piston exactly counterbalances that exerted by the fluid within the system. Since any actual machine is subject to friction, the piston would not move in either direction under the conditions assumed, and hence no work would appear in the surroundings. Even if the machine were friction-free, the piston could not move at a finite rate without giving rise to mass motions of the fluid and without causing the pressure to become nonuniform in such a manner as to reduce the internal force on the piston during an expansion and to increase that force during a compression. To permit an actual piston to move out, therefore, the external force must be less than that exerted by the fluid under static conditions, and the work appearing in the surroundings will be less than the integral of pdV. Similarly, to bring about a compression, the external force must be greater than that of the fluid, and the work done will numerically exceed the absolute value of that integral. In either event, if we take into account the sense of the process and the sign of the attending work, we can assert the following *algebraic* inequality:

$$W < \int pdV \tag{2.14}$$

This inequality arises both because of friction and because of finite speeds of expansion or compression.

It would be exceedingly difficult to describe in detail a process that is irreversible because of its being carried out at finite speed. Nonuniformities in pressure, for example, would render impossible a description of the path through any simple statement such as $p = p(V)$. If we were to stop the process at any instant, however, and allow the system to come to equilibrium, we could then specify the equilibrium pressure, which would provide a partial description of the path in the form $p = p(V)$. It should now be evident that, if a fluid is expanded along a path whose partial description reduces to $p = p(V)$, and if the fluid is subsequently compressed along a similar path with the same limiting partial description, then, because of inequality 2.14, the net total work for the cycle will be negative. If, however, the processes are carried out infinitely slowly with friction-free mechanisms, the inequality becomes an equality, and the work attending the particular cycle described above becomes equal to zero. Under such circumstances, the processes so

described, or any portions of them, are said to be *reversible* processes.*
It will readily be seen that the reversible work associated with a change
in state carried out along a path represented by $p = p(V)$ is the maximum work realizable for that path; it is not, however, a maximum for
the change in state, for, as we have already seen, no such maximum
exists (quite apart from the matter of reversibility). As long as a path
is specified, however, the reversible work has associated with it a unique-
ness arrived at through the limiting concept of a frictionless machine
operating infinitely slowly.

Reversibility also implies certain constraints with respect to heat flow
and temperature gradients. A discussion of this aspect of reversibility
will be deferred to a later chapter.

Heat

Another significant quantity in thermodynamic processes is the
amount of heat involved. We shall adopt the convention that a quantity of heat, Q, is positive when it is added to the system—in other words,
when it crosses the boundary from the surroundings into the system.
The work, W, on the other hand, is said to be positive when it appears
in the surroundings. The reason for this difference in convention is
purely historical; the difference came about because early investigators
were concerned with the conversion of heat into work and hence thought
of the heat added to an engine in relation to the work obtained from it.
We shall generally measure heat and work in calories, utilizing appropriate conversion factors as necessary.

Just as the work associated with a given change in state depends upon
the means by which the change was brought about, so the heat attending
a given change in state also depends upon the path. Although it cannot
be illustrated by simple graphical means, it is nevertheless true that the
heat associated with a given change in state can, like work, have any
value whatsoever from negative to positive infinity. It follows that Q,
as well as W, is represented by a line integral dependent upon the path,
and that dQ, which can be written in the form of expression 2.4, is not
an exact differential.

---

* The reader will note that a reversible process is not necessarily one that is actually reversed;
rather, it is a process that *can* be reversed in such a manner that both system and surroundings
would be restored to their initial conditions.

The First Law of Thermodynamics

Although Q and W depend upon the path, numerous careful experiments have shown that a restriction of fundamental importance applies to a certain combination of these quantities. This experimental conclusion, which is called the *First Law of Thermodynamics*, can be stated as follows: *When a system undergoes a change in state, the quantity* $Q - W$ *depends only upon the initial and final states of the system and is independent of the path.* This law is of tremendous importance, for it means that Q and W, which apparently could have any conceivable values for a given change in state, are not completely independent of each other. The fact that the quantity $Q - W$ does not depend on the path indicates that there exists a function of the state of the system whose change can be measured thereby. To this function we give the name *energy*, with the symbol U, and for its definition we write

$$\Delta U = Q - W \qquad (2.15)$$

Obviously, if $Q - W$ depended upon the path, it could not be used to measure the change of any function. This is so because a function of the state of a system is by definition dependent only upon the values of its independent variables. Hence a change in such a function can depend only upon the initial and final conditions and not upon the path employed to bring about the change. Equation 2.15 is usually called the First Law equation, but it is incorrect to regard it as an explicit statement of that law. The essence of the First Law is that $Q - W$ is independent of the path; because of that fact we can define a quantity, U, whose change is measured according to the equation.

The foregoing discussion can be concisely summed up by use of line-integral concepts. dQ and dW are so-called inexact differentials, whose line integrals depend upon the path. However, the line integral of $dQ - dW$, which equals $Q - W$, is independent of the path. Therefore $dQ - dW$ must be an exact differential, the differential of some function of the state of the system. To this function we give the name "energy" with the definition

$$dU = dQ - dW* \qquad (2.16)$$

which upon integration gives equation 2.15. (Once again see Appendix A.)

---

* A simple example may help the reader appreciate more fully the significance of this statement. If x and y are two independent variables, the separate expressions dy/x and ydx/x^2 are inexact. Their difference, however, is an exact differential, $d(y/x)$.

Since our definition of energy is given only in terms of its changes, we cannot specify an absolute magnitude of the energy from First Law considerations alone. Any energies arrived at through the First Law must accordingly be uncertain to the extent of an additive constant. However, this uncertainty will not prove to be a serious handicap, for the uses to which we put thermodynamics do not require a knowledge of absolute energies.

Absolute values of the energy can be specified, however, if we resort to the special theory of relativity. According to that theory, mass and energy are concomitant, the total energy of a system being equal to its mass multiplied by the square of the velocity of light:

$$U = mc^2 \qquad (2.17)$$

The change in energy attending any change in state will be given by the expression

$$\Delta U = \Delta mc^2 \qquad (2.18)$$

in which Δm is the change in mass. Since $c = 3 \times 10^{10}$ cm s^{-1}, the energy of a gram mass will equal 9×10^{20} ergs or 2.15×10^{13} calories. For ordinary chemical reactions the change in mass will be so small that it cannot be measured directly; for certain nuclear processes, however, the change in mass is appreciable, which means that the change in energy is tremendous.

Since the energy is a function of the state of the system, the change in energy attending any cyclical process must be equal to zero. (A cyclical process is any process for which the initial and final states of the system are identical.) Because $\Delta U = 0$ for such a process, the heat added to the system must exactly equal the work appearing in the surroundings. From this we conclude that any engine, process, or device operating in cycles for the purpose of converting heat into work can never produce more (or less) work than there is heat added. This rules out the possibility of constructing a so-called perpetual-motion machine of the first kind—that is, a machine that would produce work without requiring an equivalent amount of energy in the form of heat. Although this restriction is very important, it does not rule out other conceivable kinds of perpetual-motion machines, which might be equally useful from a practical point of view. (That such other perpetual-motion machines are also impossible constitutes the basis of another law of thermodynamics.)

It sometimes happens that it is difficult or impossible to evaluate directly either the heat or the work for a particular process. If, however, ΔU can be determined by indirect means for the specified change in state, then Q, for example, could be calculated from a knowledge of W by use of the First Law equation in conjunction with the indirectly established value for ΔU. This principle often provides a means for solving otherwise difficult thermodynamic problems.

EXERCISE:

Consider the differential expressions $du = xdy$ and $dv = ydx$. (A) Evaluate the line integrals $\int du$ and $\int dv$ from point 1,1 to point 3,2 along each of the three paths illustrated. Note that these line integrals depend upon the path. (B) Now consider the sums of the line integrals (namely, $\int du + dv$) for each of the paths. Are the sums independent of path? Explain.

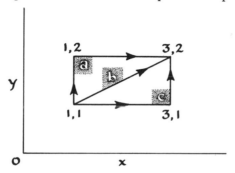

EXERCISE:

Calculate the numbers of calories and liter-atmospheres of energy equivalent to one milligram of mass.

Constant-volume and Constant-pressure Processes; Enthalpy

If a process is carried out at constant volume, the work involved (excluding electrical and other special kinds of work) will be equal to zero. With that stipulation, the heat associated with the process, which we shall now denote by Q_v, will simply equal the change in energy:

$$Q_v = \Delta U \tag{2.19}$$

For a constant-pressure process, on the other hand, the work equals $p\Delta V$, and the heat, Q_p, will be given by the expression

$$Q_p = \Delta U + p\Delta V \tag{2.20}$$

The form of equation 2.20 suggests that we might profitably define still another function of such character that its change at constant pressure will equal Q_p. Such a function, which is given by the expression $U + pV$, will be called the *enthalpy* and will be given the symbol H, with the definition

$$H = U + pV \tag{2.21}$$

Since U, p, and V are all functions of the state of the system, the enthalpy is likewise such a function, and its change will depend only upon the initial and final states of the given process. From equation 2.20, in keeping with our reason for defining enthalpy, we observe that

$$Q_p = \Delta H \tag{2.22}$$

The simplicity of equation 2.19 suggests that the energy and volume functions will tend to appear with each other, thus justifying our choice of temperature and volume as the independent variables for the energy. Likewise, the form of equation 2.22 suggests that enthalpy and pressure exhibit a similar correspondence and that temperature and pressure should be chosen as the independent variables for enthalpy. The convenience of such choices of independent variables will become apparent when we consider the differentials of energy and enthalpy.

EXERCISE:

Show that the change in enthalpy attending an isothermal change in state of an ideal gas must be equal to the change in energy.

Some Differential Expressions

Since U and H are both functions of the state of the system, we can write down their total differentials in terms of appropriate partial derivatives multiplied by the corresponding differentials of the independent variables. Thus, assuming that the energy is specified by volume and temperature, and the enthalpy by pressure and temperature, we write that

$$dU = \left(\frac{\partial U}{\partial T}\right)_V dT + \left(\frac{\partial U}{\partial V}\right)_T dV \tag{2.23}$$

and that

$$dH = \left(\frac{\partial H}{\partial T}\right)_p dT + \left(\frac{\partial H}{\partial p}\right)_T dp \tag{2.24}$$

Assuming further that the only work involved is pressure-volume work, which means that $dW = pdV$, we note that $dQ = dU + pdV = dH - Vdp$, giving rise to two expressions for dQ:

$$dQ = \left(\frac{\partial U}{\partial T}\right)_V dT + \left\{\left(\frac{\partial U}{\partial V}\right)_T + p\right\} dV \qquad (2.25)$$

$$dQ = \left(\frac{\partial H}{\partial T}\right)_p dT + \left\{\left(\frac{\partial H}{\partial p}\right)_T - V\right\} dp \qquad (2.26)$$

It is evident that dQ is an inexact differential, for the condition of exactness, equation 2.6, is not fulfilled.

EXERCISE:

By use of an equation of state such as $V = V(T,p)$, one can write that

$$dV = \left(\frac{\partial V}{\partial T}\right)_p dT + \left(\frac{\partial V}{\partial p}\right)_T dp$$

Substitute this expression for dV in equation 2.23, and show from the coefficients of dT and dp that

$$\left(\frac{\partial U}{\partial T}\right)_p = \left(\frac{\partial U}{\partial T}\right)_V + \left(\frac{\partial U}{\partial V}\right)_T \left(\frac{\partial V}{\partial T}\right)_p$$

and that

$$\left(\frac{\partial U}{\partial p}\right)_T = \left(\frac{\partial U}{\partial V}\right)_T \left(\frac{\partial V}{\partial p}\right)_T$$

By similar means, derive expressions for $(\partial H/\partial T)_V$ and $(\partial H/\partial V)_T$.

Heat Capacity

When a certain amount of material is ordinarily heated, it is usually found that the heat added is roughly proportional to the temperature change. Since the quantity of heat required to produce a definite temperature change is a matter of some importance, we shall now examine the connection between them more precisely. For this purpose we shall find it convenient to define a new quantity, the heat capacity of a system, which will be denoted by the letter C. The average heat capacity for a process is simply defined as the quotient of the heat added and the change in temperature produced by the process:

$$\langle C \rangle = \frac{Q}{\Delta T} \qquad (2.27)$$

Since Q depends upon the path, the heat capacity likewise will depend upon the path. Moreover, the average heat capacity for a process will depend upon the particular temperature interval chosen. To eliminate any ambiguity caused by the choice of interval, we shall now introduce another definition of heat capacity; this assumes the form of the limit of equation 2.27 as the temperature change approaches zero:

$$C = \frac{dQ}{dT} \qquad (2.28)$$

This expression simply means that the heat capacity at any point along any specified path will be taken as the quotient of the inexact differential dQ by the exact differential dT. Utilizing the expressions for dQ given earlier—namely, equations 2.25 and 2.26—we see that the heat capacity is

$$C = \left(\frac{\partial U}{\partial T}\right)_V + \left\{\left(\frac{\partial E}{\partial V}\right)_T + p\right\}\frac{dV}{dT} \qquad (2.29)$$

or

$$C = \left(\frac{\partial H}{\partial T}\right)_p + \left\{\left(\frac{\partial H}{\partial p}\right)_T - V\right\}\frac{dp}{dT} \qquad (2.30)$$

It is evident from equations 2.29 and 2.30 that the heat capacity must depend upon the path, as mentioned earlier, for without a specified path the *total* derivatives of V and p with respect to T are meaningless. As soon as a relationship between T and V or T and p is stipulated, the number of independent variables is reduced from two to one, and the heat capacity becomes a meaningful concept. It should be clear, therefore, that the heat capacity is not a function only of the state of the system, for it involves both the state and the slope of the path being followed. Moreover, the heat capacity can assume any real value from minus to plus infinity by proper choice of path, and its average value for a given change in state can be anything. This is in direct correspondence with the fact that Q itself might have any value whatsoever. *

The concept of heat capacity is generally used in connection with simply defined paths such as those of constant volume or constant pressure. Although the heat capacities generally employed have positive values, there are numerous paths that give rise to bizarre values. The heat capacity of saturated water vapor, for example, heated in such a way that its pressure is always equal to the vapor pressure of liquid water at the same temperature, is a negative quantity. (This comes about because a large quantity of work must be done on the vapor in such a heating process.) The heat capacity of a system undergoing an adiabatic change equals zero, of course, for an adiabatic process is defined as one for which $Q = 0$. In view of this fact, one can immediately write

---

* Since Q is path dependent, the name "heat capacity" is an unfortunate choice, for it suggests a function of the state of the system. However, no alternative name has been seriously proposed.

down an ordinary differential equation governing a reversible adiabatic process by setting a general expression for heat capacity equal to zero. The heat capacities associated with isothermal processes are infinite because heat is added or extracted without a change in temperature. The heat capacities of an ideal monatomic gas are shown for a number of paths in Figure 2-3.

As suggested earlier, the heat capacities that are commonly dealt with correspond to constant-volume or constant-pressure processes. Throughout this book, unless there is some contrary specification, "heat capacity" will refer to constant-pressure processes, such as heating in the atmosphere; this conforms to the usual practice adopted by handbooks and similar sources of information.

FIGURE 2-3   Graph showing how heat capacity depends upon path. Values next to curves indicate molar heat capacities of monatomic ideal gases heated along the indicated paths. The heat capacity of zero denotes an adiabatic path; $\frac{3}{2}R$ corresponds to constant volume, and $\frac{5}{2}R$ to constant pressure. Heat capacity becomes negatively infinite as one approaches an isothermal path from the side of positive slopes; on the other hand, heat capacity becomes positively infinite as one approaches an isothermal path from the side of negative slopes. All curves in this particular array pass through a common point, illustrating further that the heat capacity can have any real value at a given point.

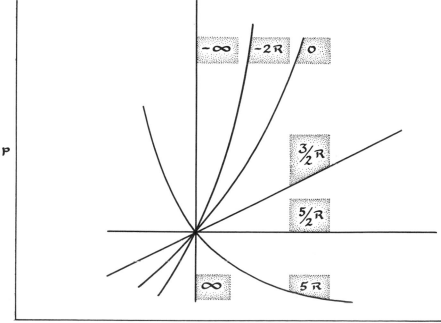

Should the volume be kept constant for a particular process, it is clear from equation 2.29 that the corresponding heat capacity, which we shall now denote with a subscript, v, will be simply given by the expression

$$C_v = \left(\frac{\partial U}{\partial T}\right)_V \tag{2.31}$$

This heat capacity will be a function of the state of the system, but, of course, its integral with respect to temperature will generally represent the heat added only for a constant-volume process.

On the other hand, if a process is carried out at constant pressure, then the slope of the path in the T-V coordinate system will be equal to the partial derivative of V with respect to T, a function that can be obtained from the equation of state of the substance involved. Hence this heat capacity, to be denoted by C_p, will be given by the expression

$$C_p = \left(\frac{\partial U}{\partial T}\right)_V + \left\{\left(\frac{\partial U}{\partial V}\right)_T + p\right\}\left(\frac{\partial V}{\partial T}\right)_p \tag{2.32}$$

Utilizing equation 2.30 in similar fashion, we can also write that

$$C_p = \left(\frac{\partial H}{\partial T}\right)_p \tag{2.33}$$

or that

$$C_v = \left(\frac{\partial H}{\partial T}\right)_p + \left\{\left(\frac{\partial H}{\partial p}\right)_T - V\right\}\left(\frac{\partial p}{\partial T}\right)_V \tag{2.34}$$

Difference Between C_p and C_v

It is often necessary, when we make thermodynamic calculations, to convert one kind of heat capacity to another; in particular, it is of some importance to know the difference between the heat capacities at constant pressure and at constant volume. It is readily seen from equations 2.31 and 2.32 that the difference between C_p and C_v will be

$$C_p - C_v = \left\{\left(\frac{\partial U}{\partial V}\right)_T + p\right\}\left(\frac{\partial V}{\partial T}\right)_p \tag{2.35}$$

Similarly, we obtain from equations 2.33 and 2.34 another expression for that difference:

$$C_p - C_v = \left\{V - \left(\frac{\partial H}{\partial p}\right)_T\right\}\left(\frac{\partial p}{\partial T}\right)_V \tag{2.36}$$

In each instance it will be noted that the difference depends, among other things, upon quantities such as $\left(\frac{\partial U}{\partial V}\right)_T$ and $\left(\frac{\partial H}{\partial p}\right)_T$. Accordingly it is now in order for us to turn our attention to evaluating these particular partial derivatives.

Energy and Enthalpy of an Ideal Gas

Near the middle of the nineteenth century, Gay-Lussac and Joule carried out experiments designed to show something about the dependence of the energy of a gas on its volume. Joule's experiment, in essence, is carried out as follows: Two containers joined by a tube provided with a stopcock are immersed in a calorimeter bath (see Fig. 2-4). One of the containers holds a gas at a specified pressure and temperature, and the other container has been evacuated. The experiment consists of opening the stopcock and allowing the gas to expand freely from one container into the other and then measuring the net temperature change attending the process after thermal equilibrium has been established. Since it is observed experimentally that considerable localized heating and cooling occur, it is important that the liquid in the calorimeter bath be stirred sufficiently to bring about uniformity in temperature.

When ordinary gases at moderate pressures are subjected to the Joule experiment, the net temperature changes observed are negligibly small. Assuming that the temperature change is actually zero, we conclude that Q, the net quantity of heat that enters or leaves the system in the course of the experiment, is zero. Moreover, from the design of the experiment, we can assert that the work attending the process is likewise zero, for obviously no weights have been lifted nor have any pistons swept out volume changes. It should be borne in mind that the system consists of the two connected containers of fixed total volume and their contents, and that nothing happening within the system can manifest itself as work in the surroundings. Since Q and W are both equal to zero, $\Delta U = 0$, from which we conclude that the energy of a gas at a given temperature is independent of its volume.

FIGURE 2-4 Schematic representation of apparatus for the Joule experiment.

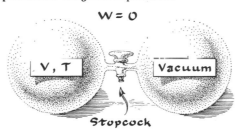

Very precise experiments of the Joule type have shown that there are actually small heat effects involved, but the heat effect per mole of gas diminishes as the initial pressure is reduced. This suggests that in the limiting state of zero pressure, in which a gas should behave ideally, the energy of the gas would cease to depend upon its volume. Accordingly we shall conclude that the energy of an ideal gas is a function of temperature only. Making use of this fact, together with the appropriate equation of state, we see from equation 2.35 that the difference between the molar heat capacities of an ideal gas at constant pressure and at constant volume is simply equal to R, the universal gas constant.

If U is a function of T only [let us say that $U = f(T)$], then C_v is likewise a function of T only, and the total differential of U will be

$$dU = C_v dT = f'(T)dT \qquad (2.37)$$

If we now make the assumption, which is not always valid, that C_v is constant, we obtain, upon integration, the expression

$$U = C_v T + \text{constant} \qquad (2.38)$$

Since the enthalpy of any substance is simply defined as the energy plus the pressure-volume product, the enthalpy of one mole of an ideal gas is equal to $U + RT$, which is also a function of temperature only. Accordingly the total differential of H will be

$$dH = C_p dT = dU + RdT$$
$$= (C_v + R)dT \qquad (2.39)$$

from which it is readily seen once more that $C_p - C_v = R$. Assuming C_v (and hence C_p) to be constant, we conclude that

$$H = C_p T + \text{constant} \qquad (2.40)$$

Just as Joule's experiment directly supplies information concerning the dependence of energy on volume, another experiment, known as the Joule-Thomson experiment, has been devised to give information about the dependence of enthalpy on pressure. The Joule-Thomson experiment, which is represented schematically in Figure 2-5, can be carried out much more conveniently and with greater accuracy than the Joule experiment. A cylindrical tube, insulated to prevent the transfer of heat to the surroundings, is provided near its middle with a porous plug, which permits gas to flow slowly through it. Initially, a certain quantity of gas at temperature T_1, volume V_1, and pressure p_1 is on the left-hand side of the plug, with no gas on the right. The gas is then allowed to effuse through the plug in such a way that its pressure on the left-hand

Q = O

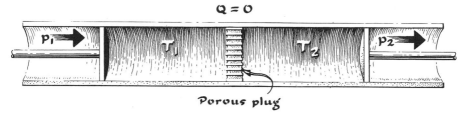

FIGURE 2-5 Schematic representation of apparatus for the Joule-Thomson experiment. The whole apparatus is insulated so that the process is adiabatic.

side is kept constant at a value p_1 by appropriate movement of a piston toward the plug. Simultaneously the piston on the right-hand side is withdrawn at precisely the rate necessary to maintain the right-hand pressure at a constant value p_2. The final volume on the right-hand side, after all the gas has passed through the plug, is denoted by V_2. It is evident that work will attend the consummation of this experiment, and that its magnitude will be $p_2V_2 - p_1V_1$, since V_2 is the change in volume occurring at constant pressure p_2 on the right-hand side and $-V_1$ is the change in volume occurring at p_1 on the left. The significant experimental quantity obtained in the course of the experiment is the change in temperature accompanying the effusion through the porous plug; we obtain this by measuring temperatures T_1 and T_2 on each side of the plug.

Since the thermodynamic system in the Joule-Thomson experiment is insulated in such a way as to render Q equal to zero, the change in energy is precisely equal to the negative of the work appearing in the surroundings. Therefore we can write that

$$\Delta U = U_2 - U_1 = -p_2V_2 + p_1V_1 \tag{2.41}$$

or that

$$U_2 + p_2V_2 = U_1 + p_1V_1 \tag{2.42}$$

from which we conclude that $H_2 = H_1$ or that $\Delta H = 0$. It appears, therefore, that the Joule-Thomson experiment is carried out under constant-enthalpy conditions. The Joule-Thomson coefficient, μ, is now defined as the change in temperature per unit change in pressure under constant-enthalpy conditions or, in terms of partial derivatives, by the expression

$$\mu = \left(\frac{\partial T}{\partial p}\right)_H \tag{2.43}$$

Evidently, if the gas cools in the process of going through the plug, the Joule-Thomson coefficient is positive, since the pressure always decreases in the experiment. On the other hand, a negative Joule-Thomson coefficient implies an increase in temperature. That a knowledge of the Joule-Thomson coefficient will give an insight into the dependence of enthalpy on pressure is clear when we recognize that μ can also be written in terms of certain partial derivatives of the enthalpy. Specifically, we note that

$$\mu = \left(\frac{\partial T}{\partial p}\right)_H = -\frac{\left(\frac{\partial H}{\partial p}\right)_T}{\left(\frac{\partial H}{\partial T}\right)_p} \tag{2.44}$$

from which we conclude that

$$\left(\frac{\partial H}{\partial p}\right)_T = -\mu C_p \tag{2.45}$$

The most convenient way of carrying out the Joule-Thomson experiment is to allow a gas coming from a reservoir of large capacity to pass through a porous plug and then escape at atmospheric pressure. Once a steady state is attained, it is easy to ascertain the temperature drop attending the effusion. Because of the convenience of this experiment, it is preferred to the Joule experiment as a way to ascertain deviations from ideal behavior.

From our earlier statement concerning the dependence of energy and enthalpy on variables other than temperature, we conclude that the Joule-Thomson coefficient should be zero for an ideal gas. Most gases at room temperature have positive Joule-Thomson coefficients; as the temperature is raised, however, the coefficient decreases to zero and then changes sign. A temperature at which the Joule-Thomson coefficient is equal to zero is called a Joule-Thomson inversion temperature. Hydrogen and helium differ from most gases in that their Joule-Thomson inversion temperatures are below room temperature.

The Joule-Thomson effect—the cooling that generally accompanies effusion through a porous plug—can be used as a basis for liquefying gases. This is not an efficient way of bringing about liquefaction, but it was one of the first methods employed. The procedure consists of allowing a gas at high pressure to be throttled from an opening and thus to be cooled, the cooled gas then being used to pre-cool incoming high-pressure gas (Fig. 2-6). As the process is continued, the gas being emitted from the throttling valve gets colder and colder, and ultimately the cooling is sufficient to bring about a certain amount of liquefaction. The method would not work, obviously, if the Joule-Thomson coefficient were negative, for then the gas would only get hotter instead of colder as the process was carried out. If one started with hydrogen and helium

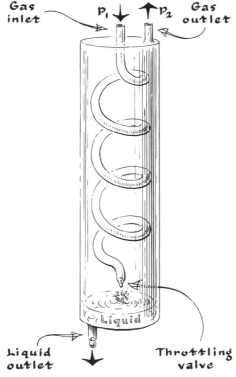

Gas inlet P_1 ↓   ↑P_2 Gas outlet

Liquid

Liquid outlet   ↓   Throttling valve

FIGURE 2-6   Apparatus for gas liquefaction through use of the Joule-Thomson effect.

at room temperature, one could not liquefy them by the Joule-Thomson effect; however, if these gases were pre-cooled by some other means to temperatures below their Joule-Thomson inversion temperatures, the effect could then be used for their liquefaction.

Kinetic Theory of Gases; Equipartition of Energy

Although classical thermodynamics can be quite fully developed without recourse to detailed molecular hypotheses, an appreciation of the subject is much enhanced by a knowledge of the kinetic theory of gases, which is concerned with the motions of molecules. Some of the important conclusions of the kinetic theory are also derivable by statistical mechanics, a topic to which we shall devote considerable attention later on. Meanwhile, we shall find it advantageous to use some of the results of the kinetic theory, particularly those that shed light on the energy of an ideal gas.

The simplest kinetic model for an ideal gaseous system involves the assumption that the molecules are point particles moving about in a perfectly random way and free of any intermolecular forces. On this assumption it can readily be shown that the pressure-volume product of such a gas equals two-thirds of the total kinetic energy of translation:

$$pV = \tfrac{2}{3}U_{\text{trans}} \tag{2.46}$$

In this expression $U_{\text{trans}} = \tfrac{1}{2}Nm\langle v^2\rangle$, N being the total number of molecules, m the mass of a molecule, and $\langle v^2\rangle$ the mean square velocity of the molecules. Combining equation 2.46 with the equation of state for an ideal gas, we conclude that for one mole

$$U_{\text{trans}} = \tfrac{3}{2}RT \tag{2.47}$$

Now let us consider, for example, a monatomic gas such as helium or argon. Apart from any energy within the individual atoms, which we can assume to be substantially constant, the only energy possessed by such molecules will be kinetic energy of translation, quantitatively expressed by equation 2.47. Since the thermal motion of such molecules appears to provide the only explanation of how energy can manifest itself in a temperature-dependent way, we shall identify the kinetic energy of translation of a monatomic gas with the thermodynamic energy defined through First Law considerations. Except for an additive constant, which will vanish when we consider changes in state, we shall now assert that the total thermodynamic energy of one mole of a monatomic ideal gas is $\tfrac{3}{2}RT$. The enthalpy, obviously, will then equal $\tfrac{5}{2}RT$, and the two most commonly used heat capacities will be given by $C_v = \tfrac{3}{2}R$ and $C_p = \tfrac{5}{2}R$.

Diatomic molecules, such as N_2 or HCl, will also possess translational energy in the amount $\tfrac{3}{2}RT$ per mole. Such molecules can also rotate and vibrate, however, and the average energies associated with these motions will likewise be temperature-dependent. At room temperatures the molar rotational energy of a diatomic molecule is roughly equal to RT, and the heat capacity is accordingly augmented by the quantity R.

As a first approximation, at least as far as heat capacity is concerned, the vibrations of diatomic molecules can be neglected at room temperatures. As the temperature is raised, however, the vibrations become important, and at very high temperatures they will also contribute an amount RT to the total energy, thus adding still another R to the heat capacity.

Polyatomic molecules likewise possess translational energy equal to $\frac{3}{2}RT$, but their rotations and vibrations are more complicated than those of diatomic molecules. Their heat capacities, accordingly, will generally be greater and more temperature-dependent than those of diatomic molecules. If we neglect vibrational excitation, linear polyatomic molecules like CO_2 and C_2H_2 can be considered equivalent to diatomic molecules as far as rotation is concerned; in other words, their molar rotational energies equal RT. A nonlinear polyatomic molecule, on the other hand, will have rotational energy equal to $\frac{3}{2}RT$ per mole and hence, except at low temperatures, a rotational heat capacity of $\frac{3}{2}R$.

The foregoing discussion can be unified through a concept known as the equipartition of energy. According to classical statistical mechanics, the average energy of a mole of gas will equal $sRT/2$, in which s equals the number of mutually independent quadratic terms in the expression for the total energy of the molecule. Thus the translational energy should be $\frac{3}{2}RT$ since the energy (all kinetic) is given by $\frac{1}{2}m(\dot{x}^2 + \dot{y}^2 + \dot{z}^2)$ in which $\dot{x}, \dot{y}$, and $\dot{z}$ are the three independent components of the velocity (in the x, y, and z directions).

A linear molecule can rotate with two degrees of freedom and the expression for its rotational energy (likewise all kinetic) will have two quadratic terms. These will contribute RT to the energy as mentioned above. On the other hand, a nonlinear polyatomic molecule can exhibit three degrees of rotational freedom, thus giving rise to $\frac{3}{2}RT$ for its energy.

Finally, a vibrating molecule, assumed to undergo simple harmonic motions, will require two quadratic terms—one for kinetic energy and one for potential energy—for each vibrational mode. Thus the vibrations of a diatomic molecule should contribute another RT to the total molar energy; for polyatomic molecules the vibrational energy would equal 2, 3, or more RT's, depending on the number of degrees of vibrational freedom. Actually, the principle of equipartition of energy is only asymptotically valid for high temperatures. Room temperature is "high" for translation, "almost high" for rotation, but generally "low" for vibrations, so the asymptotic vibrational contributions to heat capacities are not ordinarily manifested for gases.

Empirical Representation of Heat Capacities

The heat capacities of various substances can be determined experimentally by measuring the changes in temperature produced by the addition of measured quantities of heat or electrical energy. Since heat capacities are not usually constant, it is often found necessary to describe the heat capacity by some function of temperature such as

$$C_p = a + bT + cT^2 + \ldots \tag{2.48}$$

Other empirical forms can also be used, but, in any event, the constants are adjusted to fit the available data. Values of the empirical constants appearing in heat-capacity expressions for various gases are given in Table 2-1. The resulting expressions can be inserted into appropriate thermodynamic equations and subjected to the usual mathematical operations of integration, etc. Care must be exercised, however, to avoid using the expressions outside their ranges of validity. A large extrapolation outside the range for which a function was adjusted to the data may result in serious error, especially at high temperatures.

TABLE 2-1

Heat Capacities of Gases at Constant Pressure

When the parameters of the table are placed in the equation $C_p = a + bT + cT^2$, in which T is the absolute temperature, the result will be given in calories per degree per mole, valid over the range 300–1,500 K.

SUBSTANCE	a	$b \times 10^3$	$c \times 10^7$
Carbon dioxide	6.214	10.396	−35.45
Carbon monoxide	6.420	1.665	−1.96
Chlorine	7.5755	2.4244	−9.650
Ethane	2.247	38.201	−110.49
Hydrogen	6.9469	−0.1999	4.808
Hydrogen chloride	6.7319	0.4325	3.697
Methane	3.381	18.044	−43.00
Nitrogen	6.524	1.250	−0.01
Oxygen	6.148	3.102	−9.23
Water (vapor)	7.256	2.298	2.83

SOURCE: H. M. Spencer and others, *J. Am. Chem. Soc.*, *56*, 2311 (1934); *64*, 2511 (1942); *67*, 1859 (1945); *Ind. Eng. Chem.*, *40*, 2152 (1948).

EXERCISE:

Show that the ratio of heat capacities of ideal gases, $\gamma = C_p/C_v$, must lie in the range $1 < \gamma \leqslant 5/3$.

EXERCISE:

Using data from Table 2-1, calculate the reversible heat required to warm 1 mole of nitrogen from 30 to 100 °C at constant pressure.

Adiabatic Processes

An adiabatic process is one in which no heat enters or leaves the system—in other words, a process for which the heat capacity is equal to zero. As we indicated earlier, we can characterize adiabatic processes by the ordinary differential equation that results from setting $dQ = 0$ or $C = 0$. For an ideal gas, we find, in accordance with equation 2.29, that adiabatic processes are governed by the equation

$$C_v + \frac{RT}{V}\frac{dV}{dT} = 0 \qquad (2.49)$$

Assuming C_v to be constant, we obtain, upon integration, the expression

$$C_v \ln T + R \ln V = \text{constant} \qquad (2.50)$$

which represents the family of curves describing adiabatic changes in state of ideal gases. Although the assumption of constant C_v is not necessarily correct, we know that C_v must be a function of temperature only; the integration can therefore be carried out in principle, the variables being readily separable. By combining equation 2.50 with the equation of state of an ideal gas, we can write two other relationships that describe adiabatic processes equally well. One of these is

$$pV^\gamma = \text{constant} \qquad (2.51)$$

in which γ is the ratio of the heat capacity at constant pressure to that at constant volume. It should be noted that the equation of state, coupled with the conditional equation for the adiabatic process, renders the number of independent variables equal to unity.

When an adiabatic expansion is carried out for an ideal gas, the quantity of work involved will be given by the expression

$$W = -C_v(T_2 - T_1) \qquad (2.52)$$

in which the subscripts 1 and 2 denote initial and final conditions. This result can be obtained either by direct line integration of pdV along an adiabatic path or, more simply, by recognition of the fact that W must equal $-\Delta U$, the latter being simply related to the temperature change. If the heat capacity is not constant, we replace the expression in equation 2.52 by an appropriate integral, which, as long as the gas is ideal, will be valid regardless of the pressures or volumes.

Problems

1. The coefficient of cubical expansion of water at 20 °C and 1 atm is 2.1 $\times$ 10^{-4} per degree. Calculate approximately, in calories and in joules, the

work attending the heating at 1 atm pressure of 1 mole of water from 15 to 25 °C. Compare this with the work involved in heating 1 mole of an ideal gas from 15 to 25 °C.

2. The coefficient of compressibility of water at 20 °C is 49×10^{-6} per atm over the range 1–25 atm. Calculate, in calories and in joules, the work attending the compression of 1 mole of liquid water from a pressure of 1 atm to 25 atm at 20 °C. Compare this with the work involved when 1 mole of an ideal gas is compressed from 1 to 25 atm at 20 °C.

3. One mole of water is originally in the form of many droplets, each of 1 mm diameter. Each droplet is broken up into 1,000 smaller droplets of equal size. How much work, in calories, must be done to bring about this change in state at 25 °C if the surface tension of water at that temperature is 72 dyn cm^{-1}?

$\qquad\qquad$ Ans.   0.0167 cal (note the smallness of this quantity)

4. (A) If a gas that obeys the equation of state $pV = RT + \alpha p$ is heated at constant pressure from T_1 to T_2, show that the reversible work will be given by the expression

$$W = R(T_2 - T_1) + (\alpha_2 - \alpha_1)p$$

(B) If the gas is expanded isothermally, show that the reversible work will be

$$W = RT \ln \left(\frac{V_2 - \alpha}{V_1 - \alpha} \right) = RT \ln \frac{p_1}{p_2}$$

5. (A) If a gas obeys the equation of state $pV(1 - \beta p) = RT$, show that for an isothermal expansion the reversible work will be

$$W = \frac{RT}{1 - \beta p_2} - \frac{RT}{1 - \beta p_1} + RT \ln \frac{p_1(1 - \beta p_2)}{p_2(1 - \beta p_1)}$$

$$= p_2 V_2 - p_1 V_1 + RT \ln \frac{p_1^2 V_1}{p_2^2 V_2}$$

$\left(\text{Hint: Integrate by parts to evaluate } \int_{V_1}^{V_2} p\,dV.\right)$

(B) If the gas is heated at constant pressure, show that

$$W = \frac{RT_2}{1 - \beta_2 p} - \frac{RT_1}{1 - \beta_1 p}$$

6. Show, by making appropriate approximations, that if β is replaced by α/RT, the answers to Problem 5 reduce to those of Problem 4.

7. A house of constant volume, V, is warmed, the air pressure being kept constant at 1 atm. (During the heating process some of the air must be expelled through keyholes, around doorjambs, etc.)

(A) Assuming that C_v and C_p are constant for air, derive an expression for the amount of heat that must be supplied to warm the air in the house from T_1 to T_2, taking into account the continuous ejection of air.

(B) If the molar heat capacity at constant volume is $\frac{5}{2}R$, and if $V = 2,000$ cubic meters, $t_1 = 15\ °C$, and $t_2 = 20\ °C$, what is Q in kilocalories?

$$\text{Ans.}\quad \text{(A)}\quad Q = \frac{pVC_p}{R}\ln\frac{T_2}{T_1}\ \text{(why } C_p \text{ and not } C_v\text{?)}$$

8. Suppose that the house of Problem 7 were cooled from temperature T_1 to temperature T_2 with the temperature of the outside air at T_c.

   (A) How much heat must be extracted from the air within the house if you take into account the entry of additional air upon cooling? Why is the absolute value of this quantity of heat not equal to that of Problem 7?

   (B) If you assume that $t_1 = 30\ °C$, $t_2 = 25\ °C$, and $t_0 = 35\ °C$, how many kilocalories must be extracted for the house described in Problem 7?

9. One mole of an ideal gas is contained in a cylinder provided with a tightly fitting piston that is not free of friction. To cause this piston to move, one must apply a constant extra force in the direction of its movement. If this friction force is divided by the area of the piston, it reduces to a pressure equivalent of friction, p_f.

   (A) Derive expressions for the work and heat attending the isothermal expansion of the ideal gas from gas pressure p_1 to p_2 by means of the apparatus described above. If the gas is now compressed from p_2 to p_1, what are Q and W?

   (B) Calculate in calories the values of Q and W for 1 mole of the ideal gas expanded irreversibly according to the same mechanism from 1 atm to 0.5 atm if $p_f = 0.1$ atm and $t = 25\ °C$.

$$\text{Ans.}\quad \text{(B)}\quad Q = W = 351.4\ \text{cal}$$

10. (A) Equate the two general expressions for heat capacity given by equations 2.29 and 2.30, and by appropriate manipulation show that the equation so obtained is an identity.

    (B) Similarly equate the two expressions for $C_p - C_v$ given by equations 2.35 and 2.36, and show that the result is an identity.

11. One mole of an ideal monatomic gas (for which $C_v = \frac{3}{2}R$) is subjected to the following sequence of steps:

    (A) The gas is heated reversibly at a constant pressure of 1 atm from 25 to 100 °C.

    (B) Next, the gas is expanded reversibly and isothermally to double its volume.

    (C) Finally, the gas is cooled reversibly and adiabatically to 35 °C.

    Calculate in calories ΔU, ΔH, Q, and W for the over-all process A + B + C.

12. One mole of an ideal gas (not necessarily monatomic) is subjected to the following sequence of steps:

    (A) It is heated at constant volume from 25 to 100 °C.

    (B) It is expanded freely into a vacuum (Joule-type experiment) to double its volume.

(c) It is cooled reversibly at constant pressure to 25 °C.

Calculate in calories ΔU, ΔH, Q, and W for the over-all process A + B + C. (Note that it is not necessary to know the heat capacity of the gas. Explain why.)

13. How many grams of deuterium (atomic weight 2.01410) when converted into helium (atomic weight 4.00260) would liberate enough energy to lift one metric ton a height of one kilometer in the earth's gravitational field?

14. For an ideal gas, we can write for dQ the expression

$$dQ = C_v dT + \frac{RT}{V} dV$$

in which C_v is a function of temperature only. Show by direct test that dQ is not an exact differential. Now consider the quotient dQ/T, and show that it is exact. ($1/T$ is said to be an integrating factor for dQ. Actually, the proposition illustrated here is not restricted to ideal gases and will be dealt with in more detail in a later chapter.)

15. (A) Assuming that the Joule-Thomson coefficient, μ, is constant, prove that the enthalpy must be given by a function of $T - \mu p$ in the form

$$H = H(T - \mu p)$$

(B) Assuming that both μ and C_p are constant prove that

$$H = C_p T - \mu C_p p + \text{constant}$$

(c) Assuming that μ and C_p are functions of temperature but independent of pressure, show that the product μC_p must be constant, and that

$$H = \phi(T) - \mu C_p p$$

provided $\phi'(T) = C_p$.

16. Consider the following experiment, which is a hybrid between the Joule and the Joule-Thomson experiments. A tube of uniform cross-section is equipped with a porous plug and provided with a frictionless piston on one side and a fixed end on the other (see the figure). The space between the plug and the fixed end is initially evacuated, and the space between the plug and the piston contains a large quantity of an ideal gas at temperature T_1 and pressure p. The experiment consists of letting gas effuse through the plug while keeping the pressure on the left-hand side constant, the process being terminated when the gas pressure on the right-hand side of the plug equals p. Assuming that the system is insulated so that the process takes place

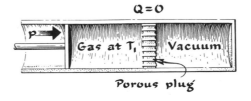

Porous plug

adiabatically, and neglecting any heat conduction through the plug, show that the final temperature of the gas on the right-hand size is given by the expression

$$T_2 = \gamma T_1$$

in which γ is the ratio C_p/C_v.

17. Consider an experiment that is somewhat the reverse of that described in Problem 16 above. The side to the right of the porous plug, of volume V_1, originally contains n_1 moles of ideal gas at pressure p_1 and temperature T_1. The piston, initially touching the plug, is withdrawn at such a rate as to maintain a constant pressure p ($<p_1$) on the left hand side of the plug. The process is carried out adiabatically and no heat is conducted from one side of the plug to the other. The experiment is over when the pressure on the right hand side equals p.

   (A) Show that, after the experiment is over, $H_2 + U_1 = U_1^o$ where H_2 is the enthalpy of the gas finally on the left hand side, U_1 the energy of the gas remaining on the right hand side, and U_1^o the original energy of the gas.

   (B) Show also that

   $$\gamma V_2 = V_1(p_1 - p)/p$$

   where V_2 is the final volume of gas on the left and γ is the ratio C_p/C_v.

18. Two containers are joined together as in the Joule experiment. One contains 0.5 mole of helium ($C_v = \frac{3}{2}R$) at 100 °C, and the other contains 1.0 mole of oxygen (for which you may assume that $C_v = \frac{5}{2}R$) at 0 °C. The gases are allowed to diffuse into each other adiabatically. What is the final temperature assuming the gases to be ideal?

19. Suppose that 1.0 mole of helium at 100 °C is on the left-hand side of the porous plug in the Joule-Thomson experiment and that 0.5 mole of oxygen at 0 °C is on the right-hand side, both pressures being 1 atm. The gases are now allowed to diffuse into each other adiabatically, the pressure being kept constant by the movement of pistons in or out. What is the final temperature if uniformity is achieved? Assume the gases to be ideal, with heat capacities given in Problem 18.

20. (A) Using the expression $C_p = a + bT + cT^2$ for the heat capacity of a substance, derive expressions for Q_p and $\langle C_p \rangle$ (the average heat capacity) for the substance heated from T_1 to T_2. What is the difference between the average heat capacity and the heat capacity at the average temperature?

   (B) Using the data of Table 2-1 calculate Q_p in calories and $\langle C_p \rangle$ for 1 mole of oxygen heated from 27 to 227 °C. What is C_p at 127 °C?

21. Prove that, if an ideal gas is subjected to a reversible change,

$$dQ = \frac{C_p p\, dV}{R} + \frac{C_v V\, dp}{R}$$

In this expression C_v and C_p are functions of the product pV and are not necessarily constant, although their difference, $C_p - C_v$, is a constant equal to R.

(A) Show directly that dQ is not an exact differential, but that dQ/pV is exact.

(B) Assuming that C_p and C_v are both constant, with $C_p = \gamma C_v$, prove that

$$\int \frac{dQ}{pV} = \frac{C_v}{R} \ln \frac{p_2 V_2^\gamma}{p_1 V_1^\gamma}$$

in which the subscripts 1 and 2 denote initial and final states. Show from the above result that, if the path followed is reversible and adiabatic, pV^γ is a constant.

22. Show that, if p and V are used as variables in place of V and T, equation 2.49 becomes

$$\frac{dp}{dV} + \gamma \frac{p}{V} = 0$$

in which $\gamma = C_p/C_v$. Assuming γ to be constant, derive equation 2.51.

23. Show that, if p and T are used as variables in place of V and T, equation 2.49 becomes

$$C_p - \frac{RT}{p} \frac{dp}{dT} = 0$$

(A) Assuming C_p to be constant, integrate this equation to obtain the p, T counterpart of equation 2.50.

(B) Assuming C_p to be given by an expression like that of equation 2.48, what does one obtain by integration?

(C) Using the result of part B, calculate the final pressure if nitrogen (assumed to be an ideal gas) is reversibly and adiabatically expanded from 1 atm and 60 °C to the point at which its temperature becomes 30 °C. Use the heat capacity parameters of Table 2-1.

24. (A) Prove equation 2.52.

(B) Show that the result can also be expressed by the equation

$$W = \frac{p_2 V_2 - p_1 V_1}{1 - \gamma}$$

(C) If C_v is not constant, how will equation 2.52 be modified if the gas remains ideal?

(D) Using data from Table 2-1, calculate W if 1 mole of nitrogen (to be considered ideal) is expanded reversibly and adiabatically from 60 to 30 °C.

25. One-tenth of a mole of an ideal monatomic gas is heated from T_1 to T_2 along a path governed by the equation $V = a \exp(bT)$, in which a and b are constants.
   (A) Derive expressions for the reversible heat and work attending this process.
   (B) If $T_1 = 300$ K, $T_2 = 400$ K, and $b = 0.01$ per degree, calculate Q, W, and ΔU in calories.

26. Prove that, if an ideal gas with a constant value for C_v is heated along a path of constant heat capacity, $C\ (-\infty < C < \infty)$, the path must belong to one of the curves in the following family:

$$(C - C_v) \ln T = R \ln V + \text{constant}$$

27. One mole of a monatomic ideal gas is heated along such a path that $C = R$. What is the nature of the path?   Ans. $V = \text{constant}/\sqrt{T}$

28. The magnetic energy of the protons in a mole of hydrogen gas is given by the equation

$$U_m = \frac{A\mathfrak{K}^2}{T}$$

in which $\mathfrak{K}$ is the strength of the magnetic field and A is a constant characteristic of hydrogen.
   (A) How much is the generalized heat capacity of hydrogen changed by the application of a magnetic field?
   (B) Derive an equation governing reversible adiabatic changes in state for hydrogen in a magnetic field.

   Ans.   (B) $R \ln V + C_v \ln T + \dfrac{A\mathfrak{K}^2}{2T^2} = \text{constant}$, if C_v is the "ordinary"

   heat capacity at constant volume.

29. An ideal gas with a constant value for C_v is expanded adiabatically but irreversibly by means of the apparatus described in Problem 9. If constant friction is the only reason for irreversibility, show that

$$\left(p \pm \frac{Rp_f}{C_p}\right) V^\gamma = \text{constant}$$

provided the plus sign holds for a compression and the minus sign for an expansion. (This problem is recommended only for students who have studied differential equations.)

3

Thermochemistry

Introduction

Heretofore our examples of changes in state have involved, for the most part, systems of fixed chemical composition existing in single phases. We shall now turn to thermochemistry, the portion of thermodynamics that is concerned with the heat effects attending processes that give rise to changes of phase and changes in composition.

In Chapter Two it was shown generally that $\Delta U = Q_v$ and $\Delta H = Q_p$ for all constant-volume or constant-pressure processes, regardless of phase or chemical changes, as long as the only work involved is pressure-volume work. It follows that those equations can be applied even to such processes as the evaporation of water, the combustion of methane, and the transition of rhombic to monoclinic sulfur.

Consider, for example, the evaporation of liquid water to water vapor, a change in state that can be symbolized by the equation

$$H_2O(liq) \longrightarrow H_2O(g) *$$

---

* The symbols (liq) and (g) denote the liquid and gaseous states. The notation (c) will be similarly used to designate the crystalline state.

It is found experimentally that, when this process takes place at constant pressure and constant temperature, say 100 °C and one atmosphere, the heat that must be supplied is 9,720 calories; this is the change in enthalpy (ΔH_{vap}) associated with the process of vaporization. The work accompanying such an evaporation, if carried out reversibly, will equal $p\Delta V$, in which expression ΔV is the difference between the volume of the vapor and that of the liquid. To obtain an excellent approximation for ΔV, we can neglect the volume of the liquid compared with that of the vapor, implying that the work is nearly equal to the pressure-volume product of the vapor. Assuming further that the vapor is an ideal gas, we can write that

$$W = RT \qquad (3.1)$$

a result that actually turns out to be in error by less than 2 percent. Recognizing the relationship between ΔU and ΔH, we conclude that the energy of vaporization will be given by the expression

$$\Delta U_{vap} = \Delta H_{vap} - p\Delta V \approx \Delta H_{vap} - RT \qquad (3.2)$$

which gives for ΔU_{vap} a numerical value of 8,980 calories at 100 °C and one atmosphere. This result means that, if a mole of liquid water at 100 °C were allowed to flash into an evacuated space of precisely the right volume to accommodate vapor at one atmosphere and 100 °C, the heat that would have to be added during the process would equal 8,980 calories. Since the process could produce no work in the surroundings, the heat added, ΔU_{vap}, would be less than ΔH_{vap}.

Let us now consider a chemical reaction, the combination of carbon with oxygen to form carbon dioxide. This combustion is known to be *exothermic*—in other words, a reaction involving the evolution of heat.* The amount of heat liberated when one mole of carbon is oxidized to carbon dioxide at 25 °C equals 94,051 calories; so the amount absorbed, or the enthalpy of the reaction, will be −94,051 calories. By definition, the enthalpy of this combustion can be written as

$$\Delta H = H_{CO_2} - H_C - H_{O_2} \qquad (3.3)$$

in which H_{CO_2} denotes the enthalpy of one mole of CO_2, etc. Rearranging the equation so that the enthalpies appear in the same order as the corresponding substances in the chemical equation describing the change in state, we note that

$$H_C + H_{O_2} = H_{CO_2} - \Delta H \qquad (3.4)$$

---

* When heat is absorbed, a reaction is said to be *endothermic*.

If we now dispense with the use of the symbol H for the enthalpy of a chemical substance and write the chemical formula instead, we arrive at the enthalpy equation

$$C(c) + O_2(g) = CO_2(g) - \Delta H \tag{3.5}$$

or, for 25 °C, at the equation

$$C(c) + O_2(g) = CO_2(g) + 94{,}051 \text{ cal} \tag{3.6}$$

In writing an equation of this type, we recognize that the chemical symbols and formulas represent the enthalpy values of the substances indicated. Since ΔH is negative for the example cited, a positive number will appear on the right-hand side of the equation; this is fully compatible with the fact that heat is liberated with the products of the reaction. Unless otherwise stated, all equations that assume the form of chemical equations combined with thermal quantities will be regarded as enthalpy equations. The enthalpy equation for the vaporization of water at 100 °C and one atmosphere, for example, can be written as

$$H_2O(\text{liq}) = H_2O(g) - 9{,}720 \text{ cal} \tag{3.7}$$

EXERCISE:

The molar enthalpy of vaporization of water at 100 °C and 1 atm is 9,717.1 cal. Under those conditions the volume of 1 g of vapor is 1,677 ml, whereas that of 1 g of liquid is 1.0 ml. (A) Calculate precisely the work attending the vaporization. Then, by subtracting that result from ΔH_{vap}, obtain a value for ΔU_{vap}. (B) Assuming the vapor to be an ideal gas, and neglecting the volume of the liquid compared with that of the vapor, repeat the calculation. (C) What is the percentage error in ΔU_{vap} calculated by method B?

EXERCISE:

A precise value for the molar enthalpy change accompanying fusion of ice at 0 °C and 1 atm is 1,436.3 cal. The volume of 1 g of ice under those conditions is 1.091 ml, whereas that of 1 g of liquid water is 1.000 ml. Calculate precisely the work involved during fusion, and then obtain the energy change. Is the correction important?

Calorimetry

Quantities of heat can be measured experimentally by means of calorimeters. Numerous kinds of calorimeters have been designed for various purposes, but we shall here mention only the salient features that must be considered from a purely thermodynamic point of view.

Some calorimeters are operated isothermally; that is, heat is either added to or extracted from the system in amounts necessary to keep the temperature constant. Heat is usually added by electrical means, and the quantity is readily calculated by the use of appropriate conversion factors. If heat is extracted (by means of a coolant, for example), an indirect calculation may be required to arrive at the proper value for Q.

Other calorimeters are operated adiabatically, either by the use of insulation or by continuously changing the temperature of the surroundings to keep the temperatures of system and surroundings equal, thus preventing heat transfer between them. The temperature change taking place under these circumstances is measured, and the heat effect that would have attended an isothermal process is computed from a knowledge of that temperature change and the appropriate heat capacities. To illustrate the calculation, let us consider a change in state involving reactants R turning into products P; to render the example more specific, we shall further assume that the process takes place at constant pressure. The following scheme will now serve to expedite our calculations:

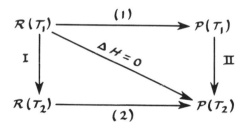

The process occurring in the adiabatic calorimeter is represented by the diagonal, for which $\Delta H = 0$. Applying the First Law, we see that ΔH_1 (the enthalpy change at T_1) can be computed from the constant-pressure heat capacity, $C_p(P)$, of the *products* by means of the following equation:

$$\Delta H_1 = -\Delta H_{II} = - \int_{T_1}^{T_2} C_p(P)\, dT \tag{3.8}$$

Similarly, the enthalpy change at T_2 (ΔH_2) can be computed from the heat capacity, $C_p(R)$, of the *reactants*:

$$\Delta H_2 = -\Delta H_I = - \int_{T_1}^{T_2} C_p(R)\, dT \tag{3.9}$$

If the process occurring in the calorimeter takes place at constant volume instead of constant pressure, we can make similar calculations from the heat capacities at constant volume to obtain the change in energy (ΔU).

It is evident that, if we wish to obtain ΔH values directly by calorimetric means, we should operate the calorimeter at constant pressure. It is often more convenient to carry out reactions, such as combustions, in so-called bomb calorimeters, which impose a constant-volume constraint. It is important, therefore, that means be developed for converting one kind of heat effect into another, in particular for converting an energy change into an enthalpy change. This type of calculation is best illustrated by the following scheme, in which we consider alternative means for bringing about equivalent isothermal changes in state:

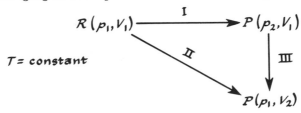

In this diagram, step I denotes a constant-volume conversion of reactants R into products P with a heat absorption $Q_v = \Delta U_I$. The same chemical reaction might take place, however, at constant pressure (process II), for which we can write $Q_p = \Delta H_{II}$. The question before us is how to use a measured value of Q_v, such as might be obtained from a bomb calorimeter, to determine Q_p. Application of the First Law to the triangular representation discloses the kind of computation that must be made. Evidently

$$\Delta H_{II} = \Delta U_I + \Delta U_{III} + p_1(V_2 - V_1) \tag{3.10}$$

or

$$Q_p = Q_v + \Delta U_{III} + p_1(V_2 - V_1) \tag{3.11}$$

Let us now examine the orders of magnitude of the correction terms that will convert ΔU_I to ΔH_{II}. For a first approximation, we can neglect the pressure-volume products of solids and liquids compared with those of gaseous materials. If we further assume that the gases behave ideally, we can assert that $p_1(V_2 - V_1)$ will equal $\Delta \nu RT$, if $\Delta \nu$ is equal to the increase in the number of moles of gases attending the change in state. Thus, for the combustion of hydrogen to liquid water, $\Delta \nu$ equals $-3/2$ for each mole of water formed; for the burning of carbon to carbon dioxide, $\Delta \nu$ equals zero. There still remains the question of the order of magnitude of ΔU_{III}. If the materials involved are all ideal gases, ΔU_{III} must be zero, for the energies of ideal gases are functions of temperature only. Since ΔU_{III} will usually be negligibly small for all substances, we shall simply set it equal to zero. Accordingly we can summarize by saying that

$$\Delta H_{II} = \Delta U_I + \Delta \nu R T \tag{3.12}$$

or that

$$Q_p = Q_v + \Delta \nu R T \tag{3.13}$$

Since even $\Delta \nu R T$ is usually small compared with ΔU_I, the several approximations introduced are fully warranted.

To illustrate the use of equations 3.12 and 3.13, let us consider the combustion of benzoic acid to carbon dioxide and water at 25 °C. When one mole of benzoic acid is burned in a bomb calorimeter at that temperature, 771.1 kilocalories of heat are liberated; this means that ΔU for the combustion is equal to -771.1 kilocalories. From the combustion reaction,

$$C_6H_5COOH(c) + \tfrac{15}{2}O_2(g) \longrightarrow 7CO_2(g) + 3H_2O(liq)$$

we note that $\Delta \nu = -1/2$. Hence we conclude that the enthalpy of combustion will be given by the expression

$$\Delta H = \Delta U - \tfrac{1}{2}RT \quad \text{or} \quad \Delta H = -771.4 \text{ kcal}$$

The small correction involved in this kind of calculation makes it clear that no appreciable error has resulted from the approximations described above.

EXERCISE:

Show that, when a carbohydrate (sugar, for example) is burned to carbon dioxide and water, the energy of combustion is almost exactly the same as the enthalpy of combustion.

"Heat of a Process" Defined

Since the heat associated with a particular change in state can have any value, depending upon the path by which the change is brought about, the phrase "heat of a process" is meaningless unless a definite path is specified or implied. It is usual, however, to assume that any process is a constant-pressure process unless stated otherwise. In accordance with this generally accepted practice, "heat of a process" will be taken to mean the heat attending the process carried out reversibly at constant pressure—in other words, the enthalpy change, ΔH. (It is tacitly assumed that no work other than pressure-volume work is involved.) When we speak of the "heat of fusion" of ice, therefore, we shall mean the change in enthalpy accompanying that fusion and shall denote it by the symbol ΔH_{fus}. Similarly, we speak of heats of vaporization, sublimation, reaction, etc., each time meaning ΔH for the corresponding process.

Heat of Formation

By "heat of formation" of a substance such as a chemical compound, we shall mean the difference between the enthalpy of the substance and the sum of the enthalpies of the elements of which it is composed, all enthalpies being measured at a specified temperature and pressure. The heat of formation is therefore simply the enthalpy change associated with the formation of the substance from its elements, as illustrated by equations 3.5 and 3.6. From these equations we see, for example, that the heat of formation of carbon dioxide is $-94,051$ calories at 25 °C.

Most compounds, of course, cannot actually be formed by direct combination of the elements; but, because the enthalpy is a function of the state of the system, the heat of formation is still a meaningful concept whose numerical value can usually be determined experimentally by indirect means. As an illustration, let us consider the substance sucrose, $C_{12}H_{22}O_{11}(c)$. Although carbon, hydrogen, and oxygen cannot simply be mixed to give sucrose, the enthalpy of formation can be indirectly obtained by recourse to combustion experiments. By burning one mole of sucrose to carbon dioxide and water, one can determine the difference between the enthalpies of sucrose and of carbon dioxide and water. Moreover, from a knowledge of the heats of formation of the latter two compounds, one can, by addition and subtraction of appropriate enthalpy equations, arrive at the following equation, valid at 25 °C:

$$12C(c) + 11H_2(g) + \tfrac{11}{2}O_2(g) = C_{12}H_{22}O_{11}(c) + 533.4 \text{ kcal} \quad (3.14)$$

EXERCISE:

When 1 mole of sucrose, $C_{12}H_{22}O_{11}(c)$, is burned in a bomb calorimeter at 25 °C, 1,346.7 kcal of heat are liberated. If the heats of formation of $CO_2(g)$ and $H_2O(liq)$ are -94.05 and -68.32 kcal, what is the heat of formation of sucrose?

Calculation of Heats of Reaction from Tables

To facilitate the calculation of heats of reaction, tables giving the heats of formation of various compounds from their elements have been prepared. If the value zero is arbitrarily assigned to the enthalpy of an element in its standard state, which is generally taken to be 25 °C and one atmosphere, we can satisfy all enthalpy equations by simply substituting the heats of formation of the compounds involved for their

TABLE 3-1

Heats of Formation of Selected Compounds

(Values given in kcal mol^{-1} relative to elements in standard states at 25 °C)

SUBSTANCE	ΔH_f	SUBSTANCE	ΔH_f
$H_2O(liq)$	−68.315	$HCOOH(liq)$	−101.51
$H_2O(g)$	−57.796	$HCOOH(g)$	−90.48
$H_2O_2(g)$	−32.58	$CH_3OH(liq)$	−57.04
$HF(g)$	−64.8	$CF_4(g)$	−221.
$HCl(g)$	−22.062	$CCl_4(g)$	−24.6
$Br_2(g)$	7.387	$CHCl_3(g)$	−24.65
$HBr(g)$	−8.70	$CBr_4(g)$	19.
$I_2(g)$	14.923	$CS_2(g)$	28.05
$HI(g)$	6.33	$HCN(g)$	32.3
$S(rhombic)$	0	$C_2H_2(g)$	54.19
$S(monoclinic)$	0.08	$C_2H_4(g)$	12.49
$S_8(g)$	24.45	$C_2H_6(g)$	−20.24
$SO_2(g)$	−70.944	$CH_3CHO(g)$	−39.72
$SO_3(g)$	−94.58	$CH_3COOH(liq)$	−115.8
$H_2S(g)$	−4.93	$CH_3COOH(g)$	−103.31
$H_2SO_4(liq)$	−194.548	$C_2H_5OH(liq)$	−66.37
$NO(g)$	21.57	$CH_3CN(g)$	20.9
$NO_2(g)$	7.93	$SiO_2(c, quartz)$	−217.72
$N_2O(g)$	19.61	$SiH_4(g)$	8.2
$N_2O_4(g)$	2.19	$SnCl_4(g)$	−112.7
$NH_3(g)$	−11.02	$ZnO(c)$	−83.24
$N_2H_4(g)$	22.80	$ZnCl_2(c)$	−99.20
$HNO_3(liq)$	−41.61	$Ag_2O(c)$	−7.42
$NH_4Cl(c)$	−75.15	$AgCl(c)$	−30.370
$P_4(\sigma)$	14.08	$FeO(c)$	−65.0
$P_4O_{10}(c)$	−713.2	$Fe_2O_3(c)$	−197.0
$PH_3(g)$	1.3	$MgO(c)$	−143.81
$H_3PO_4(c)$	−305.7	$MgCl_2(c)$	−153.28
$PF_3(g)$	−219.6	$CaO(c)$	−151.79
$PCl_3(g)$	−68.6	$CaCl_2(c)$	−190.2
$PCl_5(g)$	−89.6	$CaCO_3(c, calcite)$	−288.46
$C(c, graphite)$	0	$SrO(c)$	−141.5
$C(c, diamond)$	0.453 3	$BaO(c)$	−132.3
$CO(g)$	−26.416	$NaOH(c)$	−101.
$CO_2(g)$	−94.051	$NaCl(c)$	−98.28
$CH_4(g)$	−17.88	$KOH(c)$	−101.52
$CH_2O(g)$	−28.	$KCl(c)$	−104.33

Additional data will be found in Appendix D.

formulas in the chemically balanced enthalpy equations. By the use of this convention, the heat of any reaction can readily be computed by reference to tables. (In some instances, of course, additional information, such as a heat of solution or a heat of dilution, may be necessary, but the principles involved are the same.)

The arbitrary assignment of zero enthalpy to an element can be adopted at any temperature whatsoever as long as the heats of formation and heats of reaction all refer to the same temperature. If calculations are made for nonisothermal processes, however, the convention of assigning zero enthalpy to the elements can be adopted with respect to one temperature only. To avoid uncertainties, we shall confine our use of the convention to 25 °C whenever nonisothermal processes or cycles might be involved.

The heats of formation of a number of compounds are given in Table 3-1.

EXERCISE:

Using data from Table 3-1, calculate the heat of combustion of ethylene at 25 °C (A) when the water resulting from the combustion is in the form of liquid and (B) when the water is in the form of vapor.

Heats of Solution and Dilution

By "heat of solution" we shall mean the enthalpy change attending the combination of a certain amount of a particular solute with a specified amount of solvent to form a solution. The heat of solution at 25 °C of one mole of hydrogen chloride in ten moles of water, for example, is -16.49 kilocalories, and the process can be symbolized by the following equation:

$$HCl(g) + 10Aq = HCl \cdot 10Aq + 16.49 \text{ kcal} \qquad (3.15)$$

In this equation, water is represented by the symbol Aq, instead of $H_2O(liq)$, and for convenience it will be assigned a conventional enthalpy of zero. This notation and convention will be employed whenever the water acts as an inert solvent and is not involved in a chemical reaction. The heat of formation of $HCl \cdot 10Aq$ thus equals the heat of formation of $HCl(g)$ less 16.49 kilocalories. (The purpose of the convention with respect to Aq is to avoid exceedingly large and unwieldy numbers for heats of formation of dilute solutions, which consist mostly of solvent.)

If the solution consists of a mixture of two liquids, the heat effect attending the formation of the solution from the separate liquids is also called *heat of mixing*. When additional solvent is added to a solution so as to produce a solution of lower concentration, the heat effect for that process is called *heat of dilution*.

When a material is dissolved in a sufficiently large amount of solvent, further dilution will produce no additional heat effect. Under these circumstances the solution is said to be *infinitely dilute*, at least with respect to its enthalpy.* The concentrations below which a solution can be considered infinitely dilute will, of course, depend upon the particular solute-solvent pair. An enthalpy equation for dissolving a substance into an infinitely dilute solution (with HCl and water at 25 °C as an example) is generally written as follows:

$$HCl(g) + \infty Aq = HCl \cdot \infty Aq + 17.89 \text{ kcal} \qquad (3.16)$$

It should now be obvious why the assignment of zero enthalpy to solvent Aq is desirable for the practical working of problems.

EXERCISE:

Given the following heats of formation at 25 °C:

$BaCl_2 \cdot 100Aq$	-207.803 kcal
$BaCl_2 \cdot 300Aq$	-207.905
$BaCl_2 \cdot 500Aq$	-207.963

If $1 BaCl_2 \cdot 100Aq$ is dissolved in $1 BaCl_2 \cdot 500Aq$, what is ΔH for the process?

Heats of Reactions in Solution

Let us now consider a typical reaction carried out in aqueous solution, such as the neutralization of an acid with a base. Suppose, for example, we mix $1 NaOH \cdot \infty Aq$ with $1 HCl \cdot \infty Aq$ at 25 °C and one atmosphere. The enthalpy equation for this process is

$$NaOH \cdot \infty Aq + HCl \cdot \infty Aq = NaCl \cdot \infty Aq + H_2O(liq) + 13{,}345 \text{ cal} \quad (3.17)$$

Because one mole of water is formed in the course of the reaction, it is important to include $H_2O(liq)$ on the right-hand side of the equation. The effect of $1 H_2O(liq)$ as solvent is, of course, nil, since the solutions are infinitely dilute, but its contribution to the total enthalpy change cannot be neglected.

If the solutions of acid and base are not infinitely dilute, the representation of the neutralization reaction becomes somewhat more complicated. Specifically, let us consider the heat effect of mixing $1 NaOH \cdot xAq$

---

* Such a solution may not be infinitely dilute with respect to some other criteria, such as conformity to certain limiting solution laws.

with $1HCl \cdot yAq$, x and y being finite numbers. This process of neutralization cannot simply be represented by the equation

$$NaOH \cdot xAq + HCl \cdot yAq = NaCl \cdot (x + y)Aq + H_2O(liq) - \Delta H_1 \quad (3.18)$$

for that equation implies that the final solution consists of $1NaCl$ dissolved in $x + y$ moles of water, the additional mole of water, formed
in the course of the reaction, not serving as solvent. Evidently, to represent the desired heat effect, we must take into account the heat of dilution
of $NaCl \cdot (x + y)Aq$ with another mole of solvent:

$$NaCl \cdot (x + y)Aq + Aq = NaCl \cdot (x + y + 1)Aq - \Delta H_2 \quad (3.19)$$

The addition of equation 3.18 to equation 3.19 will then give the correct
enthalpy equation for the original process described ($\Delta H_3 = \Delta H_1 + \Delta H_2$):

$$NaOH \cdot xAq + HCl \cdot yAq + Aq = NaCl \cdot (x + y + 1)Aq + H_2O(liq) - \Delta H_3$$
$$(3.20)$$

Interpreted literally, equation 3.20 implies that an extra mole of solvent
(Aq) is added to the system while the mole of water ($H_2O(liq)$) formed
by the reaction is removed. This hypothetical process—which in reality
would produce no heat effect and actually does not occur—must be
imagined because of the different enthalpies assigned to Aq and $H_2O(liq)$.
In this connection we can write the pure formalism

$$Aq = H_2O(liq) + 68.32 \text{ kcal} \quad (3.21)$$

which is proper because the heat of formation of $H_2O(liq)$ is -68.32
kilocalories, whereas that for Aq is arbitrarily zero.

The complication of dilution accompanying a reaction does not, of
course, arise for reactions involving only infinitely dilute solutions, which,
as a matter of fact, often behave thermally in a rather simple way, as we
shall presently see.

EXERCISE:

Given the following heats of formation at 25 °C:

$HCl \cdot 12Aq$	-38.762 kcal
$NaOH \cdot 12Aq$	-112.200
$NaCl \cdot 25Aq$	-97.506
$H_2O(liq)$	-68.315

Calculate the heat attending the mixing of $1HCl \cdot 12Aq$ with $1NaOH \cdot 12Aq$ at 25 °C.

Reactions in Infinitely Dilute Solutions

When infinitely dilute solutions of two electrolytes such as NaCl and KNO_3 are mixed, it is found that the heat of such mixing is equal to zero. On the other hand, if $NaCl \cdot \infty Aq$ is mixed with $AgNO_3 \cdot \infty Aq$, an appreciable heat effect will be observed, the magnitude of which is precisely the same as that obtained if $KCl \cdot \infty Aq$ is used in place of the sodium chloride solution. Finally, if $NaOH \cdot \infty Aq$ is mixed with $HCl \cdot \infty Aq$, 13,345 calories of heat are liberated (see eq. 3.17), a quantity of heat that would also be liberated if $KOH \cdot \infty Aq$ were mixed, let us say, with $HBr \cdot \infty Aq$.

The facts just stated can all be explained by the ionic theory. The electrolytes mentioned are all assumed to be completely ionized in infinitely dilute solutions; hence any heat effects that might be observed should be attributable to reactions between ions. Since no ionic reactions occur between sodium chloride and potassium nitrate (after all, sodium nitrate and potassium chloride are also completely ionized), the heat effect of mixing this combination is zero. Silver ions, on the other hand, can react with chloride ions to form silver chloride, which will precipitate as a solid. This reaction and its enthalpy change at 25 °C can be represented by

$$Ag^+ \cdot \infty Aq + Cl^- \cdot \infty Aq = AgCl(c) + \infty Aq + 15.65 \, kcal \qquad (3.22)$$

and it does not matter what ions are originally conjugate to the silver and the chloride as long as the electrolytes are completely ionized and infinitely dilute.

The acid-base neutralization reaction can also be simply represented:

$$H^+ \cdot \infty Aq + OH^- \cdot \infty Aq = H_2O(liq) + \infty Aq + 13,345 \, cal \qquad (3.23)$$

Here again the heat effect is independent of the acids and bases used as long as no other ionization effects or reactions are involved.

Heats of Formation of Ions in Solution

From the preceding discussion we conclude that $NaCl \cdot \infty Aq$ really means $Na^+ \cdot \infty Aq + Cl^- \cdot \infty Aq$, and the heat of formation of $NaCl \cdot \infty Aq$ means the sum of the heats of formation of the separate ions in infinitely dilute solution. Since the solutions with which we are ordinarily concerned are all electrically neutral, positive and negative ions must be present in equivalent amounts; hence it is impossible by ordinary thermodynamic means to arrive at heats of formation of single ions. We can, however, by recourse to convention, prepare a table of heats of formation of ions in infinitely dilute solution, arbitrarily assigning some convenient value (zero) to the heat of formation of a convenient ion ($H^+ \cdot \infty Aq$). If we agree to say that $H^+ \cdot \infty Aq = 0$, then from equation 3.23 and the heat of formation of water we find that $OH^- \cdot \infty Aq = -54.97$ kilocalories. Combining equation 3.16 with the heat of formation of $HCl(g)$, we find that $Cl^- \cdot \infty Aq = -39.95$ kilocalories; combining the heat of formation of $NaOH \cdot \infty Aq$ with the value for the hydroxide ion, we can calculate $Na^+ \cdot \infty Aq$. By continuing this process indefinitely, we build up a complete table. By combining pairs of entries from the table, we can readily calculate heats of formation of electrolytes in infinitely dilute solution, a computation that will be valid as long as the substance is indeed completely ionized in solution. The combination of heats of formation of equivalent amounts of positive and negative ions will eliminate the element of arbitrariness introduced by the assignment of zero enthalpy to hydrogen ions. The heats of formation of a number of ions at infinite dilution are given in Table 3-2.

TABLE 3-2

Heats of Formation of Selected Ions in Aqueous Solution at Infinite Dilution

(Values given in kcal mol^{-1} relative to elements in standard states at 25 °C)

ION (∞ Aq)	ΔH_f	ION (∞ Aq)	ΔH_f
H^+	0	OH^-	-54.970
NH_4^+	-31.67	F^-	-79.50
Zn^{++}	-36.78	Cl^-	-39.952
Cu^+	17.13	Br^-	-29.05
Cu^{++}	15.48	I^-	-13.19
Ag^+	25.234	$S^=$	7.9
Fe^{++}	-21.3	$SO_3^=$	-151.9
Fe^{+++}	-11.6	$SO_4^=$	-217.32
Mg^{++}	-111.58	HS^-	-4.2
Ca^{++}	-129.74	HSO_3^-	-149.67
Sr^{++}	-130.45	HSO_4^-	-212.08
Ba^{++}	-128.50	NO_3^-	-49.56
Li^+	-66.55	$PO_4^=$	-305.3
Na^+	-57.43	$HPO_4^=$	-308.83
K^+	-60.27	$H_2PO_4^-$	-309.82
		$CO_3^=$	-161.84
		HCO_3^-	-165.39
		CH_3COO^-	-116.16

Additional data will be found in Appendix D.

EXERCISE:

The heat of formation of AgCl(c) at 25 °C is -30.37 kcal. Using heats of formation of ions from Table 3-2, calculate the heat effect of mixing $1 KCl \cdot \infty Aq$ with $1 AgNO_3 \cdot \infty Aq$.

Heats of Ionization

If one mixes equal amounts of 0.01N HCl and 0.01N NaOH, which solutions, for all practical purposes, are infinitely dilute, one observes the liberation of 13,345 calories per equivalent in accordance with equation 3.23. On the other hand, if equivalent amounts of moderately concentrated aqueous HOCl and dilute NaOH are mixed, it is found that only 10.0 kilocalories is liberated. The difference between the two figures can be attributed to the fact that hypochlorous acid is a weak electrolyte and hence is mostly in un-ionized form. Assuming that none of the hypochlorous acid is ionized, we can write the enthalpy equation representing the neutralization as

$$HOCl + OH^- = H_2O + OCl^- + 10.0 \text{ kcal} \qquad (3.24)$$

Assuming further that all ionic species appearing in equation 3.24 are at infinite dilution, we see by combining with equation 3.23 that

$$HOCl = OCl^- + H^+ - 3.3 \text{ kcal} \qquad (3.25)$$

From equation 3.25 we conclude that the heat of ionization of hypochlorous acid is approximately 3.3 kcal; this value is not exact, however, since no correction was made for the degree of ionization of the acid (which is not exactly zero), and also because the acid solution was not infinitely dilute. This type of calculation, with appropriate refinements, can be made for other un-ionized or partially ionized substances as well as for reactions involving the formation of insoluble materials.

Chemical-bond Enthalpies and Bond Energies

Let us consider a process in which a chemical molecule is completely dissociated into its separate gaseous atoms. The change in state accompanying such a process will involve a definite change in enthalpy, which can be regarded as the sum of the heats of dissociation of all of the chemical bonds that exist in the molecule. As an example, consider the molecule ethane, for which we can write the following enthalpy equation:

$$C_2H_6(g) = 2C(g) + 6H(g) - \Delta H_{bonds} \qquad (3.26)$$

The quantity ΔH_{bonds} is called the total *bond enthalpy** of ethane. From the structure of ethane we conclude that this quantity is equal to the energy of its carbon-carbon bond plus six times the average energy of its carbon-hydrogen bonds, plus $\Delta(pV)$, which in this case is $7RT$. Similarly, the heat of dissociation of gaseous water into its atoms will equal twice the bond energy of an oxygen-hydrogen bond plus $2RT$.

One can determine the total bond enthalpy of a molecule from a knowledge of the heats of formation of its gaseous atoms and the heat of formation of the molecule, using the thermochemical methods of calculation discussed earlier. Heats of formation of the gaseous atoms (relative to the elements in their standard states) can be established by various means, particularly by spectroscopic methods, which we conveniently use to evaluate heats of dissociation of diatomic molecules.

TABLE 3-3

Heats of Formation of Some Gaseous Atoms

(Values given in kcal mol^{-1} relative to elements in standard states at 25 °C)

ATOM(g)	ΔH_f	ATOM(g)	ΔH_f
O	59.553	N	112.979
H	52.095	P	75.20
F	18.88	C	171.291
Cl	29.082	Si	108.9
Br	26.741	Li	38.09
I	25.535	Na	25.60
S	66.636	K	21.33

Additional data will be found in Appendex D.

Table 3-3 gives the heats of formation of a number of the commoner atoms encountered in work on chemical structure.

By his work on the nature of the chemical bond, Pauling showed that the total bond enthalpy (or energy) of a molecule can be regarded as the sum of the values assigned to the individual chemical bonds within the molecule, such separate values being characteristic of the bond type and not of the particular molecule. Thus the enthalpy assigned to a carbon-carbon bond in an aliphatic hydrocarbon has very nearly the same value whether the bond appears in ethane or in iso-octane. Similarly, a carbon-

---

* This quantity is often called the total *bond energy* of the molecule. The enthalpy and energy would, of course, be the same except for $\Delta(pV)$. In many instances the correction is less than other errors involved.

hydrogen bond has the same average heat of dissociation whether the bond exists in benzene or in methylene chloride.*

TABLE 3-4

Enthalpies of Some Chemical Bonds

(Values given in kcal mol⁻¹ for use in enthalpy calculations for 25 °C)

BOND	ENTHALPY	BOND	ENTHALPY
O—O	34.3	F—P	117.
H—H	104.2	Cl—P	77.
O—H	110.8	Br—P	63.
F—F	37.8	C—C	82.
H—F	135.8	C=C	147.
Cl—Cl	58.2	C≡C	194.
H—Cl	103.2	O—C	84.
Br—Br	46.1	O=C	173.
H—Br	87.5	H—C	99.4
I—I	36.1	F—C	117.
H—I	71.3	Cl—C	78.
S—S	63.6	Br—C	65.
H—S	87.9	I—C	57.
N—N	38.	S—C	62.
N=N	100.	S=C	114.
N≡N	226.	N—C	70.
O—N	42.	N=C	147.
H—N	93.4	N≡C	213.
P—P	48.	Si—Si	54.4
O—P	88.	O—Si	111.4
H—P	76.7	H—Si	77.3
		F—Si	142.6

Values taken largely from books by Linus Pauling (*The Nature of the Chemical Bond* and *General Chemistry*) subject to some recalculations by the author using recent thermochemical data. The values involving silicon, in particular, and to some extent phosphorus, differ appreciably from those reported by Pauling.

Making use of Pauling's rule, we can calculate the heats of formation of compounds (approximately) from a knowledge of bond enthalpies and heats of formation of atoms. Average values of bond enthalpies, in-

---

* It should not be assumed that the successive stripping off of hydrogens from methane, for example, will involve the same heats of dissociation. The average of such individual energies, however, will be very close to what is called the carbon-hydrogen bond enthalpy.

tended for 25 °C calculations, are given in Table 3-4. Since bond energies (or enthalpies) are not necessarily the same for a given pair of atoms joined to each other in different kinds of molecules or, for that matter, different parts of the same molecule, the results of an enthalpy calculation using values from the table may be in error by perhaps 1 kcal mol^{-1}. Moreover, different molecules or molecular fragments have different thermal energies of rotation and vibration, which further complicates matters. In principle, the latter difficulty could be avoided by using bond energies valid at absolute zero and then making appropriate energy corrections for higher temperatures. Such a task can be an enormous one, however, and it is scarcely warranted for practical calculations intended to obtain molecular enthalpies at 25 °C.

In computing heats of formation from bond energies, one must provide for resonance or hybrid-bond character. From the point of view of thermodynamics, the calculations constitute a particularly interesting application of structural concepts. To the student of molecular structure, bond energies also provide data for characterizing bond types, thus contributing to a better understanding of structural problems in general.

EXERCISE:

The heat of formation of $S_8(g)$ at 25 °C is 24.45 kcal mol^{-1} and that of $S(g)$ is 66.636 kcal mol^{-1}. Calculate an enthalpy value for the S—S bond assuming S_8 to consist of a ring of eight sulfur atoms joined together by single bonds.

EXERCISE:

Estimate the heat of formation of ethane at 25 °C from values of the C—H and C—C bond enthalpies and the heats of formation of C(g) and H(g) at 25 °C. Compare the result with the value given in Table 3-1.

EXERCISE:

Estimate from bond-enthalpy values the heat of polymerization of 1 mole of ethylene to polyethylene of high molecular weight.

EXERCISE:

Using heats of formation of CH_4, H, and C from Tables 3-1 and 3-3, calculate a value for the bond enthalpy of the C—H bond.

EXERCISE:

Consider the following heats of formation at 25 °C:

N(g)	112.98 kcal
$NH_3(g)$	-11.02
$N_2H_4(g)$	22.80

Using these data only, calculate the enthalpy of the N—N bond. Note that in unnecessary to know the heat of formation of H(g) or the strength of the N—H bond.

Crystal Energies

Just as it is possible to envision the complete dissociation of a covalent chemical molecule into its separate neutral atoms, so, too, one can picture an ionic crystal undergoing dissociation into separated gaseous ions. Since many crystals, like those of the alkali halides, are known to be made up of ions arranged in definite lattice structures, it is quite reasonable and natural to inquire into the thermochemistry of dissociations like that represented by

$$NaCl(c) \longrightarrow Na^+(g) + Cl^-(g)$$

The value of ΔH for the above change in state can be determined from a knowledge of the enthalpies of formation of the ions and of the compound. (See Table 3-5 for gaseous ion enthalpies.) The connection between the various quantities is most conveniently summarized, however, by use of a thermochemical cycle known as the Born-Haber cycle. For a salt of the form MX, such as an alkali halide, the Born-Haber cycle is expressed as follows:

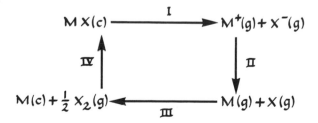

Process I involves the aforementioned dissociation of the crystal into its gaseous ions. Associated with it will be an enthalpy change, ΔH_c, and an energy change, ΔU_c, which is commonly called the lattice or crystal energy of the substance. The heat of process II will be equal to the

difference between two quantities, the electron affinity, E, of the nonmetal and the ionization potential, I, of the metal. Process III will
liberate the heat of sublimation, ΔH_{sub}, of the metal plus half of the
heat of dissociation of the nonmetal, which is equal to ΔH_X, the heat
of formation of gaseous atoms from the diatomic element. Process IV
involves the heat of formation of the crystal from the pure elements in
their standard states. By setting the total enthalpy change for the cycle
equal to zero, one obtains the following equation:

$$\Delta H_c + E - I - \Delta H_{sub} - \Delta H_X + \Delta H_{MX} = 0 \qquad (3.27)$$

Except for E and I, all of the quantities are enthalpy changes; the electron affinity and ionization potential are generally taken as the values of
the energy changes at 0 K, but no appreciable error is introduced
thereby, for the change in the kinetic energy of translation associated
with the *pair* of reactions represented by process II will be zero, and, of
course, $\Delta(pV)$ will likewise be zero.

TABLE 3-5

Heats of Formation of Some Gaseous Ions

(Values given in kcal mol^{-1} relative to elements in standard states at 25 °C)

ION(g)	ΔH_f	ION(g)	ΔH_f
H$^+$	367.16	H$^-$	33.39
Zn^{++}	665.09	OH$^-$	−33.67
Ag$^+$	243.59	F$^-$	−64.7
Mg^{++}	561.30	Cl$^-$	−58.8
Ca^{++}	460.29	Br$^-$	−55.9
Sr^{++}	427.96	I$^-$	−47.0
Ba^{++}	396.86	CN$^-$	16.0
Li$^+$	163.90		
Na$^+$	145.59		
K$^+$	122.91		
Rb$^+$	117.14		
Cs$^+$	109.45		

Additional data will be found in Appendix D.

Originally, equation 3.27 was used to estimate the electron affinities
of the halogens, since there were no good experimental methods for
getting at them directly. All of the other quantities could be determined

experimentally, except for the lattice enthalpy, which was calculated theoretically by assuming Coulomb forces between the ions and by making appropriate corrections for ultimate repulsion when the ions were brought into sufficiently close contact. In making such a calculation, one must take into account the fact that the process of dissociating the crystal does not merely involve splitting up single molecules, but entails breaking up the entire lattice, which is made up of ions that cannot individually be identified with any particular molecules. In the sodium chloride crystal, for example, each sodium ion is surrounded by six chlorides as nearest neighbors, and, similarly, each chloride ion has about it six sodiums. If we consider the interactions of nearest neighbors only, the process of dissociating a crystal of sodium chloride will involve breaking up six Coulomb attractions per mole of crystal and not just one. The second-nearest neighbors of any ion, however, will be twelve ions of the same charge, which will contribute toward a positive potential energy. The third-nearest neighbors will again have opposite charges. By summing over the entire crystal, with its infinite number of attractions and repulsions, we find the total ionic energy per molecule of sodium chloride to be greater than that expected for a single ion pair but considerably less than for six ion pairs. The factor by which one should multiply the potential energy of a single ion pair to obtain the crystal energy is known as the Madelung constant, a number whose magnitude depends upon the crystal structure and which can be calculated precisely from geometrical considerations only. If r is the distance between two univalent ions subject to Coulomb forces, the potential energy of the ions will be given by $-e^2/r$, in which e is the charge on the electron. If A denotes the Madelung constant, the energy per molecule attributable to Coulomb forces will be given by $-Ae^2/r$.

Taking into account the fact that forces of repulsion must ultimately enter into the picture to account for a stable equilibrium separation of the ions, we must modify the expression above by a factor that equals $(1 - 1/n)$ if the repulsive potential varies inversely as the nth power of the internuclear separation. The number n has values ranging from about 6 to 12, depending upon the salt. For Avogadro's number, N_A, of ion pairs, the total crystal energy, ΔU_c, of a uni-univalent salt will be given by the expression

$$\Delta U_c = \frac{N_A A e^2}{R} (1 - 1/n) \qquad (3.28)$$

in which R is the equilibrium separation between nearest-neighbor pairs of unlike ions. In writing equation 3.28, we assume that the change in thermal energy attending the dissociation of the crystal is zero. This is not strictly correct, but the error should not be serious. Neglecting the pressure-volume product of the solid and assuming the gaseous ions to be perfect gases, we find, for the dissociation, that $\Delta(pV) = 2RT$, so that

$$\Delta H_c = \frac{N_A A e^2}{R} (1 - 1/n) + 2RT \tag{3.29}$$

At room temperature, $2RT$ equals about 1.2 kilocalories, a relatively small correction. Although more refined calculations can be made, the essential features are illustrated above.

EXERCISE:

Using data from Tables 3-1 and 3-5, calculate the lattice enthalpy of one mole of KCl(c) at 25 °C.

Heats of Reaction at Different Temperatures

Consider a chemical reaction symbolized by the general equation

$$R(p,T) \longrightarrow P(p,T)$$

in which R denotes the reactants and P the products, each at pressure p and temperature T. Let us suppose that we know the enthalpy change of this reaction at temperature T_1 and that we seek to learn what the enthalpy change for the same reaction would be at some other temperature, T_2. We can do this most conveniently by application of the First Law to pairs of processes involving the reaction at each of the two temperatures of interest. Such a combination of processes is represented by the following scheme, in which we consider the reaction at each of the temperatures, T_1 and T_2, and the heating (or cooling) of the reactants and products between those temperatures:

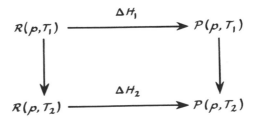

If ΔH_1 and ΔH_2 denote the enthalpies of the reactions at the two temperatures, and if $C_p(P)$ and $C_p(R)$ denote the total heat capacities of the products and the reactants at constant pressure, we note that

$$\Delta H_1 + \int_{T_1}^{T_2} C_p(P)dT = \int_{T_1}^{T_2} C_p(R)dT + \Delta H_2 \qquad (3.30)$$

Letting ΔC_p equal the difference between the heat capacities of the products and the reactants, we observe, upon rearranging equation 3.30, that

$$\Delta H_2 = \Delta H_1 + \int_{T_1}^{T_2} \Delta C_p \, dT \qquad (3.31)$$

If the change in temperature from T_1 to T_2 involves a transition, such as a fusion or an evaporation, appropriate enthalpies of transition must be incorporated into the expression. An illustration of this is seen in connection with the calculation of the heat of combustion of hydrogen and oxygen at one atmosphere pressure at a temperature greater than 100 °C. Using, for example, 25 °C for reference purposes, we can represent the problem as follows:

Application of the First Law to this set of processes gives the following expression for the heat of combustion at temperature T (assumed to be greater than 373° K):

$$\Delta H = \Delta H_{298} - \int_{298}^{T} C_p(R)dT + \int_{298}^{373} C_p(P)dT + \Delta H_{vap} + \int_{373}^{T} C_p(P)dT \quad (3.32)$$

The foregoing arguments with respect to enthalpies of reaction can be applied with equal validity to energies of reaction provided appropriate heat capacities are used for the calculations. If we are interested in the heats attending constant-volume reactions at each of two temperatures, we can establish a relationship between the ΔU's by using heat capacities at constant volume together with such energies of transition as may be required.

We can obtain an expression for the enthalpy of a reaction as a function of temperature by dropping the subscript 2 in equation 3.31 and integrating to an indefinite temperature. If we assume that ΔC_p can be represented in a form suggested by equation 2.48,

$$\Delta C_p = \Delta a + \Delta b\,T + \Delta c\,T^2 + \ldots \tag{3.33}$$

then

$$\Delta H = \Delta H_0 + \Delta a\,T + (\Delta b/2)\,T^2 + (\Delta c/3)\,T^3 + \ldots \tag{3.34}$$

in which expression

$$\Delta H_0 = \Delta H_1 - \Delta a\,T_1 - (\Delta b/2)\,T_1^2 - (\Delta c/3)\,T_1^3 - \ldots \tag{3.35}$$

It is evident that an expression of the type of equation 3.34 can always be established if all of the heat capacities are known as functions of temperature and if at least one value of the reaction enthalpy is available to determine ΔH_0, which is, in effect, a constant of integration. If the heat capacities are not known, an expression like equation 3.34 can still be established by recourse to direct measurements of ΔH at different temperatures, followed by empirical evaluation of the several coefficients.

EXERCISE:

According to a generalization known as Kopp's Law, the molar heat capacity of a solid is equal to the sum of values characteristic of the atoms that make up the material. Show that, if Kopp's Law holds, the temperature coefficient of the heat of a reaction involving solids is equal to zero.

EXERCISE:

The molar heat capacities of $Na(g)$, $Cl_2(g)$, and $NaCl(c)$ at room temperature are respectively 6.73, 8.10, and 11.98 cal deg^{-1}. The heat of formation of $NaCl(c)$ at 25 °C is -98.279 kcal. Calculate the heat of formation of $NaCl(c)$ at 35 °C.

Temperatures of Flames or Explosions

The preceding considerations can be simply modified if we wish to calculate the maximum temperature attainable from a flame at constant pressure or within an explosion at constant volume. To calculate such a maximum temperature, we shall assume that no heat is lost to the surroundings; that is, the combustion takes place adiabatically. With this hypothesis, the heat that might have been liberated by the combustion can be assumed to be used for heating the products of the reaction; hence by equating such heat quantities to appropriate integrals of the heat capacities, one can formulate the necessary equations. In particular, if one wants to determine the maximum temperature attending a combustion at constant pressure, one must equate the heat liberated under isothermal conditions $(-\Delta H)$ to an integral of the total heat capacity of the products, integrating from the initial temperature to the (unknown) maximum flame temperature. This integration will provide an equation that, at least in principle, can be solved for the maximum temperature. The total heat capacity of the products must include the heat capacities of all inert materials that might be carried along in the combustion. If one burns hydrogen in air, for example, the heat capacity of the nitrogen as well as those of the argon and other inert gases must be included in the computation since they must all be heated with the water vapor resulting from the combustion.

Similar considerations apply to calculations of the maximum temperature of an explosion within a bomb except that the total heat capacity to be used must be made up of sums of heat capacities at constant volume. Other situations, which might require other kinds of heat capacities, can be imagined. In all calculations of this type it is important to take heats of transition into account if such transitions occur in the course of heating the products of the combustion.

Problems

1. An evacuated tank of 10-liter capacity is surrounded by a thermostat at 100 °C. Inside the tank is a capsule containing exactly 1 mole of liquid water at 100 °C. The capsule is now broken, and the water is allowed to flash into vapor until the pressure is 1 atm. Neglecting the volume of the capsule and the volume of any remaining liquid water, and assuming the vapor to be an ideal gas, calculate ΔU, ΔH, Q, and W for the actual process. At 100 °C and 1 atm, $\Delta H_{vap} = 9{,}720$ cal mol^{-1}.

2. (A) Ten grams of ice at -10 °C are dropped into 25 g of water at 15 °C, and the system is allowed to reach equilibrium adiabatically. What is the final state of affairs? Assume the specific heat capacities of water and ice to be 1 and 0.5 cal deg^{-1} and the heat of fusion of ice to be 80 cal g^{-1}.

   (B) Suppose that 10 g of ice at -10 °C were dropped into 100 g of water at 15 °C. What would be the final state of affairs?

3. One-tenth (0.1000) of a mole of liquid benzene, C_6H_6(liq), is burned in a bomb calorimeter at an initial temperature of 20 °C. If the heat of combustion (ΔH) of liquid benzene at 20 °C is -782.3 kcal mol^{-1}, and if the heat capacity of the calorimeter and its contents is 5,000 cal deg^{-1}, what rise in temperature will be observed in the calorimeter if radiation losses are negligible? Make any reasonable assumptions or approximations that may appear necessary, but be sure to retain four significant figures in your answer.

4. At 25 °C the molar heat of combustion of benzoic acid, $C_6H_5COOH(c)$, is -771.4 kcal. The heats of formation of $CO_2(g)$ and $H_2O(liq)$ at 25 °C are -94.05 and -68.32 kcal respectively.

   (A) Calculate the heat of formation of $1C_6H_5COOH(c)$.

   (B) How much heat is liberated if 1 mole of benzoic acid is burned in a bomb calorimeter at 25 °C?

5. Given the following heats of formation at 25 °C:

$$
\begin{array}{ll}
Na_2SO_4 \cdot \infty \, Aq & -331.5 \text{ kcal} \\
Na_2SO_4 \cdot 10H_2O(c) & -1,033.5 \\
H_2O(liq) & -68.32
\end{array}
$$

   Calculate the heat absorbed when $1Na_2SO_4 \cdot 10H_2O(c)$ is dissolved in an infinite amount of water.

6. One mole of sulfur dioxide, $SO_2(g)$, is dissolved in an infinitely dilute solution containing two moles of sodium hydroxide, $2NaOH \cdot \infty \, Aq$, all at 25 °C. Calculate ΔH for the process, using the following enthalpy data:

$$
\begin{array}{ll}
OH^- \cdot \infty \, Aq & -55.0 \text{ kcal} \\
SO_3^= \cdot \infty \, Aq & -151.9 \\
SO_2(g) & -70.9 \\
H_2O(liq) & -68.3
\end{array}
$$

7. In the manufacture of water gas, steam is passed through hot coke, and the following reactions occur:

$$
\begin{array}{l}
C + H_2O \longrightarrow CO + H_2 \text{ (principally)} \\
CO + H_2O \longrightarrow CO_2 + H_2 \text{ (to a small extent)}
\end{array}
$$

   The water gas so obtained, when cooled to room temperature, can be considered a mixture of CO, H_2, and a little CO_2 if we neglect the water vapor.

   (A) If only the first reaction occurred during manufacture, what would be the heat of combustion at 25 °C and 1 atm per liter of water gas if all the products of combustion were in the form of vapor?

   (B) If 95 percent of the carbon should appear in CO and 5 percent in CO_2, what would be the heat of combustion per liter of gas? (Note the additional H_2 that accompanies the formation of CO_2.)

8. In the manufacture of producer gas, dry air is passed over hot coke, and the following reactions occur:

$$
\begin{array}{l}
C + \tfrac{1}{2}O_2 \longrightarrow CO \\
CO + \tfrac{1}{2}O_2 \longrightarrow CO_2
\end{array}
$$

   What is the heating value in calories per liter of producer gas at 25 °C if 95 percent of the oxygen of the air ends up in CO and 5 percent in CO_2? Assume air to consist of 20 mole percent oxygen and 80 mole percent nitrogen.

9. Some liquid water is carefully super-cooled to a temperature of -3 °C. The system is now disturbed so that the water tends to freeze. If the heat of fusion of ice at 0 °C is 80 cal g^{-1} and the specific heat capacity of water is 1.0 cal deg^{-1}, what percentage of the water will freeze when the system is brought adiabatically to a state of equilibrium at 0 °C?

10. Natural gas, which you may assume to be pure methane, is burned in a furnace employed for space-heating purposes.

   (A) If twice the minimum amount of air is used for the combustion, and if the methane and air are initially at 20 °C and the stack gases, including all of the water as vapor, are at 100 °C, how much heat is liberated per mole of methane burned? Assume that air consists of 20 mole percent oxygen and 80 mole percent nitrogen. The molar heat of combustion of methane at 20 °C is -212.91 kcal, and the heat of vaporization of water at that temperature is 10.57 kcal. For this problem you may assume the molar heat capacities in cal deg^{-1} to have the following constant values: $O_2(g)$, 7.02; $N_2(g)$, 6.96; $H_2O(g)$, 8.03; $CO_2(g)$, 8.87.

   (B) How much heat is liberated if the stack gases are cooled just to the point where they become saturated with water vapor, but without condensation? (See a handbook for vapor pressures of water.)

11. Calculate the heat of mixing of $2(HCl \cdot 50Aq)$ with $1(HCl \cdot 200Aq)$ at 25 °C. If $3HCl(g)$ is now dissolved in the resulting solution, what is the additional heat effect? Use the following enthalpy data, which are valid at 25 °C:

$HCl(g)$	-22.06 kcal
$HCl \cdot 50Aq$	-39.52
$HCl \cdot 100Aq$	-39.66
$HCl \cdot 200Aq$	-39.74

12. Estimate the heat of the following reaction at 25 °C by the use of bond enthalpies:

   $$CH_4(g) + Cl_2(g) \longrightarrow CH_3Cl(g) + HCl(g)$$

13. Suppose that the potential energy of a pair of ions with opposite unit charges (like Na^+ and Cl^-) is given by the equation

   $$V = -\frac{e^2}{r} + \frac{C}{r^n}$$

   in which e is the electronic charge, C is a constant, and r is the internuclear separation of the ions. (The first term is the well known Coulomb attraction and the second is a repulsive potential.) If V is at a mimimum when $r = R$, show that the value of that minimum is given by

   $$V_{min} = -\frac{e^2}{R}(1 - 1/n)$$

   This constitutes a part of the derivation of equation 3.28.

14. Calculate the lattice enthalpy of NaCl(c), using equation 3.29. The Madelung constant for the sodium chloride structure is 1.748 when used in conjunction with R, the distance between nearest neighbors ($R = 2.81$ A for NaCl). The repulsive exponent, n, appearing in equation 3.29, can be taken as equal to 8.

15. At 25 °C and 1 atm the heat of formation of HCl(g) is -22.06 kcal mol^{-1}. Deduce an expression for the heat of formation as a function of temperature, using heat capacities from Table 2-1.

16. One-tenth mole of carbon monoxide is mixed with 1/20 mole of oxygen and 1/5 mole of nitrogen and exploded in a bomb calorimeter. What is the maximum possible temperature resulting from the explosion if the original temperature is 25 °C? Use heat-capacity expressions from Table 2-1. (Use successive approximations to solve the resulting cubic equation.)

17. The Deacon process, formerly used for making chlorine, involves the oxidation of hydrogen chloride by oxygen as follows:

$$2HCl(g) + \tfrac{1}{2}O_2(g) \longrightarrow H_2O(g) + Cl_2(g)$$

Using heat capacities expressed by data in Table 2-1 and heats of formation at 25 °C from Table 3-1, derive expressions for the heat of the Deacon reaction as a function of temperature.

18. A number of gases, including the "inert" gases, curiously form solid hydrates with heats of formation (at 25 °C) as follows:

$Ar \cdot 5H_2O(c)$	-357.2 kcal
$Kr \cdot 5H_2O(c)$	-357.1
$Xe \cdot 6H_2O(c)$	$-428.$
$Cl_2 \cdot 8H_2O(c)$	-564.4
$CH_4 \cdot 6H_2O(c)$	$-445.$

Calculate the heats of hydration of Ar, Kr, Xe, Cl$_2$, and CH$_4$ with liquid water, and note how similar the numbers are.

19. Xenon forms a hexafluoride, XeF$_6$, the solid form of which has an enthalpy of formation equal to -88 kcal mol^{-1}. If the molar heat of sublimation of XeF$_6$ is 14.9 kcal, what is the average strength of the xenon-fluoride bond?

20. (A) Calculate the strength of the C—Cl bond using no more data than the following enthalpies of formation at 25 °C:

C(g)	171.291
Cl(g)	29.082
CH$_4$(g)	-17.88
CHCl$_3$(g)	-24.65

(B) Estimate the heat of formation of CH$_2$Cl$_2$(g) using only the heats of formation of CH$_4$(g) and CHCl$_3$(g). Compare the answer so obtained with the value reported in the literature, namely, -22.10 kcal. Note that it is unnecessary to know any heats of formation of atoms or any bond enthalpies to solve this part of the problem.

4

The Second Law
of Thermodynamics

Conversion of Heat into Work

It has already been shown that certain systems can be carried through processes that will produce work while absorbing heat. For example, during the isothermal expansion of an ideal gas, which is a constant-energy process, a certain amount of heat is added to the system, and an exactly equivalent amount of work appears in the surroundings. For such a process, since the energy of an ideal gas is a function of temperature only, we can say that heat is converted into work.

Although heat can be converted into work by isothermal processes of the kind just described, such processes by themselves are of little use for constructing heat engines, since for practical reasons heat engines must operate in cycles. A cycle necessarily calls for restoration of the system to its initial condition. If one directly reverses the isothermal expansion of an ideal gas, a quantity of work at least equal to that which was obtained in the expansion will be required for the compression. It is obvious, therefore, that such a cycle, involving direct reversal, would be of no use as a basis for making an engine.

Nevertheless, in accordance with the First Law of Thermodynamics, it is conceivable that cycles for converting heat into work can be designed. That such cycles exist, subject to certain limitations with respect to their efficiencies, can be readily demonstrated by considering a particular kind of cycle, the Carnot cycle.

Work Attending a Reversible Carnot Cycle for an Ideal Gas

A Carnot cycle is a cycle consisting of two isothermal and two adiabatic steps. We shall now calculate the work appearing in the surroundings when one mole of an ideal gas is carried through the reversible Carnot cycle illustrated in Figure 4-1. Starting at pressure p_1, volume V_1, and temperature T_1, the gas is first expanded reversibly and isothermally

FIGURE 4-1 p-V representation of a reversible Carnot cycle for an ideal gas. Steps I and III represent isothermal processes at T_1 and T_2. Steps II and IV represent adiabatic processes. (The adiabatics for this cycle were calculated for $\gamma = 5/3$— that is, for a monatomic gas.)

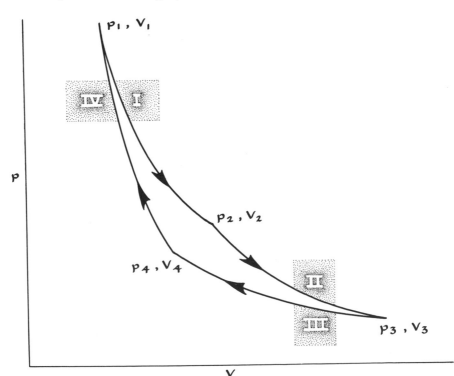

to a state characterized by p_2, V_2, and T_1. After this expansion the gas is further expanded adiabatically to a condition represented by p_3, V_3, and T_2; this is followed by an isothermal compression to p_4, V_4, and T_2, which point is so chosen that it can be joined to the initial point of the cycle by an adiabatic compression that will constitute the last stage of the cycle. Utilizing equations previously developed for reversible isothermal and adiabatic changes of ideal gases, we now write for each of the steps, denoted by I, II, III, and IV, the energy changes, the work, and the heat as follows:*

$$\Delta U_I = 0; \; Q_1 = W_I = RT_1 \ln \frac{V_2}{V_1} \tag{4.1}$$

$$\Delta U_{II} = -W_{II} = C_v(T_2 - T_1); \; Q_{II} = 0 \tag{4.2}$$

$$\Delta U_{III} = 0; \; Q_{III} = W_{III} = RT_2 \ln \frac{V_4}{V_3} \tag{4.3}$$

$$\Delta U_{IV} = -W_{IV} = C_v(T_1 - T_2); \; Q_{IV} = 0 \tag{4.4}$$

In writing equations 4.2 and 4.4 we have assumed that C_v is constant. This assumption is not really necessary, for all we need to establish is that $W_{II} + W_{IV} = 0$, a conclusion derivable from the fact that the energy of an ideal gas is a function of temperature only. By virtue of equation 2.50 we can further assert that

$$\left(\frac{T_1}{T_2}\right)^{C_v/R} = \frac{V_3}{V_2} = \frac{V_4}{V_1} \tag{4.5}$$

Here again we have assumed, for convenience, not from necessity, that C_v is constant. Upon rearrangement of equation 4.5 we observe that

$$\frac{V_4}{V_3} = \frac{V_1}{V_2} \tag{4.6}$$

This equation enables us to write W_{III} in the following form:

$$W_{III} = -RT_2 \ln \frac{V_2}{V_1} \tag{4.7}$$

Upon adding together all of the work terms for the cycle, we conclude that the net work is equal to

$$W = R(T_1 - T_2) \ln \frac{V_2}{V_1} \tag{4.8}$$

Assuming that $T_2 < T_1$ and $V_1 < V_2$, we see that Q_I is a positive quantity equal to the heat added to the system at T_1 whereas $-Q_{III}$ equals the heat extracted from (or rejected by) the system at T_2. From the point

---

* The reader will also find it helpful to refer to Fig. 4-2.

of view of the machine operator, who is concerned with how much work he can get from a given amount of heat, the ratio W/Q_{I} becomes of prime importance. This ratio, which is called the efficiency of the cycle, simply turns out to be

$$\frac{W}{Q_{\mathrm{I}}} = \frac{T_1 - T_2}{T_1} \tag{4.9}$$

Evidently this efficiency depends only upon the temperatures of the isothermal processes of the cycle. Since we assumed our working substance was an ideal gas, the calculated efficiency, so far as we now know, may be characteristic of such a gas. Indeed, so far as the First Law is concerned, it is conceivable that working substances other than ideal gases can give rise to other efficiencies; this important possibility will be considered presently.

EXERCISE:

A general trend in the construction of engines for the generation of power is toward higher temperatures of internal operation and as low temperatures as possible for the discharged products. What is the thermodynamic justification for this?

EXERCISE:

Draw a diagram depicting a reversible Carnot cycle for an ideal gas, using temperature and volume as variables.

EXERCISE:

Draw a diagram for a reversible Carnot cycle for an ideal gas, using $\ln p$ and $\ln V$ as coordinates.

The Second Law of Thermodynamics

We shall now consider another basic principle that can be regarded as a second postulate for our thermodynamic development. This principle, known as the Second Law of Thermodynamics, can be formulated in many ways, but we shall here use a statement similar to that employed by Planck: *It is impossible to construct a machine operating in cycles that will convert heat into work without producing any other changes in the surroundings.* The ultimate justification for the Second Law, like that for the First Law, rests with the fact that conclusions derived from it, either directly or indirectly, are in accord with experience and observation. In short, nobody has been able to violate the law.

Let us now examine Planck's statement, in the light of examples already considered, to see precisely what it means and what kinds of restrictions it imposes. At the beginning of the chapter we pointed out that heat could be converted into work by the isothermal expansion of an ideal gas. At first glance this might appear to be a violation of the Second Law, but further examination discloses that it is not. In the isothermal expansion of an ideal gas the system undergoes a net change and a cycle is not completed. The Second Law does not rule out conversion of heat into work if other changes take place, although it does rule out conversion of heat into work if everything else remains the same. Hence the example just cited does not constitute a violation of the Second Law.

Let us now consider the operation of a Carnot cycle, which, at least for an ideal gas, brings about the appearance of work without producing a net change in the working material subjected to the cycle. In this case (see Fig. 4-2), although a quantity of heat, Q_I, has been removed from a reservoir at temperature T_1 and a net quantity of work, W, has appeared in the surroundings, another very important change has accompanied the operation—namely, the transfer of another quantity of heat, $-Q_{III}$, to another heat reservoir at temperature T_2. Thus the operation of the cycle does give rise to work, but simultaneously some heat is discharged

FIGURE 4-2 Schematic representation of work and heat quantities involved in a Carnot cycle. "Engine" denotes the mechanism, including the gaseous working substance, that undergoes the cycle. An arrow marked W, pointing away from the engine, denotes work appearing in the surroundings, and so on.

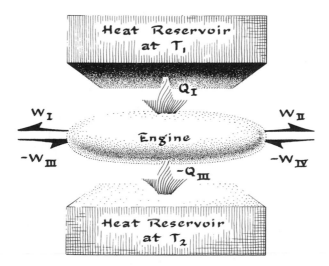

into a low-temperature reservoir to produce the other change in the surroundings required by the Second Law. Hence a cycle involving extraction of heat at one temperature and discharge of a portion thereof at a lower temperature can be accompanied by the appearance of work subject, of course, to the First Law of Thermodynamics, which requires equality of the net total heat and work.

A machine operating in cycles in accordance with the First Law, but otherwise intended to convert heat into work without producing any other changes in the surroundings, is said to be a *perpetual-motion machine of the second kind*. If it could be constructed, such a perpetual-motion machine, which is ruled out by the Second Law, might be fully as useful as one of the first kind. Numerous attempts have been made to design perpetual-motion machines, the more naive of which are in violation of the First Law. The more sophisticated attempts take cognizance of the First Law but try to get round the Second Law, often by subtle methods that are not readily detected. This is an outgrowth of the fact that certain aspects of the First Law are more generally understood than the Second Law, the significance of which often escapes comprehension.

Efficiency of Reversible Carnot Cycles for Any Substance

Near the beginning of this chapter, we considered, for illustrative purposes, an ideal gas subjected to a reversible Carnot cycle. Actually any gas, or indeed any substance whatsoever, might conceivably be put through a Carnot cycle, which simply calls for two isothermals and two adiabatics. Taking cognizance of this possibility, we shall now prove an important theorem, based on the Second Law, to the effect that any two substances carried through reversible Carnot cycles operating between the same pair of temperatures must exhibit the same thermodynamic efficiency. To avoid the risk of inadvertently bringing in conclusions dependent upon the properties or existence of ideal gases, we shall now temporarily forgo any reference to ideal gases or even to the ideal-gas temperature scale.

Let us consider two substances, A and B, each capable of being carried through the reversible cycles illustrated in Figure 4-3. In each instance the first isothermal will be carried out at the same temperature, which we shall denote by t_1, using an arbitrary scale. (The actual value of t_1 is unimportant, for we need only assure ourselves that the temperature is the same for both substances.) We shall further stipulate that the quan-

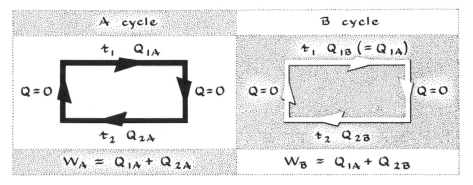

FIGURE 4-3 Schematic representation of reversible Carnot cycles for two different substances, A and B, operating between the same pair of temperatures, t_1 and t_2, with $Q_{1A} = Q_{1B}$.

tity of heat, Q_{1A}, associated with the first isothermal for substance A shall be exactly equal to Q_{1B}, the corresponding quantity for substance B. Each substance, after its first isothermal, is subjected to an adiabatic cooling step of such character that they both end at some convenient temperature, t_2, that must be the same for both substances, though we need not know its actual value. Next we shall isothermally compress each substance to those points which, when followed by adiabatics, will restore them to their initial conditions. The isothermals at t_2 will involve heat quantities Q_{2A} and Q_{2B}, which on an *a priori* basis might be unequal. The work quantities attending the two cycles, W_A and W_B, will then be given, according to the First Law, by the following equations:

$$W_A = Q_{1A} + Q_{2A} \qquad (4.10)$$
$$W_B = Q_{1A} + Q_{2B} \qquad (4.11)$$

We shall now prove that these work quantities must be equal, else the Second Law of Thermodynamics be violated. To do so, let us first consider the possibility that W_A is greater than W_B, which would mean that the efficiency of cycle A is greater than that of cycle B. If W_A is greater than W_B, let us construct a composite engine consisting of engines A and B constrained to operate in such a way that cycle A is carried out in a forward direction while cycle B is carried out backward (see Fig. 4-4). For this combination of cycles a net quantity of work, $W_A - W_B$, will appear in the surroundings, and simultaneously a net quantity of heat, $Q_{2A} - Q_{2B}$, will have been extracted from the low-temperature reservoir. No net change in the high-temperature

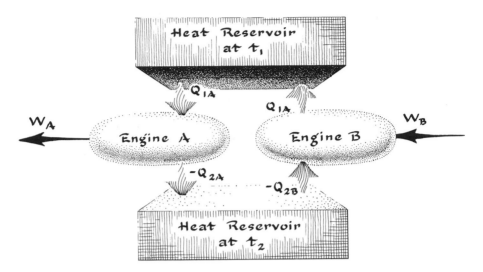

FIGURE 4-4   Diagram representing operation of engine A in forward direction and of engine B in reverse. Note that, if the two engines are consolidated, the heat reservoir at temperature t_1 can be eliminated. If W_A were greater than W_B, a quantity of work, $W_A - W_B$, would appear in the surroundings while an equivalent amount of heat would be extracted from the heat reservoir at temperature t_2; this would violate the Second Law.

reservoir shall have occurred, for the heat extracted from it by substance A is exactly replaced by an equivalent amount of heat discharged in the reverse of cycle B. The net result of the operation of the composite engine is the conversion of a quantity of heat, $Q_{2A} - Q_{2B}$, into an equivalent amount of work, $W_A - W_B$, without other changes in the surroundings. Since this is clearly a violation of the Second Law of Thermodynamics, our original premise must be in error, and hence we conclude that W_A cannot be greater than W_B. Similarly, if we assume that W_B is greater than W_A, we can imagine the construction of a composite engine involving the forward operation of cycle B and the reverse operation of cycle A. This would also give rise to the conversion of a certain quantity of heat, $Q_{2B} - Q_{2A}$, extracted from the low-temperature reservoir, into a net quantity of work, $W_B - W_A$, appearing in the surroundings. This, too, would violate the Second Law of Thermodynamics, so we conclude that W_B cannot be greater than W_A. The work quantities must, accordingly, be the same, and, as a corollary thereto, Q_{2A} must equal Q_{2B}. Recognizing this fact, we see that the operation of the composite engine just envisioned will produce zero work and likewise bring about no changes in the surroundings.

Thermodynamic Scale of Temperature

In the proof of the preceding section, no direct use was made of the temperatures, provisionally represented by t_1 and t_2, that served to characterize the two heat reservoirs. It was shown, however, that the efficiency of a reversible Carnot cycle—and hence the ratios Q_{1A}/Q_{2A} and Q_{1B}/Q_{2B}—must depend only on these temperatures and not upon the working substances. The fact that the ratio Q_1/Q_2 (we can now drop the subscripts A and B) depends only on temperatures t_1 and t_2, and not on the substance used, suggests a logical definition of absolute temperature, one not requiring the concept of an ideal gas. If we simply agree to define the ratios of the absolute temperatures as the ratios of the absolute values of the corresponding heat quantities in a reversible Carnot cycle, we arrive at a scale that is fully compatible with the ideal gas scale without assuming the existence of such a gas. Specifically, we shall now agree to define absolute temperature in the following way:

$$\frac{T_1}{T_2} = -\frac{Q_1}{Q_2} = \frac{|Q_1|}{|Q_2|} \tag{4.12}$$

In accordance with this definition of absolute temperature, the efficiency of any reversible Carnot cycle is given by the equation

$$\frac{W}{Q_1} = \frac{Q_1 + Q_2}{Q_1} = \frac{T_1 - T_2}{T_1} \tag{4.13}$$

which is, of course, precisely the same as that obtained for an ideal gas using the ideal-gas temperature scale. We conclude, therefore, that the thermodynamic scale just defined is equivalent to the ideal-gas scale, but we draw logical satisfaction from the fact that the concept of an ideal gas is not necessary, although it will continue to remain convenient in our theoretical deductions.

The above definition of absolute temperature provides only a relative scale since nothing has been said about the size of a degree. It was formerly stipulated that 100 units would represent the difference between the temperatures of melting ice and boiling water at one atmosphere pressure. If the two qualitative temperatures, t_1 and t_2, correspond to boiling water and melting ice, their absolute temperatures would then be given by the equations

$$T_2 = -\frac{100Q_2}{Q_1 + Q_2} = -\frac{100}{(Q_1/Q_2) + 1} \tag{4.14}$$

and

$$T_1 = \frac{100Q_1}{Q_1 + Q_2} = \frac{100}{1 + Q_2/Q_1} \tag{4.15}$$

Present practice, however, is to define the size of the degree by setting the absolute ice point equal to 273.15 K. This has the advantage that improved measurements of relative temperatures cannot affect the absolute ice point and will not significantly change temperatures near absolute zero. On the other hand, the boiling point of water under this system is no longer necessarily equal to 100 °C, although departures from it will be quite inconsequential. (See Prob. 5, Chap. One.) The newer practice is particularly important in connection with cryogenic work, but it makes little practical difference in thermodynamic measurements around room temperature.*

Definition of Entropy

Just as the First Law of Thermodynamics suggested a definition of a function of the state of the system, to which we gave the name "energy," so, too, the Second Law suggests, less obviously, the definition of still another function. The procedure we shall employ to prove the existence of this new function is not complicated in principle, but it is somewhat involved and does require careful analysis. Our proof will consist of showing that for a reversible process a certain differential expression, to be denoted by dQ_{rev}/T, is an exact differential—that is, the differential of some function. Since it is by no means obvious that dQ_{rev}/T will be exact, a proof is definitely called for.

FIGURE 4-5 Reversible cycle (in V-T diagram) consisting of three isothermals and three adiabatics. The broken line shows how the complete cycle can be regarded as a composite of two Carnot cycles.

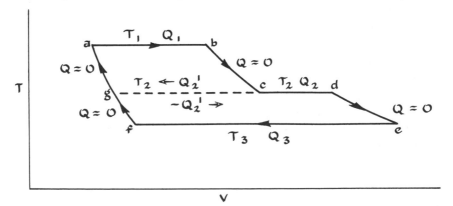

---

*W. F. Giauque, *Nature*, *143*, 623 (1939).

At this point the reader might properly ask why anyone would even suspect dQ_{rev}/T to be exact, quite apart from attempting to prove it. The suggestion comes from a rearrangement of equation 4.12 in the form

$$\frac{Q_1}{T_1} + \frac{Q_2}{T_2} = 0 \tag{4.16}$$

which is really nothing more than a restatement of the definition of absolute temperature. It will be observed that the left-hand side of equation 4.16 is precisely equal to the line integral of dQ/T for a reversible Carnot cycle. Now, if the line integral of dQ/T about a reversible Carnot cycle is equal to zero, maybe the same would be true about any reversible cycle, in which case dQ_{rev}/T would be an exact differential.[*] With this suggestion in mind, we now turn to proving the correctness of this conjecture.

To carry out the proof, let us first consider a particular kind of cycle, one made up of alternating isothermals and adiabatics. Specifically, let us examine the cycle illustrated in Figure 4-5, which is made up of three isothermals and three adiabatics. The isothermals consist of ab at T_1, cd at T_2, and ef at T_3, accompanied by heat quantities Q_1, Q_2, and Q_3; the adiabatics consist of bc, de, and fa. For the sake of the proof, let us extend the isothermal cd backward along the broken line until it cuts adiabatic fa at g; finally, let us assign a value Q_2' to the heat along cg. Now the over-all cycle can be regarded as a composite of two Carnot cycles, $abcg$ and $gdef$. Following our definition of absolute temperature, we write

$$\frac{Q_1}{T_1} + \frac{Q_2'}{T_2} = 0 \tag{4.17}$$

and

$$\frac{-Q_2' + Q_2}{T_2} + \frac{Q_3}{T_3} = 0 \tag{4.18}$$

The addition of equation 4.17 to equation 4.18 then yields

$$\frac{Q_1}{T_1} + \frac{Q_2}{T_2} + \frac{Q_3}{T_3} = 0 \tag{4.19}$$

This demonstration can obviously be generalized for any reversible cycle made up of n alternating isothermals and n adiabatics to give

$$\frac{Q_1}{T_1} + \frac{Q_2}{T_2} + \cdots \frac{Q_n}{T_n} = 0 \tag{4.20}$$

---

[*] Problem 14, Chapter Two, also suggests that $1/T$ might generally be an integrating factor for dQ_{rev}.

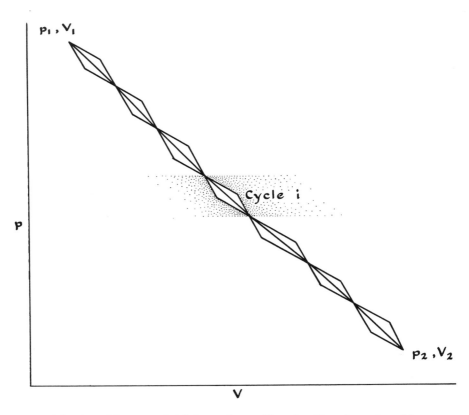

FIGURE 4-6   An arbitrary path joining points p_1, V_1 and p_2, V_2 and enveloped by upper and lower paths made up of alternating isothermals and adiabatics.

or

$$\sum_{i=1}^{n} \frac{Q_i}{T_i} = 0 \qquad (4.21)$$

As a corollary to equations 4.20 and 4.21 we can assert that, for non-cyclical reversible processes made up exclusively of isothermals and adiabatics, the summation of Q_i / T_i will depend only upon the initial and final states. With this background we shall now turn to the general theorem, applicable to paths not necessarily made up of isothermals and adiabatics.

Consider an arbitrary reversible path represented by the smooth line joining p_1, V_1 to p_2, V_2 in Figure 4-6. Approximations to this path can be realized either by a series of alternating isothermals and adiabatics above it or by a similar series below it. Properly chosen, the upper and lower isothermal-adiabatic paths can be made to meet each other on the chosen path, as shown in the figure; thus the given path can be regarded as enveloped by a number of Carnot cycles. Let us now examine the ith group of steps, which for convenience is enlarged in Figure 4-7. For the upper path there will be a quantity of heat, Q_i, associated with an isothermal at T_i. For the lower path the heat will be Q_i' at temperature $T_i - \Delta T_i$ $(= T_{i+1})$. The segment of the given path being investigated involves a quantity of heat, Q_i'', but, since it is nonisothermal, we cannot specify a temperature beyond saying that for each point it will lie in the range T_i to $T_i - \Delta T_i$. Now let W_i, W_i', and W_i'' represent the quantities of work associated with the upper, lower, and middle paths from the upper left-hand corner of the four-sided figure to the lower right-hand corner. Then, using the First Law, we can assert that

$$Q_i - W_i = Q_i' - W_i' = Q_i'' - W_i'' \tag{4.22}$$

From the construction of the diagram it is seen that

$$W_i \geqslant W_i'' \geqslant W_i' \tag{4.23}$$

Therefore

$$Q_i \geqslant Q_i'' \geqslant Q_i' \tag{4.24}$$

But, according to the definition of absolute temperature, we can also

FIGURE 4-7 An enlarged view of one section of Figure 4-6.

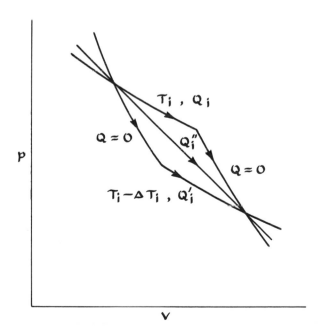

write that

$$\frac{Q_i}{T_i} = \frac{Q_i'}{T_i - \Delta T_i} \tag{4.25}$$

from which it follows, in the light of equation 4.24, that we can always find a θ_i such that

$$\frac{Q_i}{T_i} = \frac{Q_i''}{T_i - \theta_i \Delta T_i} \tag{4.26}$$

provided $0 \lessgtr \theta_i \lessgtr 1$. If we now sum up the terms in equation 4.26 from the point p_1, V_1 to p_2, V_2, we obtain

$$\sum_{i=1}^{n} \frac{Q_i}{T_i} = \sum_{i=1}^{n} \frac{Q_i''}{T_i - \theta_i \Delta T_i} \tag{4.27}$$

But we have already seen that $\displaystyle\sum_{i=1}^{n} \frac{Q_i}{T_i}$ depends only on the initial and final states, regardless of n, for it represents a sum along alternating isothermals and adiabatics. Therefore the limit of that sum as n becomes infinite and as each ΔT_i approaches zero is the same as the finite sum and hence also a function only of the initial and final states. Accordingly we can write that

$$\lim_{\Delta T_i \to 0} \sum_{i=1}^{n} \frac{Q_i}{T_i} = \lim_{\Delta T_i \to 0} \sum_{i=1}^{n} \frac{Q_i''}{T_i - \theta_i \Delta T_i} = \phi(p_2, V_2) - \phi(p_1, V_1) \tag{4.28}$$

$\phi(p, V)$ being some function of pressure and volume. Let us now consider what is meant by the line integral of dQ/T along the arbitrarily chosen path. By definition that integral is precisely equal to

$$\lim_{\Delta T_i \to 0} \sum_{i=1}^{n} \frac{Q_i''}{T_i - \theta_i \Delta T_i}$$

Hence

$$\int \frac{dQ}{T} = \phi(p_2, V_2) - \phi(p_1, V_1) \tag{4.29}$$

In other words, the line integral of dQ/T for any reversible path depends only upon the initial and final states, which is equivalent to saying that dQ_{rev}/T is an exact differential—that is, the differential of some function of the state of the system. This function will be given the name *entropy* and will be denoted by the symbol S. Therefore

$$dS = \frac{dQ_{\text{rev}}}{T} \tag{4.30}$$

The subscript "rev" is inserted to emphasize that the relationship is a valid one only for reversible processes. Equation 4.30 is equivalent to saying that $1/T$ is an integrating factor for the inexact differential dQ_{rev}.* The change in entropy attending a given change in state can therefore be written as

$$\Delta S = \int \frac{dQ_{rev}}{T} \tag{4.31}$$

In this connection it is important to note that one computes an entropy change by evaluating the line integral of dQ/T along a *reversible* path even if the process of interest is an irreversible one. To calculate an actual entropy change, therefore, one must conceive of reversible means of bringing about the change of state that may actually occur irreversibly. In the next section we shall inquire into a possible relationship between dS and dQ_{irrev}/T.

Entropy Changes Attending Irreversible Processes

In order to learn something about entropy changes associated with irreversible processes, let us again consider a Carnot cycle consisting of two isothermals (with heats Q_1 and Q_2 at temperatures T_1 and T_2) and two adiabatics. Now what possible effect can irreversibility have on the efficiency of such a cycle? In accordance with the Second Law, no cycle operating between two definite temperatures can be more efficient than a reversible one working between the same two temperatures. (If such were possible, an appropriate combination of cycles, reversible and irreversible, would constitute a perpetual-motion machine of the second kind.) We can also rule out the possibility that an irreversible cycle possesses the *same* efficiency as a reversible one, for under this hypothesis the irreversible portions of the cycle could be thermodynamically reversed—a contradiction of terms. The Second Law does not, however, rule out the possibility of an irreversible cycle being *less* efficient than a reversible one, a conclusion that is entirely reasonable and is, indeed, observed to be true. We can therefore write, in general, for any Carnot cycle, not necessarily reversible, that

$$W \leqslant \left(\frac{T_1 - T_2}{T_1}\right) Q_1 \tag{4.32}$$

---

* As an illustration of an integrating factor, consider the (inexact) differential expression $ydx/x - dy$. If this is divided by y, the expression becomes exact: $d(\ln x/y)$. Accordingly we say that $1/y$ is an integrating factor for that differential expression.

Q_1 being the actual heat associated with the isothermal at T_1, whether the cycle is reversible or not, and T_1 and T_2 being the absolute temperatures of the heat reservoirs, defined, of course, in terms of reversible cycles. But, since $W = Q_1 + Q_2$ (from the First Law),

$$\frac{Q_1}{T_1} + \frac{Q_2}{T_2} \leqslant 0 \qquad (4.33)$$

the equality holding for a reversible Carnot cycle and the inequality for irreversible ones. Now let us suppose that the isothermal at T_2 and the two adiabatic steps are each reversible and that any irreversibility in the cycle can be attributed solely to the isothermal at T_1. Since the entropy is a function of the state of the system, ΔS for the cycle must be equal to zero. But ΔS for the cycle is also equal to the sum $(\Delta S_1 + \Delta S_2)$ of the entropy changes associated with the two isothermal processes, since the entropy changes attending the reversible adiabatics must be zero. By hypothesis,

$$\Delta S_2 = \frac{Q_2}{T_2} \qquad (4.34)$$

and, since $\Delta S_1 + \Delta S_2 = 0$, we see from statements 4.33 and 4.34 that

$$\Delta S_1 \geqslant \frac{Q_1}{T_1} \qquad (4.35)$$

the inequality holding for an irreversible process and the equality for a reversible one.

EXERCISE:

Consider a Carnot cycle for which the two isothermals are carried out reversibly, one or both of the adiabatics being irreversible. Prove by methods similar to those employed above that $\Delta S \geqslant 0$ for the adiabatic steps.

By recourse to methods similar to those employed in the preceding section, but involving the concept of inequalities introduced by irreversibility, it can be shown in general that

$$\Delta S \geqslant \int \frac{dQ}{T} \qquad (4.36)$$

the equality holding for reversible paths and the inequality for all other paths.* Expression 4.36 assumes a particularly simple form if the

---

* Expression 4.36 can be regarded as another statement of the Second Law provided the temperature is first defined as the reciprocal of an integrating factor for dQ_{rev}. That such system under consideration is thermally insulated from its surroundings. Under such circumstances, $dQ = 0$, and the inequality becomes

$$\Delta S \geqslant 0 \qquad (4.37)$$

From expression 4.37 we conclude that the entropy of an adiabatic system tends to increase unless it is at its maximum value, in which case it must remain constant. This suggests that the entropy function can be regarded as a measure of the degradation of the system; for, when the system is completely degraded or "run down," its entropy can no longer increase. Assuming that the laws of thermodynamics as we know them are valid everywhere, and supposing further that the entire universe can be regarded as an adiabatic system, we conclude that the total entropy of the universe tends to increase with time.

Calculation of Entropy Changes

Since by definition $dS = dQ_{rev}/T$, it is obvious that, for a reversible isothermal process, $\Delta S = Q/T$. For the isothermal expansion of n moles of an ideal gas, it follows that

$$\Delta S = nR \ln \frac{V_2}{V_1} = nR \ln \frac{p_1}{p_2} \qquad (4.38)$$

Equation 4.38 is valid even for irreversible processes if the initial and final temperatures are the same. This is true because ΔS must be computed not from $\int \frac{dQ}{T}$ but from $\int \frac{dQ_{rev}}{T}$. Thus, in the Joule expansion of an ideal gas, $Q = 0$, but ΔS will still be given by equation 4.38, in full agreement with expression 4.36.

For the reversible heating or cooling of a system, we note that

$$\Delta S = \int_{T_1}^{T_2} \frac{C \, dT}{T} \qquad (4.39)$$

C being the (reversible) heat capacity. Equation 4.39 is perfectly general and can be used with the concept of generalized heat capacity expressed by equations 2.29 and 2.30. If C is constant, then

$$\Delta S = C \ln \frac{T_2}{T_1} \qquad (4.40)$$

---

an integrating factor exists must, of course, be demonstrated in such an approach to the proposition. This has been done by C. Carathéodory, *Math. Ann.* *67*, 355 (1909). See also Margenau and Murphy, *The Mathematics of Physics and Chemistry*, 2nd edition, D. Van Nostrand Co., New York, 1956, p. 26.

EXERCISE:

Using the concept of generalized heat capacity in eqs. 2.29 and 2.30, derive general expressions for the change in entropy for reversible changes in state. Hence show for one mole of an ideal gas, for which C_v is constant, that

$$\Delta S = C_v \ln \frac{T_2}{T_1} + R \ln \frac{V_2}{V_1}$$

$$= C_p \ln \frac{T_2}{T_1} - R \ln \frac{p_2}{p_1}$$

EXERCISE:

Calculate the entropy change that occurs when one mole of helium (to be regarded as an ideal gas) is subjected to the Joule experiment, both vessels being of the same volume.

If the heat capacity of a system is given as a function of temperature (for example, in the form $C = a + bT + cT^2 + \ldots$), the entropy change for heating such a system from T_1 to T_2 will be given by the expression

$$\Delta S = a \ln \frac{T_2}{T_1} + b(T_2 - T_1) + \frac{c}{2}(T_2^2 - T_1^2) + \ldots \qquad (4.41)$$

If a system undergoes a reversible transition, such as a fusion or an evaporation, the entropy change for the process will equal the heat of transition divided by the absolute transition temperature. Since the heat of such a transition carried out at constant pressure is just the enthalpy change, we can write

$$\Delta S_{\text{trans}} = \frac{\Delta H_{\text{trans}}}{T} \qquad (4.42)$$

EXERCISE:

Calculate ΔS for the following change in state:

$$\begin{array}{cc} \text{H}_2\text{O(c)} & \longrightarrow \text{H}_2\text{O(g)} \\ 0\ ^\circ\text{C, 1 atm} & 100\ ^\circ\text{C, 1 atm} \end{array}$$

Assume that $\Delta H_{\text{fus}} = 1{,}436$ cal mol^{-1} at $0\ ^\circ\text{C}$, $\Delta H_{\text{vap}} = 9{,}720$ cal mol^{-1} at $100\ ^\circ\text{C}$, and the molar heat capacity of $\text{H}_2\text{O(liq)}$ is 18 cal deg^{-1}.

If a transition occurs under nonequilibrium conditions, the entropy change will not be given by equation 4.42; one must compute it by integrating dQ/T for some reversible means of bringing about the same change in state. To illustrate such a calculation, let us consider the isothermal freezing of super-cooled water at $-5\ ^\circ\text{C}$, for which the over-all change in state can be represented by

$$H_2O(\text{liq}) \longrightarrow H_2O(c)$$
$$-5\,°C \qquad\qquad -5\,°C$$

Since super-cooled liquid water cannot, by direct freezing, be reversibly converted into solid at the same temperature, we must conceive of some indirect reversible means of bringing about the specified change in state.

One series of processes that might be employed for this purpose consists of heating the liquid to 0 °C, reversibly freezing it to solid at that temperature, and finally cooling the solid back to -5 °C:

I $\quad H_2O(\text{liq}) \longrightarrow H_2O(\text{liq})$
$$\quad\;\; -5\,°C \qquad\qquad 0\,°C$$

II $\quad H_2O(\text{liq}) \longrightarrow H_2O(c)$
$$\quad\;\; 0\,°C \qquad\qquad 0\,°C$$

III $\quad H_2O(c) \;\; \longrightarrow H_2O(c)$
$$\quad\;\; 0\,°C \qquad\qquad -5\,°C$$

The over-all entropy change will then be given by the expression

$$\Delta S = \int_{268}^{273} \frac{C(\text{liq})}{T}\,dT - \frac{\Delta H_{\text{fus}}}{273} + \int_{273}^{268} \frac{C(c)}{T}\,dT$$

$$= \int_{268}^{273} \frac{C(\text{liq}) - C(c)}{T}\,dT - \frac{\Delta H_{\text{fus}}}{273} \tag{4.43}$$

in which $C(\text{liq})$ and $C(c)$ are the heat capacities of the liquid and solid and ΔH_{fus} is the heat of fusion at 0 °C.

We can produce the same change in state isothermally by reversibly evaporating the liquid, expanding the vapor until its pressure is equal to the vapor pressure of the solid, and then condensing the vapor to solid:

I $\quad H_2O(\text{liq}) \longrightarrow H_2O(g)$
$$\qquad\qquad p_1$$

II $\quad H_2O(g) \;\; \longrightarrow H_2O(g)$
$$\quad\;\; p_1 \qquad\qquad p_s$$

III $\quad H_2O(g) \;\; \longrightarrow H_2O(c)$
$$\quad\;\; p_s$$

In this representation, p_s and p_1 equal the vapor pressures of the solid and liquid at -5 °C. We can again compute the over-all entropy change by summing up the entropy changes for the three steps. If we assume the vapor to be an ideal gas, the entropy change for the second stage will be given by equation 4.38. Using primed quantities to represent heats of transition at -5 °C (as distinguished from the unprimed ones at 0° C), we can now write that

$$\Delta S = \frac{\Delta H'_{vap}}{268} + R \ln \frac{p_1}{p_s} - \frac{\Delta H'_{sub}}{268} \tag{4.44}$$

Recognizing, further, that the heat of sublimation, $\Delta H'_{sub}$, less the heat of vaporization, $\Delta H'_{vap}$, is precisely equal to the heat of fusion, we can also write that

$$\Delta S = - \frac{\Delta H'_{fus}}{268} + R \ln \frac{p_1}{p_s} \tag{4.45}$$

The value of $\Delta H'_{fus}$ appearing in equation 4.45 must correspond to -5 and will not be exactly the same as ΔH_{fus} at 0, which is used in equation 4.43.

The answers obtained by the two methods—expressed by equations 4.43 and 4.45—are not identical in form, but they must be equal if the Second Law is correct. Equating the two answers will therefore give us an equation relating a quotient of certain vapor pressures to calorimetric quantities. This is just one example of the kind of thing that can be done with the Second Law. We shall systematically explore such matters in later chapters. At this stage, however, we can properly reflect on the far-reaching nature of the Second Law and on the extraordinary utility of the entropy concept.

EXERCISE:

Neglecting the difference between the heat capacities of liquid water and ice, calculate the ratio of vapor pressures, p_1/p_s, for water and ice at $-5\,°C$. The heat of fusion of ice is 1,436 cal mol^{-1}.

Entropy Changes Attending Irreversible Heat Flow

If two heat reservoirs of infinite heat capacity are placed in contact with each other, heat will flow from the hotter reservoir to the colder one. If the net quantity of heat so transferred is Q, and if the temperatures of the hot and cold reservoirs are T_2 and T_1, the net entropy change attending this process will be

$$\Delta S = Q \left(\frac{1}{T_1} - \frac{1}{T_2} \right) > 0 \tag{4.46}$$

If the reservoirs have finite heat capacities, the net entropy change will be given by appropriate integrals, but will still be positive. If we now regard the *pair* of reservoirs as our system, and stipulate further that no heat shall enter or leave that system (this does not rule out heat flow between the reservoirs within the system), the net increase in entropy implies irreversibility in accordance with expression 4.37. It is evident, therefore, that heat flowing through a finite drop in temperature does so irreversibly, and that under no circumstances can heat flow spontaneously from one reservoir to another at a higher temperature. All this raises the question of how heat might be transferred reversibly, a topic to which we shall now turn.

Reversible Heating and Cooling

In Chapter Two we considered what was meant by a "reversible change in state," with particular reference to friction, differences in pressure, and work obtained. Let us now examine the matter of reversibility of heating and cooling, particularly in connection with entropy changes. We have already asserted that a reversible process is a process after the completion of which both system and surroundings can be restored to their initial conditions. In the light of this statement and what we know about the Second Law, how can an object (of constant heat capacity C, let us say) be heated reversibly from temperature T_1 to T_2? Simply placing the object at temperature T_1 next to a very large heat reservoir at temperature T_2 will not suffice; for, although the object will be heated, there is no conceivable way of restoring the initial conditions without somehow leaving a net change in the surroundings.

There are ways, however, of bringing about a reversible heating. One method would involve the operation of an appropriate reversible engine, which would extract heat from a large reservoir at T_2 and discharge a portion of it into the object to warm it, any remainder of energy appearing as work in the surroundings. The original condition could then be restored if we used the work that appeared in the surroundings to operate the same engine in reverse.

Another method of approaching reversible heating involves the use of a large number of heat reservoirs, each of heat capacity C, at temperatures T_1, $T_1 + \Delta T$, $T_1 + 2\Delta T$, ... $T_2 - \Delta T$, and T_2. We can now heat the object stepwise, first by placing it in contact with the reservoir at temperature $T_1 + \Delta T$, then in contact with that at $T_1 + 2\Delta T$, etc. To cool the object back to T_1, the process is reversed by stepwise cooling. If the heating and cooling are carried out in this way, it is evident that each of the intermediate reservoirs will be unchanged, for each will have lost a quantity of heat, $C\Delta T$, during the heating of the object and gained a like amount during the cooling. There will, however, be a net change in the two extreme reservoirs, a loss of heat amounting to $C\Delta T$ by the reservoir at T_2 and a gain of an equal amount by the reservoir at T_1. The net total entropy change will then be equal to $C\Delta T\left(\dfrac{1}{T_1} - \dfrac{1}{T_2}\right)$, which quantity can be made vanishingly small if we increase the number of reservoirs indefinitely, simultaneously causing ΔT to approach zero. In summary we can say that, for reversible processes, all direct heat flows must occur between vanishingly small temperature differences, just as all pistons sweeping out volume changes must have only infinitesimal differences in pressure on opposite sides.

Some Modifications of Carnot-cycle Equations

Equation 4.13, which relates the efficiency of a reversible Carnot cycle to the temperatures involved, can be rewritten in several different ways according to the ultimate use in mind. If the quantity of heat Q_1 is infinitesimal, we can replace Q_1 by dQ_1 and W by dW to obtain

$$dW = \frac{T_1 - T_2}{T_1} dQ_1 \qquad (4.47)$$

This equation is particularly useful under certain circumstances when T_1 or T_2 is variable (see Prob. 8).

Another form of equation 4.13 can be written if $T_2 - T_1$ is regarded as infinitesimal—say dT. Then the work associated with such a cycle will also be infinitesimal:

$$dW = -Q\frac{dT}{T} \qquad (4.48)$$

The subscripts have now been omitted. Equation 4.48 is useful for deriving certain Second Law corollary equations. Equations 4.47 and 4.48 are not to be regarded as equivalent, for the infinitesimal character of W comes about for different reasons in the two cases.

Refrigerating Machines

We have seen that heat can be converted into work by the operation of an appropriate cycle provided a certain fraction of the heat is discharged into a low-temperature reservoir. By operating such a cycle in reverse, we can evidently take heat from a low-temperature reservoir and discharge it at a higher temperature, although this will require expending a certain amount of work. Such reverse operation of an engine constitutes what is called a refrigeration cycle. The only qualitative difference between a refrigerator and a heat engine is that the signs of the net work and heat are changed; the same equations obviously apply to reversible operations.

The foregoing discussion means that upon reversible expenditure of a quantity of work, W, an amount of heat, Q_1, can be removed from a reservoir at temperature T_1 and discharged at a higher temperature, T_2, according to the equation

$$Q_1 = \frac{T_1 W}{T_2 - T_1} \qquad (4.49)$$

Moreover, the total heat, Q_2, *discharged* at the higher temperature will equal $Q_1 + W$, a quantity given by the expression

$$Q_2 = \frac{T_2}{T_2 - T_1} W \qquad (4.50)$$

Q_1 is of particular interest in refrigeration since it equals the heat removed—for example, from a refrigerator. Since Q_2 equals the heat discharged, it is an item of prime importance in connection with "heat pumps," which are mechanical devices designed for heating purposes. It is evident that Q_2 can be much greater than W under reversible conditions of operation. It would therefore be more efficient to heat a house by pumping heat from outside into the house through appropriate operation of a refrigeration cycle as a "heat pump" than it would be to dissipate the same amount of work energy through electrical resistance heaters within the house.

It is generally wasteful of fuel to use electrical resistance heaters, since about twice as much heat is discharged at the power plant as is finally utilized for heating. If large quantities of heat are required, such as for space and water heating, it is better from the point of view of fossil-fuel conservation to burn the fuel directly where the heat is required than to use electrical resistors. Nevertheless, vast quantities of increasingly precious natural gas are burned to generate electricity that is often marketed for heating purposes.

EXERCISE:

If a refrigerator operates reversibly between 0 and 20 °C, how much heat is extracted from the refrigerator per calorie expended as work?

EXERCISE:

How many calories of heat can be expected per calorie of work expended during the reversible operation of a heat pump if the temperature of the system to be heated is 20 °C and the temperature of the heat reservoir is

(A) 10 °C (B) 0 °C (C) -20 °C? Show that the use of a heat pump does not appear promising for regions lacking warm heat reservoirs.

EXERCISE:

Draw a diagram, similar to Figure 4-2, to represent schematically the operation of a refrigerator. Note carefully which quantities are positive and which are negative.

Problems

1. Consider the entropy change attending the heating of a substance at constant pressure from T_1 to T_2 when the heat capacity is given by the expression $C_p = a + bT + cT^2$. What average constant value should the heat capacity have over the same temperature range to give the same entropy change? Calculate ΔS and such average heat capacity for 1 mole of oxygen heated from 27 to 227 °C. (Compare with the results of Prob. 20, Chap. Two.)

2. It was proved in the text that $1/T$ is an integrating factor for dQ_{rev}. Show that there is an infinite number of integrating factors for dQ_{rev} of the form $f(S)/T$, in which $f(S)$ is any integrable function of the entropy, S.

3. Assuming the heat capacities of super-cooled water and ice to be constant (but not equal to each other), calculate the vapor-pressure ratio, p_l/p_s, for water and ice at -5 °C. At 0 °C the specific heat of fusion of ice is 79.7 cal. The specific heat capacities of water and ice can be taken as 1.0 and 0.5 cal deg^{-1}. (Compare with the exercise on page 94.)

4. Consider the following nonisothermal cycle, in which p is the vapor pressure of water at absolute temperature T:

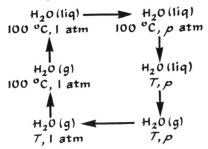

(A) Set up an expression for the entropy change ($\Delta S = 0$) of the entire cycle by adding together the terms for all the steps. Neglect the entropy change attending the compression of the liquid, and assume the vapor to be an ideal gas.

(B) Neglecting the difference between the heat capacities of liquid and vapor, and hence assuming ΔH_{vap} to be constant, derive an expression for the vapor pressure of water at temperature T.

(C) If $\Delta H_{vap} = 9,720$ cal mol^{-1}, calculate the vapor pressure in Torr of water at 105 °C.

(Note: This problem illustrates again the utility of the entropy concept. Vapor pressures will be dealt with in greater detail in later chapters, but the underlying features are already disclosed.)

5. Consider two different steam engines, one operating by use of steam from a boiler at 20 atm pressure and the other using steam at 10 atm. Using the reversible-Carnot-cycle formula as a guide to the relative efficiencies, calculate the ratio of the efficiencies of the two engines. Refer to a handbook for steam pressure-temperature data, and assume in each instance that the steam is discharged at 1 atm.

6. A brick of heat capacity C_1, originally at temperature T_1, is placed upon a brick of heat capacity C_2, originally at temperature T_2. Heat flows from one brick to the other until thermal equilibrium is established.

(A) What is the total entropy change of the bricks?

(B) If $C_1 = 2,000$ cal deg^{-1}, $t_1 = 0$ °C, $C_2 = 1,000$ cal deg^{-1}, and $t_2 = 100$ °C, what is the final temperature, and what is ΔS in cal deg^{-1}?

7. Consider again the two bricks described in Problem 6. Suppose that the two bricks can be brought to a common temperature, T_0, so determined that the sum of the entropies of the two bricks remains unchanged.

(A) Derive an expression for T_0, and compare it with the final temperature in Problem 6.

(B) For the bricks described in Problem 6, calculate a numerical value for T_0.

8. Two heat reservoirs of constant heat capacities C_1 and C_2 are initially at temperatures T_1 and T_2 respectively. What is the maximum amount of work attainable by use of only those reservoirs in conjunction with reversible machines operating in cycles? Solve this problem by using eq. 4.47 in conjunction with $dQ_1 = -C_1 dT_1$ and $dQ_2 = -C_2 dT_2$. (Compare with Prob. 7.)

9. Consider two heat reservoirs, one of finite heat capacity C_1 at an initial temperature T_1 and the other of infinite heat capacity at a temperature T_2. Calculate the maximum amount of work obtainable from this pair of heat reservoirs with reversible machines operating in cycles. You can solve this problem by taking the solution of Problem 8 and passing to the limit as $C_2 \longrightarrow \infty$, or, directly, by the method suggested for Problem 8, bearing in mind that $dT_2 = 0$.

10. Consider a three-step reversible cycle consisting of an isothermal expansion at temperature T_1, a constant-volume cooling to temperature T_2, and an adiabatic compression to the initial state. If one mole of an ideal gas is

used, what is the total work obtained from operation of this cycle? Show
that the efficiency, calculated on the basis of the heat supplied at T_1, is
equal to $1 - (T_1 - T_2)/[T_1 \ln (T_1/T_2)]$.

11. Consider a three-stage reversible cycle consisting of an isothermal expansion
at T_1, a constant-pressure cooling to T_2, and an adiabatic compression back
to the starting point. If one uses 1 mole of an ideal gas, what is the work
obtained from the cycle? What is the efficiency, calculated on the basis
of the heat supplied during the isothermal stage?

12. An air-conditioned house is kept at a temperature of 20 °C while the outside
air is at 30 °C. The cooling is done by a refrigerating machine whose compression cylinders, located outside the house, operate at 50 °C, and whose
expansion coils, within the house, are at 10 °C. How much work must be
done for every kilocalorie of heat removed from the house if the machine
operates reversibly? What entropy changes occur within and outside the
house for this amount of refrigeration?

13. One mole of an ideal monatomic gas is heated reversibly along a path
defined by the expression $T = AV^2$, in which A is a constant. If the initial
temperature is 0 °C, what must the final temperature be if the entropy
change is to equal 2.5 cal deg^{-1}?

14. An object of constant heat capacity C is cooled by the reversible operation
of a refrigerator whose expansion coils are kept exactly at the temperature
of the object and whose compressor is kept uniformly at temperature T_0.
If the object is cooled from T_1 to T_2, how much work must be expended in
operating the refrigerator?

$$\text{Ans. } -W(\text{work done}) = CT_0 \ln (T_1/T_2) - C(T_1 - T_2)$$

15. The air in a house of volume V is cooled from temperature T_1 to temperature T_2 when the outside air is at temperature T_0. The pressure is kept
constant at 1 atm so that some air enters the house during the cooling
process. (See Prob. 8, Chap. Two.)

(A) If the house is cooled by a refrigerating mechanism that operates reversibly between the outside air temperature (T_0) and the inside air
temperature (variable from T_1 to T_2), how much work must be expended to bring about the cooling?

(B) If $V = 2,000$ cubic meters, $t_1 = 30$ °C, $t_2 = 25$ °C, and $t_0 = 35$ °C,
what is W in kilocalories? Assume C_v for air to be constant and equal
to $\frac{5}{2}R$ per mole.

$$\text{Ans. } (A) \quad -W \text{ (work done)} = \frac{pVC_pT_0}{R} \left[\frac{1}{T_1} - \frac{1}{T_2} - \frac{T_0}{2T_1^2} + \frac{T_0}{2T_2^2} \right]$$

16. The air in the house described above is heated from temperature T_1 to
T_2 by means of a heat pump operating reversibly between the inside temperature (variable between T_1 and T_2) and a constant outside temperature
(T_0).

(A) How much work must be done to produce the heating if air is expelled during the heating process? (See Prob. 7, Chap. Two.)

(B) If $t_1 = 15$ °C, $t_2 = 20$ °C, $t_0 = 10$ °C, and $V = 2,000$ cubic meters, what is W in kilocalories if the molar heat capacity of air is $\frac{5}{2}R$?

$$\text{Ans. (A)} \quad -W \text{ (work done)} = \frac{C_p pV}{R}\left\{\ln\frac{T_2}{T_1} + \frac{T_0}{T_2} - \frac{T_0}{T_1}\right\}$$

17. (A) A mass, m, is raised reversibly and isothermally from the earth's surface to height h. (The work required for this change will equal mgh, in which expression g is the acceleration of gravity.) Show that the entropy change for the process equals zero.

(B) The raised mass, at temperature T, is now permitted to fall freely until it collides inelastically with the earth. The heat generated by the impact can be presumed to be transmitted to the earth (of infinite heat capacity), which is at temperature T, the same as that of the falling mass. Show that the entropy of the earth must increase by mgh/T.

18. A thermodynamic system consists of a rigid box of volume V containing materials that might undergo a change in state. The box is also insulated so that no heat can enter or leave it. If something should happen within the box to produce a pressure change, Δp, what if anything can be said about ΔU, ΔH, Q, W and ΔS?

$$\text{Ans.} \quad \Delta U = 0; \; \Delta H = V\Delta p; \; Q = 0; \; W = 0; \; \Delta S \gtrless 0$$

19. (A) A cylinder containing materials that might undergo a change in state is provided with a frictionless piston that maintains a constant pressure, p, on the contents of the cylinder. The cylinder is also insulated so that no heat can enter or leave it. If something happens in the cylinder to produce a change in volume equal to ΔV, what if anything can be said about ΔU, ΔH, Q, W, and ΔS?

$$\text{Ans.} \quad \Delta U = -p\Delta V; \; \Delta H = 0; \; Q = 0; \; W = p\Delta V; \; \Delta S \gtrless 0$$

(B) If the system described in part A had a piston subject to friction, then what can be said about ΔU, ΔH, Q, W and ΔS? Consider the cylinder and its contents to be the system and assume that any heat generated by reason of friction is not transferred to the surroundings.

$$\text{Ans.} \quad \Delta U > -p\Delta V; \; \Delta H > 0; \; Q = 0; \; W < p\Delta V; \; \Delta S > 0$$

5

Entropy and Probability;
the Third Law
of Thermodynamics

The concept of entropy, which was developed in the preceding chapter, is a logical outgrowth of the postulates of thermodynamics, particularly of the Second Law. Now it should be borne in mind that the existence of the entropy function was demonstrated without recourse to any chemical or molecular model; hence, from the point of view of classical, phenomenological thermodynamics, the concept of entropy would be meaningful even if we knew nothing about atoms and molecules. The reader will recall, however, that the energy function, which was suggested by the First Law, and which likewise required no molecular model, becomes more fully appreciated and understood when the kinetic point of view is introduced. Thus it became expedient to ascribe the energy of an ideal gas to the molecular motions of translation, rotation, and vibration. For nonideal systems there are other contributions to the energy, attributable, for example, to interactions between different molecules.

The molecular picture, having helped us understand more fully the nature of the energy function, will contribute even more strikingly to an understanding of entropy. The connection, however, is not obvious, for it involves the concept of probability, which at the outset may appear far removed from entropy as introduced in Chapter Four. The connection exists, nevertheless, as we shall now illustrate by considering the entropy of an ideal gas and its dependence on volume.

Entropy of an Ideal Gas

We have already seen that the increase in entropy accompanying the free expansion of one mole of an ideal gas (Joule experiment) is given by the expression

$$\Delta S = R \ln \frac{V_2}{V_1} \tag{5.1}$$

This increase is fully compatible with inequality 4.37, which asserts that the entropy of an adiabatic system increases for spontaneous processes. Now we might also expect, intuitively, that an isolated system made up of a large number of molecules will strive for a state of maximum probability. Utilizing the free expansion of an ideal gas as an example, we shall now inquire into the precise connection between entropy and probability.

Consider two vessels, of volumes V and V', connected by a tube as shown in Figure 5-1. Let us put into this system N_A molecules of ideal gas, N_A being Avogadro's number. (By "ideal gas" we mean, of course, a gas made up of point particles, free of intermolecular forces.) Now the most probable distribution of molecules between the two containers will correspond to an equal concentration in both. (Minor fluctuations from a state of uniform concentration may occur, but they will be insignificant

FIGURE 5-1 Two connected vessels of volumes V and V'. When N_A molecules are put into the system, the most probable distribution will involve $N_A V/(V + V')$ molecules in vessel V and $N_A V'/(V + V')$ molecules in vessel V'.

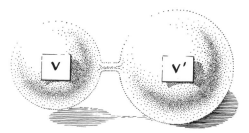

Total Volume = V + V'

and of little consequence in the subsequent argument.) Let us now ask what the probability is of finding all N_A molecules in the container of volume V, bearing in mind the fact that the two containers are connected and are both accessible. The probability of finding any one molecule in the volume V is given by the expression $V/(V + V')$, which is simply the fraction of the total volume represented by the container of interest. The probability, P, of having N_A noninteracting molecules all present in volume V is given by the expression

$$P = \left(\frac{V}{V + V'}\right)^{N_A} \tag{5.2}$$

Now by hypothesis it is certain that all N_A molecules are located somewhere in the volume $V + V'$. If we consider the ratio of the probability of finding all the molecules somewhere in the system to that of finding all the molecules in volume V, we obtain the reciprocal of expression 5.2—namely, $[(V + V')/V]^{N_A}$. Furthermore, if we take the logarithm of that ratio and multiply by Boltzmann's constant, k, we obtain

$$k \ln (1/P) = k \ln [(V + V')/V]^{N_A} = N_A k \ln [(V + V')/V] \tag{5.3}$$

Examination of the quantity appearing in this equation discloses that it is precisely equal to the entropy change (eq. 5.1) attending the free expansion of one mole of an ideal gas from volume V ($=V_1$) to volume $V + V'$ ($=V_2$). This demonstration suggests, therefore, that the entropy is simply related to the logarithm of the probability, a conclusion we shall amplify further.

That the logarithm of the probability is the function that connects entropy to probability can be shown in a more general way. Consider a homogeneous system arbitrarily divided into the two parts illustrated in Figure 5-2. The total entropy of that system can be regarded as the sum of the separate entropies; that is, $S = S_1 + S_2$. Let us assume fur-

FIGURE 5-2   Representation of a homogeneous system divided (by dotted lines) into two parts with entropies S_1 and S_2.

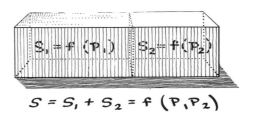

ther that the entropy is given by some function, $f(P)$, of the probability of finding the molecules in their state of existence. Then we can write that the individual entropies of the two parts of the system will be given by similar functions of the individual probabilities, as follows:

$$S_1 = f(P_1)$$

$$S_2 = f(P_2) \tag{5.4}$$

We shall now introduce an important property of probabilities—namely, that the probability of realizing a combination of systems is equal to the product of the probabilities of realizing the separate systems (that is, $P = P_1P_2$). This is equivalent to assuming that the probability of experiencing two separate events of a noninteracting character is precisely equal to the product of the probabilities of experiencing the separate events. We now assert, in other words, that the entropy is such a function of probability that the function of the product of the probabilities is equal to the sum of the separate functions. * As we shall presently see, the one function that fulfills this condition is the logarithmic function. Let us write the following as our basic equation:

$$S = f(P) = f(P_1P_2) = f(P_1) + f(P_2) \tag{5.5}$$

Differentiating equation 5.5 partially with respect to P_1 and then again with respect to P_2, we obtain the following two equations:

$$P_2f'(P_1P_2) = f'(P_1)$$

$$P_1f'(P_1P_2) = f'(P_2) \tag{5.6}$$

Upon appropriate manipulation we observe that

$$P_1P_2f'(P_1P_2) = Pf'(P) = P_1f'(P_1) = P_2f'(P_2) \tag{5.7}$$

Now if P_1 is independent of P_2 (and this must surely be so, since there is no constraint on the sizes of the two parts of the whole system), we can assert that the products $P_1f'(P_1)$ and $P_2f'(P_2)$ must each be equal to some constant, k. This means that

$$f'(P_1) = \frac{k}{P_1}$$

$$f'(P_2) = \frac{k}{P_2}$$

$$f'(P) = \frac{k}{P} \tag{5.8}$$

---

* This discussion should be considered a pedagogical demonstration and not a proof or derivation of the connection between entropy and probability. The assumptions quite obviously determine the form of equation to be derived.

Integration of these differential equations yields

$$f(P_1) = k \ln P_1 + C_1$$
$$f(P_2) = k \ln P_2 + C_2$$
$$f(P) = k \ln P + C_3 \tag{5.9}$$

in which C_1 and C_2 are constants of integration and $C_3 = C_1 + C_2$. Hence we can write for the entropy

$$S = k \ln P + \text{constant} \tag{5.10}$$

in which k must be Boltzmann's constant to agree with equation 5.3, which in turn is compared with equation 5.1. There still remains, however, the question of the value of the constant of integration, a subject we shall discuss later.

Entropy and Disorder

It appears reasonable from the arguments employed to arrive at equation 5.3 that the entropy of a system can be regarded as a monotonically increasing function of the probability of attaining the system. Now a system having a high degree of order is in an improbable state

FIGURE 5-3 Two-dimensional model of a crystal with atoms undergoing random vibrations.

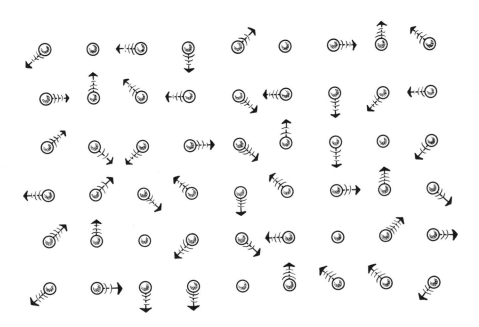

and hence of low entropy. Conversely, a state of high entropy or high probability will appear to be disorganized or random. Hence the increase in entropy attending a spontaneous adiabatic process can be regarded as the direct counterpart of the tendency for a system left to itself to become more random.

To illustrate the connection between entropy and disorder more precisely, let us consider a model of a crystal, which for convenience is depicted in two dimensions in Figure 5-3. To keep the argument simple, we shall further assume that each atom (or ion) in this crystalline lattice can be regarded as an isotropic harmonic oscillator possessing certain well-defined quantized energy levels.* Now, if all the little oscillators were in precisely the same quantum state, we could say that the crystal had complete order; however, if some of the particles were in low quantum states and others in high quantum states and we knew only the average energy, an element of randomness would exist, for there would be numerous ways of realizing the statistical state corresponding to the specified average energy. The latter state would therefore have a higher probability and higher entropy than the former. In practice it is difficult, if not impossible, to get all the oscillators into the same quantum state unless that state is the lowest one. We can approach the lowest quantum levels by cooling the crystal toward absolute zero, where it will have a minimum amount of energy. Under such circumstances all the atoms will be in the same state, and, according to the above discussion, the crystal will have maximum order, minimum probability, and minimum entropy. If, by some means, we could get all the atoms into some other quantum state, each characterized by the same set of quantum numbers and the same energy, the system would again have low probability and low entropy. Unfortunately, we cannot as yet do this by altering the ordinary thermodynamic variables available to us. The underlying idea can be further illustrated by analogy. A crack drill squad of soldiers exhibits perfect order whether at attention or marching; in either event the "entropy" is low. In the crystal, however, we can attain order only by putting the atoms at attention, and this we can do by sufficient cooling.† No method is currently available for getting all the atoms of a crystal to vibrate simultaneously in identical excited states.

---

* This rather artificial picture is the basis of the Einstein theory of heat capacities of crystals. Although somewhat deficient for the calculation of heat capacities, this model will suffice for our present purpose.

†Strictly speaking, the atoms will never be at "attention" since they will always exhibit motion associated with the zero-point energy. The important point is that they are all alike, quantum-mechanically speaking.

The Third Law of Thermodynamics

In the light of the preceding discussion, we are now in a position to state the Third Law of Thermodynamics, which provides a basis for determining absolute entropies. According to the Third Law, as formulated by Planck, *the entropy of a perfect crystal at absolute zero is equal to zero.* This law, like the two preceding it, is a postulate, the validity of which ultimately rests on experiment. Actually, we cannot completely test the Third Law by experiment, for we have no means available for direct measurement of absolute entropies. Nevertheless, the measured entropy changes associated with chemical reactions are compatible with the law. Moreover, the Third Law is consistent with theoretical considerations, which we shall discuss later under "Statistical Mechanics."

Before the development of the Third Law, Nernst proposed a heat theorem, bearing his name, that was a predecessor to the law and quite as useful as far as chemical reactions are concerned. Nernst asserted that the entropy change for reactions between pure crystalline substances at absolute zero is equal to zero. This is fully compatible with Planck's statement but leaves the possibility that all pure crystalline compounds at absolute zero possess entropies equal to the sum of entropies of their pure crystalline elements under like conditions.

Thermodynamic Probability

In view of the connection between entropy and probability, the Third Law is certainly reasonable. Practically, it enables us to determine the constant of integration in equation 5.10—in other words, to ascertain the kind of probability that should be used. To establish the quantitative connection, we shall now introduce the concept of "thermodynamic probability," which is a quantity proportional to ordinary probability but endowed with the characteristic that its logarithm multiplied by Boltzmann's constant will equal the total entropy. By "thermodynamic probability," which will be represented by the symbol Ω, we shall mean the number of different ways in which a thermodynamic state can be realized. A state of perfect order, one in which all the atoms of a crystal, for example, are in the same vibrational level, has a thermodynamic probability of unity. This is so because there is only one way in which all the atoms can be vibrationally alike. On the other hand, if one has a disorganized state, in which some atoms are in high vibrational levels while others are in lower levels, then—as long as one specifies only the average energy—there are numerous ways of realizing the state of affairs.

Thermodynamic probability can be readily illustrated in connection with the dependence of entropy on volume. Returning to Figure 5-1, let us suppose that the two volumes V and V' are equal. Under that stipulation the probability of finding any particular molecule in a specified half of the total system will be precisely one-half, and the most probable distribution will consist of an equal division of molecules between the two containers. Now there are many different ways in which a large number of molecules can be divided into two equal-sized groups. Hence the thermodynamic probability associated with the uniform distribution of molecules is large. On the other hand, there is only one way of selecting molecules so that all are in one container and none in the other. Thus the thermodynamic probability of the state characterized by having all of the molecules exhibit one of just two possible aspects will be unity.

Entropies of Systems Other than Perfect Crystals at Absolute Zero

Our statement of the Third Law, which concerns the entropy of perfect crystals at absolute zero, leaves open the question of what the entropy might be for other systems under similar circumstances. To answer this question, let us consider, for example, a solid crystalline solution at absolute zero. Suppose that this crystal has perfect order, so far as occupancy of lattice sites is concerned, but consists of two different kinds of atoms that are distinguishable from each other either by chemical or by physical means. If the two kinds of atoms are distributed throughout the lattice in a perfectly random way, then, it is clear, the thermodynamic probability will not be unity, and hence we can expect a positive entropy. This means that the formation of a mixed crystal from perfect crystals of the separate kinds of atoms is accompanied by an increase in entropy, a conclusion that is entirely reasonable.*

A solid solution will not, of course, have a positive entropy unless it possesses some degree of randomness. If each atom of one kind should bear an ordered spatial relationship to atoms of the other kind, the system would have a low thermodynamic probability even if it were, in composition, a mixture. As an artificial illustration, consider a crystal of sodium chloride, which is made up of sodium ions and chloride ions arranged in a definite lattice structure. Such a crystal is not to be regarded as a solid solution of sodium ions and chloride ions, for the ions

---

* See also Problem 1 at the end of this chapter.

occur in a well-defined repeating pattern. Hence a perfect crystal of sodium chloride at absolute zero will have a thermodynamic probability of unity even though two ionic species are present.

A kind of solid solution that is nearly perfect but still has a positive entropy, even at absolute zero, is a crystalline mixture of chemical species differing only in their isotopic composition. Under most circumstances, the chemical reactivities of different isotopes of the same element can be regarded as equivalent, and hence two or more isotopes, in either elemental or compound form, will tend to be randomly distributed among the sites available. Taking cognizance of the existence of the isotopes thus requires assigning a positive entropy to the system. Practically speaking, however, one can disregard the entropy of mixing of isotopes provided the entropies so calculated are used only in connection with reactions in which no separation of isotopes occurs. Since the separation of isotopes is difficult to attain, especially through chemical reactions, one can in practice forget about the existence of isotopes without introducing appreciable error into the thermodynamic calculations. For processes that do give rise to separations, however, full cognizance must be taken of the entropy changes attending isotopic mixing. Ordinarily, chemists do not bother to take into account the effects of isotope mixing but proceed on the assumption of uniform atomic species.

For super-cooled liquids or glasses at absolute zero one can also expect entropies greater than zero. Since a liquid does not possess the order that is characteristic of a crystal, it will have a positive entropy, which can be regarded as "frozen in" when the liquid is subjected to super-cooling. The subject of the entropies of mixtures and disordered systems will be dealt with quantitatively in a later chapter.

Use of the Third Law

By using the Third Law for reference purposes, we can compute absolute entropies of pure substances at temperatures other than 0 K from their heat capacities and heats of transition. A solid material at temperature T, for example, will have an entropy given by the expression

$$S = \int_0^T \frac{C_p}{T} \, dT \tag{5.11}$$

A liquid material will have an entropy given by the expression

$$S = \int_0^{T_m} \frac{C_p(c)}{T} \, dT + \frac{\Delta H_{fus}}{T_m} + \int_{T_m}^T \frac{C_p(liq)}{T} \, dT \tag{5.12}$$

in which T_m is the melting point of the solid, ΔH_{fus} is the heat of fusion at that melting point, and $C_p(c)$ and $C_p(liq)$ are the heat capacities of solid and liquid respectively. Under some circumstances, other transitions, such as those between different forms of solids, may also be involved, but they can be introduced in an obvious way. The extension to gases is also obvious; this will, of course, involve entropies of evaporation or sublimation, depending upon the route used to reach the gaseous state.

The heat capacities appearing in equations 5.11 and 5.12 can be determined by calorimetric means, although at low temperatures recourse is usually taken to theoretical formulas for the heat capacities of crystals. For temperatures approaching absolute zero, the atomic heat capacity of a crystalline substance is given quite accurately by Debye's formula:

$$C_v = \frac{12\pi^4 R}{5}\left(\frac{T}{\theta}\right)^3 = 464.6 \left(\frac{T}{\theta}\right)^3 \text{ cal deg}^{-1} \text{ mol}^{-1} \tag{5.13}$$

The quantity θ, which can be determined by low-temperature heat-capacity measurements, is called the "characteristic temperature" and has values depending upon the substance. Since equation 5.13 is supposed to be valid only for low temperatures, its use is generally restricted to the range $0 < T < \theta/10$. In applying the Third Law, one integrates the Debye equation from absolute zero to some convenient low temperature above which experimental heat capacity data can be used. The integrations at higher temperatures are carried out either by numerical methods or by analytical operations on empirical equations.

Debye's equation was intended to be valid only for elemental substances, but it can also be applied with some success to compounds, provided the heat capacity given by equation 5.13 is multiplied by the number of atoms in the molecular formula. Since the difference between C_p and C_v is negligible for solids, especially at low temperatures, equation 5.13 can be used for C_p as well. In very precise work, corrections of an empirical nature can be applied.

Problems

1. (A) Consider a crystal of a diatomic molecule, AA′, in which A and A′ are distinguishable in principle but so similar (like isotopes of heavy elements) that they behave alike in intermolecular interactions. If each molecule has two possible orientations (AA′ and A′A) in a crystal,

what is the total number of ways in which N such molecules can be arranged in the crystal? If we assume that the molecular orientation is completely random even at absolute zero, what would the molar entropy of the crystal be at absolute zero? Compare the answer with the entropy change attending the free expansion of an ideal gas to double its volume. Explain.

Ans. $S = R \ln 2$

(B) Clayton and Giauque* showed indirectly that the molar entropy of carbon monoxide at absolute zero is 1.2 cal deg^{-1}. Compare this with the answer to part A; what does this suggest concerning the nature of crystalline CO at absolute zero?

2. Using the Debye formula for heat capacities of solids, show that the entropy of a solid at low temperatures is equal to one-third of its heat capacity.

3. The Debye characteristic temperatures for lead and aluminum are 96.3 and 375 respectively. (Observe that the softer metal has the lower characteristic temperature; this is generally true.) Calculate and plot C_v for lead and aluminum for temperatures up to $T = \theta/10$.

4. (A) Suppose that an object of constant heat capacity C were to be cooled from temperature T_1 to absolute zero by use of a reversible refrigerating machine whose expansion coils are always at the temperature of the object and whose compressor is at constant temperature T_0. Show, in view of the answer to Problem 14, Chapter Four, that an infinite amount of work would have to be done.

(B) Actually, all heat capacities approach zero at absolute zero. Taking into account the variability of heat capacity, and assuming that the object to be cooled is a perfect crystal and is cooled at constant volume, show, in view of the Third Law, that the reversible work expended in operating the refrigerator would equal

$$-W \text{ (work done)} = -U(T_1,V) + U(0,V) + T_0 S(T_1,V)$$

U being the energy of the object and S the entropy.

(Note: Although the reversible work required for cooling an object to absolute zero is finite, an infinite number of cycles of ever-diminishing effect would be necessary to do the job; hence absolute zero is actually unattainable.)

5. The heat capacities of cadmium and magnesium between 12 and 298.61 K are given in the table on page 113.† Assuming the T^3 law to be valid below 12 K, and neglecting the difference between C_p and C_v at those low temperatures, calculate the absolute entropies of Cd and Mg at 25 °C. (Use a method of graphical integration.)

---

* J. O. Clayton and W F. Giauque, *J. Am. Chem. Soc.*, *54*, 2610 (1932).

† R. S. Craig, C. A. Krier, L. W. Coffer, E. A. Bates, and W. E. Wallace, *J. Am. Chem. Soc.*, *76*, 238 (1954).

T, K	C_p, cal deg^{-1} mol^{-1}		T, K	C_p, cal deg^{-1} mol^{-1}	
	magnesium	cadmium		magnesium	cadmium
12	0.016	0.392	130	4.527	5.608
14	.026	.592	140	4.718	5.684
16	.042	.804	150	4.876	5.746
18	.065	1.020	160	5.013	5.799
20	.086	1.240	170	5.133	5.844
25	.188	1.803	180	5.236	5.884
30	.341	2.306	190	5.331	5.922
35	.550	2.760	200	5.418	5.956
40	.803	3.158	210	5.487	5.988
45	1.076	3.503	220	5.550	6.018
50	1.367	3.803	230	5.611	6.047
60	1.953	4.283	240	5.667	6.073
70	2.498	4.647	250	5.719	6.096
80	2.981	4.920	260	5.766	6.119
90	3.404	5.138	270	5.811	6.144
100	3.753	5.284	280	5.853	6.171
110	4.052	5.413	290	5.896	6.201
120	4.307	5.518	298.16	5.929	6.224

6

Applications of the Entropy Concept

Introduction

Although the Second Law of Thermodynamics was stated with respect to the impossibility of constructing a certain kind of machine (the so-called perpetual-motion machine of the second kind), the most useful applications of the law, from the point of view of the chemist, deal with chemical reactions and changes in state. The relationship between heat engines and chemical reactions is not obvious, but a fundamental connection does exist, and that connection is best established through the concept of entropy, which is an outgrowth of Second Law considerations. Once it has been shown that dQ_{rev}/T is an exact differential, which we call dS, it becomes possible to derive innumerable additional relationships, involving energy, work, heat capacities, etc. Moreover, by modification of the statement that $dS \geqslant dQ/T$, we can formulate still other inequalities governing spontaneous processes and can establish upper limits to the useful work obtainable under certain conditions. An appreciable effort will be required if we are to accomplish all these things; nevertheless, we shall undertake this task by considering, first of all, several applications of the Second Law that are illustrative of its broad generality.

Dependence of Energy on Volume and of Enthalpy on Pressure

In Chapter Two we discussed an approach to a knowledge of $(\partial U/\partial V)_T$ and $(\partial H/\partial p)_T$ through the Gay-Lussac, Joule, and Joule-Thomson experiments. By use of Second Law considerations, however, we can now establish practical formulas for those partial derivatives —formulas that will require only a knowledge of the equations of state of the materials involved.

Combining the general expression for dQ, equation 2.25, with the definition of entropy, equation 4.30, we obtain

$$dS = \frac{1}{T}\left(\frac{\partial U}{\partial T}\right)_V dT + \frac{1}{T}\left\{\left(\frac{\partial U}{\partial V}\right)_T + p\right\} dV \qquad (6.1)$$

Since dS is an exact differential, examination of the coefficients of dT and dV shows that

$$\left(\frac{\partial S}{\partial T}\right)_V = \frac{1}{T}\left(\frac{\partial U}{\partial T}\right)_V = \frac{C_v}{T} \qquad (6.2)$$

and

$$\left(\frac{\partial S}{\partial V}\right)_T = \frac{1}{T}\left\{\left(\frac{\partial U}{\partial V}\right)_T + p\right\} \qquad (6.3)$$

But we also know that $\partial^2 S/\partial T\partial V = \partial^2 S/\partial V\partial T$, since the order of differentiation is immaterial. Hence by forming the indicated second partial derivatives from equations 6.2 and 6.3, we obtain, upon simplification and rearrangement, the following result:

$$\left(\frac{\partial U}{\partial V}\right)_T = T\left(\frac{\partial p}{\partial T}\right)_V - p \qquad (6.4)$$

Also, by substituting equation 6.4 back into equation 6.3, we deduce that

$$\left(\frac{\partial S}{\partial V}\right)_T = \left(\frac{\partial p}{\partial T}\right)_V \qquad (6.5)$$

EXERCISE:

Show from equation 6.2 that $\partial^2 S/\partial V\partial T = (1/T)(\partial^2 U/\partial V\partial T)$. Also obtain another expression for $\partial^2 S/\partial V\partial T$ by differentiating equation 6.3 partially with respect to T. By equating the second partial derivatives so obtained, derive equation 6.4.

Since the right-hand sides of equations 6.4 and 6.5 depend only upon the equation of state of the system under consideration, it is evident that those equations should be particularly useful for evaluating quantities that might otherwise be exceedingly difficult to determine experimentally. For example, if we are concerned with an ideal gas, for which $p = nRT/V$, we can immediately show that

$$\left(\frac{\partial U}{\partial V}\right)_T = 0 \tag{6.6}$$

This is the same conclusion that was reached from the experiments of Gay-Lussac and Joule. The present indirect approach is more satisfying, however, than that which depended upon the observations of a not very precise experiment.

We can also show, from equation 6.5 and the equation of state for an ideal gas, that

$$\left(\frac{\partial S}{\partial V}\right)_T = \frac{nR}{V} \tag{6.7}$$

Integration of this equation immediately yields equation 4.38, which was previously obtained for the entropy change attending the isothermal expansion of an ideal gas.

The method just employed for deriving expressions for $(\partial U/\partial V)_T$ and $(\partial S/\partial V)_T$ will work equally well to establish equations for $(\partial H/\partial p)_T$ and $(\partial S/\partial p)_T$. To accomplish this end, one should start with a combination of equations 2.26 and 4.30 and proceed as before. It will first be recognized that

$$\left(\frac{\partial S}{\partial T}\right)_p = \frac{1}{T}\left(\frac{\partial H}{\partial T}\right)_p = \frac{C_p}{T} \tag{6.8}$$

and that

$$\left(\frac{\partial S}{\partial p}\right)_T = \frac{1}{T}\left\{\left(\frac{\partial H}{\partial p}\right)_T - V\right\} \tag{6.9}$$

Upon forming appropriate second partial derivatives in each of two ways, one obtains, after simplification, the following equations:

$$\left(\frac{\partial H}{\partial p}\right)_T = V - T\left(\frac{\partial V}{\partial T}\right)_p \tag{6.10}$$

$$\left(\frac{\partial S}{\partial p}\right)_T = -\left(\frac{\partial V}{\partial T}\right)_p \tag{6.11}$$

For ideal gases it can readily be shown that

$$\left(\frac{\partial H}{\partial p}\right)_T = 0 \tag{6.12}$$

and that

$$\left(\frac{\partial S}{\partial p}\right)_T = -\frac{nR}{p} \tag{6.13}$$

The last two equations are likewise consistent with earlier conclusions.

EXERCISE:

Derive equation 6.10 by the method described in the text and illustrated in the exercise on page 115.

EXERCISE:

Show, in the light of the Third Law, that $(\partial S/\partial V)_T$ and $(\partial S/\partial p)_T$ must both equal zero for a perfect crystal at absolute zero. Hence, in view of equations 6.5 and 6.11, show that $(\partial p/\partial T)_V$ and $(\partial V/\partial T_p)$ must both approach zero at $T \longrightarrow 0$.

Equations 6.4 and 6.10 also enable us to rewrite in more convenient form the generalized heat-capacity expressions. Substituting those equations into equations 2.29 and 2.30, and taking cognizance of what is meant by C_v and C_p, we obtain the equations

$$C = C_v + T\left(\frac{\partial p}{\partial T}\right)_V \frac{dV}{dT} \tag{6.14}$$

$$C = C_p - T\left(\frac{\partial V}{\partial T}\right)_p \frac{dp}{dT} \tag{6.15}$$

Finally, we obtain from either equation 2.35 or 2.36 another expression for $C_p - C_v$:

$$C_p - C_v = T\left(\frac{\partial p}{\partial T}\right)_V \left(\frac{\partial V}{\partial T}\right)_p \tag{6.16}$$

From this result we can readily establish that $C_p - C_v = R$ for one mole of an ideal gas. If we introduce into equation 6.16 the coefficient of compressibility, $\kappa = -\frac{1}{V}\left(\frac{\partial V}{\partial p}\right)_T$, and the coefficient of thermal expansion, $\alpha = \frac{1}{V}\left(\frac{\partial V}{\partial T}\right)_p$, we shall see that

$$C_p - C_v = \frac{TV\alpha^2}{\kappa} \tag{6.17}$$

Equation 6.17 is, of course, valid for liquids and solids as well as for gases. Although that equation is not among the most important conclusions derivable from the Second Law, it is nevertheless a striking example of the kind of result that can be obtained from such a mundane proposition as the impossibility of constructing a perpetual-motion machine.

EXERCISE:

Calculate the difference $C_p - C_v$ for 1 mole of water at 20 °C. At that temperature the coefficient of compressibility is 49×10^{-6} per atm and the coefficient of thermal expansion is 2.1×10^{-4} per degree.

Entropy of an Ideal Gas

It has already been shown that the energy and enthalpy of an ideal gas are functions of temperature only, and that their differentials must accordingly be given by the expressions $dU = C_v dT$ and $dH = C_p dT$. C_v and C_p are, of course, also functions of T alone and are subject to the further condition that $C_p - C_v = R$ for one mole. Let us now consider the total differential of S for a mole of ideal gas, utilizing first V and T as independent variables and then p and T. From equations 6.2 and 6.7 we see that

$$dS = \frac{C_v}{T} dT + \frac{R}{V} dV \tag{6.18}$$

Integration of equation 6.18 immediately yields the expression

$$S = \int \frac{C_v}{T} dT + R \ln V + k \tag{6.19}$$

in which k is a constant of integration whose value could be determined by Third Law calculations. If we assume C_v to be constant, equation 6.19 reduces to the equation

$$S = C_v \ln T + R \ln V + k \tag{6.20}$$

Evidently, if one considers a change in state from T_1, V_1 to T_2, V_2, the change in entropy will be given by the expression

$$\Delta S = C_v \ln \frac{T_2}{T_1} + R \ln \frac{V_2}{V_1} \tag{6.21}$$

Operating in a manner similar to that employed above, but using p and T as the independent variables, one finds that

$$S = \int \frac{C_p}{T} dT - R \ln p + k' \tag{6.22}$$

Furthermore, if C_p can be regarded as constant,

$$S = C_p \ln T - R \ln p + k' \tag{6.23}$$

Finally, for a change in state from T_1, p_1 to T_2, p_2, we obtain the equation

$$\Delta S = C_p \ln \frac{T_2}{T_1} - R \ln \frac{p_2}{p_1} \tag{6.24}$$

In view of the equation of state for an ideal gas, the equations for S involving V and T are fully compatible with those expressed in terms of p and T. The choice of equations to use for any particular set of circumstances is usually determined by convenience and the form in which the data are supplied.

Thermodynamic Properties of a van der Waals Fluid

We have just seen that a number of important thermodynamic quantities can be deduced from a knowledge of the equation of state of the substance involved. To illustrate these calculations further, let us consider one mole of a van der Waals fluid, for which the equation of state is

$$(p + a/V^2)(V - b) = RT \tag{6.25}$$

Solving for p and substituting into equation 6.4, we obtain the result

$$\left(\frac{\partial U}{\partial V}\right)_T = \frac{a}{V^2} \tag{6.26}$$

Upon integration this yields the equation

$$U = -\frac{a}{V} + f(T) \tag{6.27}$$

in which $f(T)$ is some function of the temperature only, a function whose derivative is precisely the heat capacity at constant volume:

$$\left(\frac{\partial U}{\partial T}\right)_V = C_v = f'(T) \tag{6.28}$$

Assuming C_v to be constant, we conclude that

$$U = -\frac{a}{V} + C_v T + \text{constant} \tag{6.29}$$

It is evident from equations 6.27 and 6.29 that, as the volume becomes infinite, the energy approaches that of an ideal gas. This is in keeping with the fact that at low enough pressures all gases tend to behave ideally. To obtain the entropy of a van der Waals fluid, we should start with its total differential, making use of equations 6.2 and 6.5 in conjunction with the equation of state:

$$dS = \frac{C_v}{T} dT + \frac{R}{V - b} dV \tag{6.30}$$

Upon integration, once again assuming C_v to be constant, we obtain the equation

$$S = C_v \ln T + R \ln (V - b) + k \tag{6.31}$$

Equation 6.31 is quite similar to that applicable to an ideal gas, equation 6.20. Once again, as with the energy, we observe that, as V becomes infinite, the entropy approaches that of the ideal gas, since b becomes negligible compared with V under such circumstances. Although van der Waals' equation has two constants that give rise to deviations from ideality, only one of them, a, affects the energy, the other, b, modifying the entropy, provided the independent variables are chosen as V and T.

The heat added during an *isothermal* reversible expansion of a van der Waals gas will evidently be given by the expression

$$Q_{rev} = T\Delta S = RT \ln \frac{V_2 - b}{V_1 - b} \tag{6.32}$$

From equation 6.27 we also conclude that the isothermal energy change will be

$$\Delta U = -a \left(\frac{1}{V_2} - \frac{1}{V_1} \right) \tag{6.33}$$

Combining these results with the First Law expression, we find that

$$W_{rev} = RT \ln \frac{V_2 - b}{V_1 - b} + a \left(\frac{1}{V_2} - \frac{1}{V_1} \right) \tag{6.34}$$

This result is identical with that obtained by direct integration of pdV.

The preceding equations relating to a van der Waals system have all been simple in form because we have employed V and T as independent variables and have adhered to the U,S,V,T scheme of representation. If p and T are chosen as independent variables, so as to carry out the formulation in terms of H, S, p, and T, the treatment becomes less tractable, and approximations must be made to avoid unduly complicated expressions. The underlying reason is that, although p can readily be eliminated as a function of V and T, one cannot write V as a function of p and T without solving a cubic equation. To avoid the complication of solving a cubic equation, we shall accordingly write van der Waals' equation in an approximate form, valid for low pressures:

$$pV = RT + (b - a/RT)p \tag{6.35}$$

It is evident from this equation* that the Boyle temperature—that is, the temperature at which pV is constant—is given by the expression

$$T_{Boyle} = \frac{a}{Rb} \tag{6.36}$$

Solving equation 6.35 for V and substituting into equation 6.10, we can show that

---

* Equations 6.35 and 6.36 were the subject of derivations in Problem 8, Chapter One.

$$\left(\frac{\partial H}{\partial p}\right)_T = b - \frac{2a}{RT} \qquad (6.37)$$

From this we conclude that

$$H = \left(b - \frac{2a}{RT}\right)p + g(T) \qquad (6.38)$$

in which expression $g(T)$ is some function of T. The heat capacity at constant pressure will then be given by

$$C_p = \left(\frac{\partial H}{\partial T}\right)_p = \frac{2ap}{RT^2} + g'(T) \qquad (6.39)$$

From equation 6.11, moreover, we deduce that

$$\left(\frac{\partial S}{\partial p}\right)_T = -\frac{R}{p} - \frac{a}{RT^2} \qquad (6.40)$$

Hence, for an isothermal expansion of a van der Waals fluid between moderate pressures, we obtain the expression

$$\Delta S = -R \ln\frac{p_2}{p_1} - \frac{a}{RT^2}(p_2 - p_1) \qquad (6.41)$$

It can be shown, finally, that at low pressures

$$C_p - C_v = R + \frac{2ap}{RT^2} \qquad (6.42)$$

In Chapter Two we saw that the Joule-Thomson coefficient can be written as

$$\mu = -\frac{\left(\frac{\partial H}{\partial p}\right)_T}{C_p} \qquad (6.43)$$

In the light of equation 6.10 this expression can now be rewritten as

$$\mu = \frac{T\left(\frac{\partial V}{\partial T}\right)_p - V}{C_p} \qquad (6.44)$$

It is clear, from its equation of state, that an ideal gas will have a Joule-Thomson coefficient equal to zero. In view of equation 6.37, however, that coefficient for a dilute van der Waals gas will be given by the expression

$$\mu = \frac{\frac{2a}{RT} - b}{C_p} \qquad (6.45)$$

Although equation 6.45 does not agree exactly with experimental observations, it nevertheless enables one to predict qualitatively a behavior that is actually observed for gases at low pressures. At low temperatures the Joule-Thomson coefficient is positive; as the temperature rises, it passes through an inversion point and becomes negative. Equation 6.45 suggests this behavior and predicts such an inversion temperature to be

$$T_{\text{inv}} = \frac{2a}{Rb} \tag{6.46}$$

The Joule-Thomson inversion temperature, according to the van der Waals equation, is therefore twice the Boyle temperature, a conclusion that is only roughly correct.

EXERCISE:

One mole of a van der Waals gas expands freely into a vacuum from volume V_1 at temperature T_1 to a new total volume, V_2. If the process is adiabatic, what is the final temperature? Assume that C_v is constant.

$$\text{Ans.} \quad T_2 = T_1 + \frac{a}{C_v}\left(\frac{1}{V_2} - \frac{1}{V_1}\right)$$

EXERCISE:

Show that $g'(T)$ and $f'(T)$, appearing in equations 6.39 and 6.28, are so related that $g'(T) - f'(T) = R$.

Entropy Change on Mixing of Ideal Gases

Up to the present, our treatment of chemical thermodynamics has been concerned for the most part with pure substances, and little

FIGURE 6-1   Diffusion of ideal gases into each other.

Before diffusion   **After diffusion**

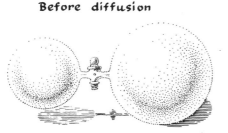
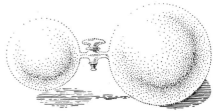

n_A, V_1, p, T n_B, V_2, p, T $n_A + n_B, V_1 + V_2, p, T$

reference has been made to systems of variable composition. At this point we shall make an inquiry into composition effects; specifically, we shall investigate the change in entropy attending the mixing of ideal gases.

The process upon which we shall focus our attention is the following: n_A moles of gas A and n_B moles of gas B, each at pressure p and temperature T, are brought together and allowed to diffuse into each other at constant total volume and temperature (Fig. 6-1). This process closely resembles the Joule experiment, the difference being that we allow two gases to diffuse into each other instead of allowing one gas to flow into a vacuum.

If V_1 and V_2 are the original volumes of the gases, then, in accordance with the ideal-gas equation of state, we can assert that

$$pV_1 = n_A RT$$

and that

$$pV_2 = n_B RT \tag{6.47}$$

Moreover, if the temperature is kept constant, the final pressure must be the same as the original pressure, since

$$p(V_1 + V_2) = (n_A + n_B)RT \tag{6.48}$$

in which expression $V_1 + V_2$ is the total volume of the mixture. The final partial pressures will be given by the expressions

$$p_A = x_A p$$

and

$$p_B = x_B p \tag{6.49}$$

x_A and x_B being the mole fractions of A and B in the mixture.

Now the energy of an ideal gas is a function of temperature only, and the energy of a mixture of such gases is the sum of their separate energies. The latter is true because no intermolecular forces operate in the ideal gaseous state, and the average energy of a molecule is accordingly independent of its environment. The change in energy associated with the isothermal mixing of ideal gases must therefore be zero. Because the experiment described above is so constructed that $W = 0$, it follows from the First Law that Q also is equal to zero. Since the process is clearly an irreversible one, we conclude from inequality 4.37 that $\Delta S > 0$.

To evaluate ΔS precisely, we must conceive of some reversible means of bringing about the same change in state and must sum up Q_{rev}/T for that path. A reversible and isothermal method of mixing the two gases can be carried out in three steps as follows:

1. Expand gas A from volume V_1 to volume $V_1 + V_2$ at constant T. For this process we can assert that $Q_{rev} = n_A R T \ln (V_1 + V_2)/V_1$ and that

$$\Delta S_1 = n_A R \ln \frac{V_1 + V_2}{V_1} = -n_A R \ln x_A \qquad (6.50)$$

x_A being the mole fraction of A in the final mixture.

2. Likewise expand gas B from volume V_2 to volume $V_1 + V_2$ at constant T. Then, similarly,

$$\Delta S_2 = -n_B R \ln x_B \qquad (6.51)$$

3. Mix the gases reversibly by means of an apparatus illustrated in Figure 6-2. This apparatus consists of two cylindrical boxes, aa' and $b'b$, each of volume $V_1 + V_2$ and of such character that one can be slid into the other until they enclose the same volume. Gas A is placed in box aa' and gas B in box $b'b$. End a' of box aa' consists of a semi-permeable membrane that will permit molecules of gas B to pass through but is impermeable to molecules of A. Similarly, end b' of box $b'b$ is a semi-permeable membrane that permits A molecules to pass through but not B. As the boxes are pushed together, the pressure in the interior region, $a'b'$, will equal $p_A + p_B = p$, whereas the pressures in ab' and $a'b$ will equal p_A and p_B respectively.

FIGURE 6-2 Apparatus for reversible mixing of gases. Membranes impermeable to gases A and B are represented at a' and b'. Volumes a–a' and b–b' are equal and remain constant as the sections slide together, one into the other. Initially, a' and b' are in contact; at the end of the operation, a' is in contact with b and b' in contact with a. Axial rod holds b and b' at fixed distance from each other; rod slides through close-fitting hole in membrane a'.

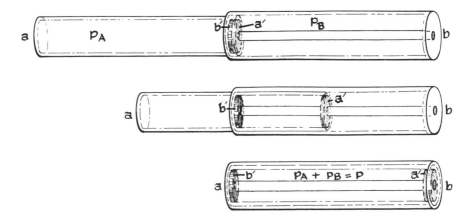

The pressure effective on the two ends of aa' is always equal to p_A because the B molecules exert no net pressure on wall a'. Likewise, the net pressure on each end of box $b'b$ is equal to p_B. The work, W, accompanying the process of bringing the two boxes into coincidence must therefore equal zero, for no net forces are operating through any distance. Because the temperature is unchanged, the energy is likewise unchanged; so from the First Law it follows that $Q_{rev} = 0$. Hence

$$\Delta S_3 = 0 \tag{6.52}$$

Since the state attained after completion of process 3 is precisely that obtained by the original mixing experiment, we conclude that

$$\Delta S_{mix} = \Delta S_1 + \Delta S_2 + \Delta S_3$$

$$= -n_A R \ln x_A - n_B R \ln x_B \tag{6.53}$$

In view of Dalton's Law of Partial Pressures, we can equally well write that

$$\Delta S_{mix} = -n_A R \ln \frac{p_A}{p} - n_B R \ln \frac{p_B}{p} \tag{6.54}$$

In other words, the entropy of mixing is exactly equal to the sum of the entropy changes for the expansion of the separate gases, provided their final pressures are the same as their partial pressures would be in the mixture.

The total entropy of a mixture will, of course, be equal to the original entropies of the pure gases plus the entropy of mixing. Employing expressions like equation 6.23 for the individual entropies, we see that for a mixture

$$S_{total} = (n_A C_{pA} + n_B C_{pB}) \ln T - (n_A + n_B) R \ln p$$

$$- n_A R \ln x_A - n_B R \ln x_B + n_A k'_A + n_B k'_B \tag{6.55}$$

k'_A and k'_B being Third Law entropy constants for the two gases. An equally good expression involving partial pressures is

$$S_{total} = (n_A C_{pA} + n_B C_{pB}) \ln T - n_A R \ln p_A$$

$$- n_B R \ln p_B + n_A k'_A + n_B k'_B \tag{6.56}$$

from which it is clear that the total entropy is the sum of the separate entropies, each calculated as if the other gas were not present. Using volume as a variable, we can write

$$S_{total} = (n_A C_{vA} + n_B C_{vB}) \ln T$$

$$+ (n_A + n_B) R \ln V + n_A k_A + n_B k_B \tag{6.57}$$

in which V is the total volume.

EXERCISE:

Consider two vessels, each of 1-liter capacity, connected as in the Joule experiment. Originally, one of the vessels contains 0.1 mole of ideal gas A at 25 °C, and the other contains 0.1 mole of ideal gas B, also at 25 °C. The gases are now allowed to diffuse into each other isothermally. What is ΔS in cal deg^{-1} for the process?

Concluding Remarks

The applications of the entropy concept dealt with in this chapter are only illustrative of the many things that can be done. Further developments will be carried through in later chapters, but the significant groundwork has already been laid. For convenience we shall define some new functions in terms of energy and entropy, and when we use those functions we may occasionally be inclined to lose sight of entropy. Actually, the entropy concept and the Second Law of Thermodynamics will be ever present in our subsequent developments, although they may be disguised in various forms. What follows really amounts to application, for the basis of thermodynamics has now been formulated.

Problems

1. Show that, if $(\partial U/\partial V)_T = 0$, the equation of state of the substance must be of the form
$$p = T\phi(V)$$
in which $\phi(V)$ is an arbitrary function of V.

2. Show that, if $(\partial H/\partial p)_T = 0$, the equation of state of the substance must be of the form
$$V = T\psi(p)$$
in which $\psi(p)$ is an arbitrary function of p.

3. Show, in view of the results of Problems 1 and 2, that, if $(\partial U/\partial V)_T = 0$ and $(\partial H/\partial p)_T = 0$, then pV/T is constant.

4. Along what kind of path would a van der Waals gas possess a heat capacity equal to $C_v + R$?

   Ans. $p + a/V^2 = $ constant

5. A cylinder with a frictionless piston contains 1 mole of an ideal gas. In addition, the closed end of the cylinder is joined to the piston by means of a spring that obeys Hooke's Law with a potential energy of $\frac{1}{2}k(L - L_0)^2$, k being the Hooke's Law constant, L and length of the spring, and L_0 the equilibrium length of the spring. (See the following figure.) An external observer studies the p-V-T relations of the gas without knowing about the presence of the spring.

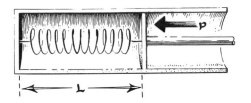

(A) Assuming that k is independent of the temperature, what will the observer find for the apparent equation of state? Let V and V_0 be the volumes corresponding to L and L_0, and let A be the cross-sectional area of the cylinder.

(B) From the apparent equation of state, in conjunction with equation 6.4, show how the energy of this "gas" depends upon its volume. Explain.

$$\text{Ans.} \quad (\text{B}) \quad U = \frac{k(V - V_0)^2}{2A^2} + C_v T + \text{constant}$$

(This problem shows how hidden forces operating between molecules might also be measured, at least to some extent, by thermodynamic analysis of the equation of state.)

6. A certain liquid obeys the equation of state

$$V = V_0[1 + \alpha t - \kappa (p - 1)]$$

in which α and κ are constants and V_0 is the volume at 0 °C and 1 atm pressure, p being expressed in atmospheres.

(A) Prove that

$$\left(\frac{\partial S}{\partial p}\right)_T = - V_0 \alpha$$

(B) Prove that

$$\left(\frac{\partial S}{\partial V}\right)_T = \frac{\alpha}{\kappa}$$

(c) Prove that

$$\left(\frac{\partial H}{\partial p}\right)_T = V_0[1 - 273\alpha - \kappa(p - 1)]$$

(D) For water in the neighborhood of 20 °C and 1 atm, $\alpha = 2.1 \times 10^{-4}$ per degree and $\kappa = 49 \times 10^{-6}$ per atm. Calculate ΔS in cal deg^{-1} and ΔH in calories when 1 mole of water is compressed at 20 °C from 1 to 25 atm. (Compare with Prob. 2, Chap. Two.)

7. Show that, if a gas obeys the equation of state

$$pV = RT + \alpha p$$

its energy will be expressed by an equation of the form

$$U = -pT\frac{d\alpha}{dT} + \phi(T)$$

8. (A) Show that, if a gas obeys the equation of state

$$pV = RT + \alpha p$$

then

$$C_p - C_v = R \left(1 + \frac{p}{R} \frac{d\alpha}{dT}\right)^2$$

$$= R + 2p \frac{d\alpha}{dT} \text{ (approximately)}$$

Compare the approximate form of the answer with equation 6.42 of the text, and discuss.

(B) For hydrogen, α equals 0.0142 liter at 0 °C and 0.0148 liter at 20 °C. What is $C_p - C_v$ for hydrogen at 1 atm pressure in the neighborhood of 10 °C?

9. Show that, to the order of approximation implicit in equation 6.35, $g(T) - f(T) = RT$, where $g(T)$ and $f(T)$ are the functions introduced in equations 6.38 and 6.27. (In other words, upon integrating the result of the second exercise on page 122, the constant of integration must be zero.)

10. (A) Prove, for all substances, that

$$\left(\frac{\partial p}{\partial T}\right)_S = \left(\frac{\gamma}{\gamma - 1}\right)\left(\frac{\partial p}{\partial T}\right)_V$$

if $\gamma = C_p/C_v$. Suggestion: Start with the expression $(\partial p/\partial T)_S = -(\partial S/\partial T)_p/(\partial S/\partial p)_T$, make use of equations 6.8, 6.11, and 6.16, and rearrange to final answer.

(B) From the result of A, show, for an ideal gas with constant γ, that a reversible adiabatic process is subject to the equation

$$p = \text{constant } T^{\gamma/(\gamma-1)}$$

11. (A) Prove the following:

$$\left(\frac{\partial V}{\partial T}\right)_S = -\frac{1}{(\gamma - 1)}\left(\frac{\partial V}{\partial T}\right)_p$$

Proceed by a method similar to that suggested for Problem 10.

(B) Show, as a consequence, that for an ideal gas a reversible adiabatic process is subject to the equation $VT^{1/(\gamma-1)} = \text{constant.}$

12. (A) Prove the following:

$$\left(\frac{\partial p}{\partial V}\right)_S = \gamma \left(\frac{\partial p}{\partial V}\right)_T$$

This is somewhat more difficult than Problems 10 and 11. Proceed by replacing $(\partial p/\partial V)_S$ by $-(\partial S/\partial V)_p/(\partial S/\partial p)_V$ and $(\partial p/\partial V)_T$ by $-(\partial T/\partial V)_p/(\partial T/\partial p)_V$. Then observe that $(\partial S/\partial V)_p = (\partial S/\partial T)_V(\partial T/\partial V)_p + (\partial S/\partial V)_T$ and that a similar equation holds for $(\partial S/\partial p)_V$. By appropriate manipulation based on equations derived in the text, complete the proof.

(B) Show, as a consequence, that for an ideal gas a reversible adiabatic process is subject to the equation $pV^\gamma = \text{constant.}$

13. Prove that

$$C_p = T \left(\frac{\partial p}{\partial T}\right)_S \left(\frac{\partial V}{\partial T}\right)_p$$

14. A 1-liter vessel containing hydrogen at 25 °C and 1 atm is connected to a 2-liter vessel containing oxygen at 25 °C and 0.5 atm. If the gases are allowed to diffuse into each other isothermally, what is ΔS in cal deg^{-1}? (Assume gaseous ideality.)

15. A 1-liter vessel containing helium at 100 °C and 1 atm is connected to a 2-liter vessel containing neon at 0 °C and 0.4 atm. The gases are allowed to mix with each other and reach thermal equilibrium adiabatically. What is the final state of affairs, and what is ΔS for the process in cal deg^{-1}?

16. Prove, for any homogeneous substance, that

$$\left(\frac{\partial U}{\partial p}\right)_T = -T \left(\frac{\partial V}{\partial T}\right)_p - p \left(\frac{\partial V}{\partial p}\right)_T$$

and that

$$\left(\frac{\partial H}{\partial V}\right)_T = T \left(\frac{\partial p}{\partial T}\right)_V + V \left(\frac{\partial p}{\partial V}\right)_T$$

7

Isothermal Processes; Helmholtz and Gibbs Free Energies

Introduction

The Second Law of Thermodynamics is a very general proposition and is applicable to all kinds of individual processes or combinations of processes. It will be of interest, however, to examine it under certain restricted conditions to gain a fuller insight into some of its more detailed implications. Since isothermal processes are particularly important, it appears worth while to see what additional conclusions might be obtained, and what new functions might profitably be defined by application of the Second Law to such processes.

If we restrict ourselves to isothermal processes, the Second Law can be restated as follows: *It is impossible to construct a machine operating in isothermal cycles that will convert heat into work.* This statement is, of course, a corollary of the general Second Law, for we showed from Carnot-cycle considerations that at least two heat reservoirs at different temperatures are necessary before any work can be produced by a cycle. The restricted statement, however, is well suited to our immediate needs, for it enables us to come to an important conclusion regarding work attending non-cyclical isothermal processes.

Consider a change in state that can be brought about along either of two different reversible isothermal paths, and let W_1 and W_2 represent the work quantities associated with those paths. In accordance with our modified statement of the Second Law, we now conclude that W_1 must equal W_2. This proposition is proved as follows:

Suppose that the two work quantities were not equal and that W_1 were greater than W_2. Under such circumstances we could construct a perpetual-motion machine by simply coupling process 1 with the reverse of process 2. Evidently process 1 would yield quantity of work W_1, and the reverse of process 2 would require expenditure of work W_2; hence the net work appearing in the surroundings for each cycle of operation would equal $W_1 - W_2$, a positive quantity. Since this would be in violation of the Second Law, we conclude that W_1 cannot be greater than W_2. By a similar argument, involving a reversal of the sense of the cycle, we can also show that W_2 cannot be greater than W_1, thus leaving only the possibility that they are equal. This is equivalent to saying that the reversible work associated with an isothermal change in state must be independent of the path followed; the isothermal reversible work must, therefore, be a function only of the initial and final states, and hence can be measured by the change in some function of the state of the system. We shall now seek to identify that function.

Helmholtz Free Energy and Gibbs Free Energy

For isothermal processes, we have seen in accordance with expression 4.36 that $T\Delta S \geqslant Q$, the equality holding for reversible processes and the inequality for all others. If we combine this with the First Law equation, it follows that

$$W \leqslant -\Delta U + T\Delta S \tag{7.1}$$

In Chapter Two we saw, in general, that W could assume any real value whatsoever for a given change in state, depending upon the path followed. However, it now appears that, if the change in state is carried out isothermally, there is an upper limit to W—namely, $-\Delta U + T\Delta S$. This upper limit is precisely equal to the isothermal *reversible* work and, in agreement with the conclusion of the preceding section, it must be a function of the initial and final states of the system, for ΔU and ΔS are both such functions. It is precisely this proposition that suggests that we define a new function of the state of a system, a function whose change will measure the reversible (or maximum) work associated with an isothermal process. To this function we shall give the symbol A and the name *Helmholtz free energy,*[*] with the definition

$$A = U - TS \qquad (7.2)$$

In terms of this new function, expression 7.1 can be rewritten as

$$W \leqslant -\Delta A \qquad (7.3)$$

which means that the decrease in Helmholtz free energy is an upper limit to the work obtainable in an isothermal process.

It will now be evident that, if an isothermal system is so constrained that $W = 0$ (which we can realize, for example, by keeping the volume constant and by excluding electrical and other special kinds of work), then

$$\Delta A \leqslant 0 \qquad (7.4)$$

We conclude, therefore, that the Helmholtz free energy of an isothermal system constrained to constant volume will spontaneously tend to decrease until it reaches a minimum, and that, once the minimum is attained, the system will be in a state of equilibrium. This conclusion is comparable to the statement (expression 4.37) that the entropy of an adiabatic system tends to increase until it reaches its maximum (equilibrium) value.

Let us now turn to processes that occur both isothermally and at constant pressure. Under such circumstances, the reversible pressure-volume work will equal $p\Delta V$, and the quantity $W - p\Delta V$ will equal all non-pV work, such as electrical work, that might appear in the surroundings. If we denote $W - p\Delta V$ by W', it follows from expression 7.3 that

$$W' \leqslant -\Delta A - p\Delta V \qquad (7.5)$$

If the system under consideration consists of several parts at different pressures, and if these pressures are all kept constant, expression 7.5 can be readily generalized to cover such a situation. If we let p_i be the pressure of the ith portion and ΔV_i the change in volume of that portion, the total pressure-volume work will be $\sum_i p_i \Delta V_i$; so we can write that

$$W' \leqslant -\Delta A - \sum_i p_i \Delta V_i \qquad (7.6)$$

in which expression W' once again represents all work in excess of pressure-volume work. The forms exhibited by expressions 7.5 and 7.6

---

[*] This important quantity is also called the *work content,* the *Helmholtz energy,* and the *Helmholtz function.*

suggest that still another function might be defined, one bearing the same relationship to the Helmholtz free energy as the enthalpy bears to the energy. This function will be called the *Gibbs free energy** and will be denoted by G, with the definition

$$
\begin{aligned}
G &= A + pV \\
  &= E - TS + pV \\
  &= H - TS
\end{aligned}
\tag{7.7}
$$

Rewriting expression 7.5 or 7.6, we observe that

$$W' \leqslant -\Delta G \tag{7.8}$$

from which we conclude that the decrease in Gibbs free energy attending an isothermal, constant-pressure process is an upper limit to all work, other than pressure-volume work, that might appear in the surroundings. If an isothermal system is so constrained that the only possible work is pressure-volume work, then

$$\Delta G \leqslant 0 \tag{7.9}$$

which is another condition for spontaneity and equilibrium. In other words, at constant T and p the Gibbs free energy tends to decrease until it reaches a minimum, at which point the system will be in equilibrium.

EXERCISE:

Consider a general chemical reaction carried out at constant temperature:

$$aA + bB + \cdots \longrightarrow eE + fF + \cdots$$

Show that, if the gases involved can be considered ideal and if the volumes of solids and liquids, compared with those of gases, can be neglected, then

$$\Delta G = \Delta A + \Delta \nu RT$$

provided $\Delta \nu$ measures the increase in the number of moles of gases for the reaction.

Application to Electrical Cells

Before we established expression 7.8, all of our explicit textual references to work involved work of the pressure-volume type only. We are now in a position to deal with other kinds of work—in particular, with electrical work, such as might accompany a change in state brought about through the operation of an electrical cell. Specifically, let us

---

* The *Gibbs free energy* is also known as the *free energy*, the *Gibbs energy*, and the *Gibbs function*.

consider a cell consisting of two hydrogen electrodes (at pressures p_1 and p_2) immersed in a conducting solution such as dilute aqueous sulfuric acid, all at a uniform temperature, T:

$$\text{Pt, } H_2(p_1), H_2SO_4 \cdot x\text{Aq}, H_2(p_2), \text{ Pt}$$

In this representation, Pt denotes platinized platinum, a metallic surface that is capable of catalyzing, and rendering reversible, the dissociation of hydrogen into hydrogen ions and electrons. We shall now adopt the convention that the electromotive force of a cell (to be denoted by E) is positive if, during spontaneous operation, a positive current flows from left to right within the cell. A negative electromotive force will then imply a tendency for current to flow in the opposite direction. The electrical work associated with z faradays moving through the cell will be given by the expression

$$W_{el} = zEF \tag{7.10}$$

in which F is the value of the faraday,* the equivalent of one mole of electricity.

Let us now evaluate the total work associated with the disappearance of one mole of hydrogen (at p_1) and the appearance of one mole of hydrogen (at p_2) brought about by the reversible operation of the cell. The reaction that occurs at the left-hand electrode can be written as

$$H_2(p_1) \longrightarrow 2H^+ + 2e^-$$

whereas the reaction at the right-hand electrode will be represented by the expression

$$2H^+ + 2e^- \longrightarrow H_2(p_2)$$

The net result of the operation, so far as hydrogen is concerned, is the sum of the two reactions:

$$H_2(p_1) \longrightarrow H_2(p_2)$$

The pressure-volume work at the left-hand electrode will be simply equal to $-p_1V_1$ since V_1, the volume of one mole of hydrogen at pressure p_1, is exactly equal to the decrease in volume of gas at that electrode. Similarly, the work at the other end will equal p_2V_2 since V_2, the volume of one mole of hydrogen at pressure p_2, equals the increase in volume at the right-hand electrode.† The total work will then equal the sum of

---

* One faraday equals 96,487 coulombs, a quantity equivalent to the charge on Avogadro's number of protons or electrons. It is also equivalent to 23,061 calories per volt-equivalent.
† The reader will note that the pressure-volume work associated with the operation of the cell is identical in mathematical form to that of the Joule-Thomson experiment; in this respect the aqueous solution of sulfuric acid fills a role akin to that of the porous plug. The shift from Helmholtz free energy to Gibbs free energy indicated in equation 7.13 is quite analogous to the shift from energy to enthalpy suggested by equation 2.42.

the electrical and pressure-volume works:

$$W = zEF + p_2V_2 - p_1V_1 \qquad (7.11)$$

In this expression z, the number of faradays, equals 2 for each mole of hydrogen. According to equation 7.3, however, the total work for a reversible process is precisely equal to the decrease in Helmholtz free energy $(A_1 - A_2)$; so we further write that

$$-\Delta A = A_1 - A_2 = zEF + p_2V_2 - p_1V_1 \qquad (7.12)$$

Rearranging equation 7.12, we observe that

$$(A_1 + p_1V_1) - (A_2 + p_2V_2) = G_1 - G_2 = zEF \qquad (7.13)$$

or that

$$-\Delta G = zEF = W_{el} \qquad (7.14)$$

Equations 7.14 are, of course, fully compatible with expression 7.8; in the present instance we are concerned with the reversible operation of a cell.

The net change in state attending the passage of two faradays of electricity through the cell is the conversion of $1H_2(g)$ at p_1, T to $1H_2(g)$ at p_2, T. Such a change in state can also be brought about directly by the isothermal expansion (or compression) of one mole of hydrogen, and the reversible work for this alternative process must be exactly equal to that given by equation 7.11. If the work quantities were not equal, one could violate the Second Law through construction of a perpetual-motion machine by using the cell for one part of a cycle and the direct expansion (or compression) for another part, the sense of the cycle to be determined so as to leave a net positive quantity of work. Since this is impossible, we conclude, by equating the two kinds of work, that

$$\int_{V_1}^{V_2} p\,dV = zEF + p_2V_2 - p_1V_1 \qquad (7.15)$$

One can readily integrate the left-hand side of equation 7.15 by parts; upon doing so and rearranging, one obtains

$$\int_{p_1}^{p_2} V\,dp = -zEF \qquad (7.16)$$

Evidently equation 7.16 will enable us to calculate E from a knowledge of the equation of state for hydrogen. In particular, assuming hydrogen to be an ideal gas and noting that $z = 2$, we find upon integration that

$$E = \frac{RT}{2F} \ln \frac{p_1}{p_2} \qquad (7.17)$$

Equation 7.17 shows that E will be positive as long as $p_1 > p_2$, a conclusion that the physical picture makes entirely reasonable. Obviously, if $p_2 > p_1$, hydrogen will tend to disappear from the right-hand electrode and appear on the left side; this would make E negative, for a positive current would flow from right to left within the cell, which direction is opposite to that agreed upon for a positive electromotive force. If hydrogen cannot be regarded as an ideal gas, a more precise equation of state should be utilized for evaluating the integral appearing in equation 7.16. Comparing equation 7.16 with equations 7.14, we conclude that the integral referred to must be equal to the Gibbs free-energy change of the hydrogen gas. The generality of this relationship will be one of the next topics for discussion.

EXERCISE:

Assuming hydrogen to be an ideal gas, calculate the electromotive force at 25 °C for the cell described on page 134 if $p_1 = 10$ atm and $p_2 = 1$ atm.

EXERCISE:

Show that at 25 °C the expression $(\ln 10) RT/F$ equals 0.05916 volt. Show, as a consequence, that at 25 °C equation 7.17 becomes

$$E = \frac{0.05916}{2} \lg \frac{p_1}{p_2} \text{ volt}$$

(Note: The factor 0.05916 will recur frequently in problems involving electromotive force at 25 °C.)

The Differentials of U, H, A, and G

In accordance with the First Law and our definition of energy, we have been able to write $dU = dQ - dW$, an equation valid for all processes, reversible or not. If, in particular, we consider a reversible process, we can write that $dQ = TdS$ and, assuming all work to be pressure-volume work, that $dW = pdV$. Substituting these expressions into that for dU, we obtain the equation

$$dU = TdS - pdV \qquad (7.18)$$

Equation 7.18 can be used for all processes, reversible or not, provided one uses for T and p the values they would have at equilibrium when the independent variables, S and V, are specified.

At this point the reader may properly ask, "How, after reversible paths were assumed for its derivation, can equation 7.18 be used for irreversible processes?" The reason for this becomes clear when we realize that, for an irreversible process, both $TdS > dQ$ and $pdV > dW$, and that the differences giving rise to the two inequalities must of necessity be exactly the same. Thus, even though TdS does not always equal dQ, the differential expression $TdS - pdV$ must always equal $dQ - dW$ since dU, the principal part of the energy increment, is an exact differential. In other words, $dU_{rev} = dU_{irrev}$ for any given infinitesimal change of state since dU depends only on the partial derivatives of the energy and the increments of its independent variables, and not upon how the process occurs.

Equation 7.18 suggests that the energy might properly be regarded as a function of S and V, such that

$$\left(\frac{\partial U}{\partial S}\right)_V = T \quad \text{and} \quad \left(\frac{\partial U}{\partial V}\right)_S = -p \tag{7.19}$$

Using equation 7.18 and the definition of enthalpy ($H = E + pV$), we can also write that

$$dH = TdS + Vdp \tag{7.20}$$

from which it appears that H can be regarded as a function of S and p, such that

$$\left(\frac{\partial H}{\partial S}\right)_p = T \quad \text{and} \quad \left(\frac{\partial H}{\partial p}\right)_S = V \tag{7.21}$$

Moreover, from the definition of Helmholtz free energy ($A = U - TS$), it is seen that

$$dA = -SdT - pdV \tag{7.22}$$

from which it is evident that the natural independent variables to choose for A are T and V and that

$$\left(\frac{\partial A}{\partial T}\right)_V = -S \quad \text{and} \quad \left(\frac{\partial A}{\partial V}\right)_T = -p \tag{7.23}$$

Finally, if we introduce the Gibbs free energy ($G = A + pV$), we conclude that

$$dG = -SdT + Vdp \tag{7.24}$$

and, since T and p appear to be the natural independent variables for G, that

$$\left(\frac{\partial G}{\partial T}\right)_p = -S \quad \text{and} \quad \left(\frac{\partial G}{\partial p}\right)_T = V \tag{7.25}$$

Maxwell's Relations

From the equations given in the preceding section we can now conveniently derive four additional equations, known as Maxwell's thermodynamic relations. We obtain these equations by taking mixed second partial derivatives of the type $\partial^2 U/\partial S\partial V$. Since the order of differentiation is immaterial, it follows from equation 7.19 that

$$\left(\frac{\partial T}{\partial V}\right)_S = -\left(\frac{\partial p}{\partial S}\right)_V \tag{7.26}$$

Similarly, by considering $\partial^2 H/\partial S \partial p$, we obtain from equation 7.21 the result

$$\left(\frac{\partial T}{\partial p}\right)_S = \left(\frac{\partial V}{\partial S}\right)_p \tag{7.27}$$

Likewise, from equation 7.23, we readily see that

$$\left(\frac{\partial S}{\partial V}\right)_T = \left(\frac{\partial p}{\partial T}\right)_V \tag{7.28}$$

And finally, from equation 7.25, we find that

$$\left(\frac{\partial S}{\partial p}\right)_T = -\left(\frac{\partial V}{\partial T}\right)_p \tag{7.29}$$

The last two equations are identical with equations 6.5 and 6.11 but are here derived more simply than before. In this connection it should become even more evident that the essence of the Second Law and its implications reside in the entropy concept, although the two free-energy functions will suggest convenient means of deducing further conclusions.

Calculation of Changes in Helmholtz and Gibbs Free Energies

It is clear from equations 7.23 and 7.25 that changes in the Helmholtz free energy, ΔA, and the Gibbs free energy, ΔG, for isothermal processes can readily be calculated from the equation of state of the material involved. Evidently

$$\Delta A = -\int_{V_1}^{V_2} p\,dV = -W_{\text{rev}} \tag{7.30}$$

and

$$\Delta G = \int_{p_1}^{p_2} V\,dp = \Delta A + \Delta(pV) \tag{7.31}$$

For an ideal gas, $\Delta G = \Delta A$, in compliance with Boyle's Law, and the precise expressions for one mole of such a gas will be

$$\Delta G = \Delta A = -RT \ln \frac{V_2}{V_1} = -RT \ln \frac{p_1}{p_2} \tag{7.32}$$

For a gas obeying the equation of state $pV = RT + \alpha p$, ΔG will not equal ΔA except at the Boyle temperature; the appropriate expressions follow:

$$\Delta A = -RT \ln \frac{V_2 - \alpha}{V_1 - \alpha} = RT \ln \frac{p_2}{p_1} \tag{7.33}$$

$$\Delta G = RT \ln \frac{p_2}{p_1} + \alpha(p_2 - p_1) \tag{7.34}$$

The difference between ΔG and ΔA [namely, $\alpha(p_2 - p_1)$] is precisely equal to $\Delta(pV)$ for the gas. Calculations of ΔA and ΔG for other equations of state are left as exercises for the student.

EXERCISE:

(A) If a gas obeys the equation of state $pV(1 - \beta p) = RT$, show by direct integration of $V dp$ that

$$\Delta G = RT \ln \frac{p_2(1 - \beta p_1)}{p_1(1 - \beta p_2)}$$

(B) By subtracting $\Delta(pV)$ from ΔG, show also that

$$\Delta A = RT \ln \frac{p_2(1 - \beta p_1)}{p_1(1 - \beta p_2)} - \frac{RT}{1 - \beta p_2} + \frac{RT}{1 - \beta p_1}$$

EXERCISE:

Show for a van der Waals gas that

$$\Delta A = -RT \ln \frac{V_2 - b}{V_1 - b} - a \left(\frac{1}{V_2} - \frac{1}{V_1} \right)$$

and that

$$\Delta G = -RT \ln \frac{V_2 - b}{V_1 - b} - 2a \left(\frac{1}{V_2} - \frac{1}{V_1} \right) + bRT \left[\frac{1}{V_2 - b} - \frac{1}{V_1 - b} \right]$$

Free Energies of Ideal Gases

The functional dependence of Helmholtz free energy on volume and of Gibbs free energy on pressure can readily be established for ideal gases by indefinite integration of the equations $\partial A / \partial V = -p$ and $\partial G / \partial p = V$. Making use of the equation of state for one mole of an ideal gas and performing the necessary integrations, one obtains the equations

$$A = -RT \ln V + \phi(T) \tag{7.35}$$

$$G = RT \ln p + \psi(T) \tag{7.36}$$

in which $\phi(T)$ and $\psi(T)$ are functions of temperature only. Equations 7.35 and 7.36 are, of course, fully compatible with equations 7.32, which apply to changes for isothermal processes.

More complete expressions for A and G can be established by recourse to the definitions of the Helmholtz and Gibbs free energies ($A = U - TS$ and $G = H - TS$) in combination with expressions already obtained for U, H, and S for ideal gases. (See eqs. 2.38, 2.40, 6.20, and 6.23.) In particular, assuming that C_v and C_p are constant, we see that

$$A = C_v T + K - C_v T \ln T - RT \ln V - kT \tag{7.37}$$

and that

$$G = C_p T + K - C_p T \ln T + RT \ln p - k'T \tag{7.38}$$

in which expressions K is the constant of integration associated with the energy function and k and k' are the entropy constants. It will be observed that both A and G are uncertain to the extent of a constant plus another constant multiplied by the temperature. By comparing equations 7.37 and 7.38 with equations 7.35 and 7.36 one can readily ascertain the nature of $\phi(T)$ and $\psi(T)$.

The total Helmholtz free energy and total Gibbs free energy of a mixture of n_A moles of gas A and n_B moles of gas B can be similarly written by the use of equations 6.56 and 6.57 if one recognizes that the energy of a mixture of ideal gases is equal to the sum of the separate energies. For the Helmholtz function we can write

$$A = n_A C_{vA} T + n_B C_{vB} T + n_A K_A + n_B K_B - (n_A C_{vA} + n_B C_{vB}) T \ln T$$
$$- (n_A + n_B) RT \ln V - n_A k_A T - n_B k_B T \tag{7.39}$$

For the Gibbs function we can write

$$G = n_A C_{pA} T + n_B C_{pB} T + n_A K_A + n_B K_B - (n_A C_{pA} + n_B C_{pB}) T \ln T$$
$$+ n_A RT \ln p_A + n_B RT \ln p_B - n_A k'_A T - n_B k'_B T \tag{7.40}$$

It should be clear that the total Helmholtz or Gibbs free energy of a mixture of ideal gases is the sum of the separate functions calculated as if each gas occupied the total volume by itself.

Changes in Free Energies for Nonisothermal Processes

Although this chapter is concerned primarily with isothermal processes and related functions, it is appropriate at this time to comment briefly on changes in Helmholtz free energy and Gibbs free energy for non-isothermal processes. First of all, we should recall that the inequalities 7.3 and 7.8, relating work to ΔA and ΔG, are valid only for isothermal processes, since no limits can be set upon the nonisothermal work attending a change in state. It appears, therefore, that a change in A or G brought about by a change in temperature would be of little direct thermodynamic significance.* Accordingly, whenever we write ΔA or

---

* In this connection it will be observed that it is impossible to evaluate ΔA or ΔG for a process in which the initial and final temperatures are different unless one knows absolute values of the entropy. To demonstrate this, consider two particular states with variables A_1, U_1, S_1, T_1 and A_2, U_2, S_2, T_2. From the definition of the Helmholtz free energy we conclude that

ΔG, we shall imply a change for which the initial and final temperatures are the same; moreover, we shall never relate ΔA or ΔG to work quantities unless the corresponding processes are isothermal.

The foregoing arguments do not, however, preclude comparing ΔG's (or ΔA's) at different temperatures if each ΔG (or ΔA) refers to a definite temperature. We can, for example, compare a ΔG at T_1 with the corresponding ΔG at T_2 even though we shall not attempt to make use of the difference between G at T_1 and G at T_2.

The Gibbs-Helmholtz Equations

Although changes in the free-energy functions are evaluated only for isothermal processes, those changes are nevertheless functions of the temperature at which such processes occur. It is therefore entirely proper to ask how ΔA or ΔG depends on the temperature, and with that question in mind we shall now establish equations for $\partial \Delta A/\partial T$ and $\partial \Delta G/\partial T$.

Let us consider a perfectly general isothermal change in state, such as a chemical reaction represented as follows:

$$R(V_1, T) \longrightarrow P(V_2, T)$$

In this representation $R(V_1, T)$ denotes reactants occupying volume V_1 at temperature T, and $P(V_2, T)$ denotes products of volume V_2 at the same temperature. Associated with this process will be a change in Helmholtz free energy; $\Delta A = A_2 - A_1$, if A_2 denotes the Helmholtz energy of the products and A_1 that of the reactants. Evidently we can now write that

$$\left(\frac{\partial \Delta A}{\partial T}\right)_{V_1, V_2} = \left(\frac{\partial A_2}{\partial T}\right)_{V_2} - \left(\frac{\partial A_1}{\partial T}\right)_{V_1} \tag{7.41}$$

This equation, if we apply the first of equations 7.23, becomes

$$\left(\frac{\partial \Delta A}{\partial T}\right)_{V_1, V_2} = -S_2 + S_1 = -\Delta S \tag{7.42}$$

in which ΔS is the entropy change for the specified isothermal process. In view of the definition of A, we can further write that

$$\Delta S = \frac{\Delta U - \Delta A}{T} \tag{7.43}$$

$$\Delta A = \Delta U - T_1 \Delta S - S_2 \Delta T$$

Before we established the Third Law, the necessity of knowing an absolute entropy would have ruled out use of ΔG or ΔA for nonisothermal processes. Even though we might now have absolute entropy values, our conclusions about the possible significance of ΔA and ΔG for nonisothermal processes remain unaltered.

So

$$\left(\frac{\partial \Delta A}{\partial T}\right)_{V_1,V_2} = \frac{\Delta A - \Delta U}{T} \tag{7.44}$$

This equation is one of those known as the Gibbs-Helmholtz equations. It is valid even when the initial and final volumes are not the same; however, it is implicit that both volumes are constant when forming the partial derivative of ΔA with respect to temperature. If the change in state takes place at constant volume, so that $V_2 = V_1$, the equation, of course, is still valid, but the derivative is then simply written as $(\partial \Delta A/\partial T)_V$.

EXERCISE:

Consider the following reversible cycle, in which R and P represent reactants and products of a general change in state:

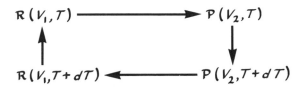

(A) If ΔA is the change in Helmholtz free energy at T, the change in Helmholtz free energy for the forward reaction at $T + dT$ will be $\Delta A + (\partial \Delta A/\partial T)_{V_1,V_2} dT$. Show, as a consequence, that the total work for the cycle will equal $(\partial \Delta A/\partial T)_{V_1,V_2} dT$.

(B) If ΔS is the entropy change at T, then $T\Delta S$ is the reversible heat, $Q_{\rm rev}$, at T. By the use of equation 4.48 and the result of part A, deduce, as a consequence, the Gibbs-Helmholtz equation (7.44).

Employing methods quite analogous to those used above, we can derive a second Gibbs-Helmholtz equation, applicable to changes in Gibbs free energy. Consider a change in state represented by the expression

$$R(p_1, T) \longrightarrow P(p_2, T)$$

in which the symbols are to be interpreted as before. The change in Gibbs energy, ΔG, will, of course, equal $G_2 - G_1$, and its partial derivative with respect to temperature will be

$$\left(\frac{\partial \Delta G}{\partial T}\right)_{p_1,p_2} = \left(\frac{\partial G_2}{\partial T}\right)_{p_2} - \left(\frac{\partial G_1}{\partial T}\right)_{p_1}$$

$$= -S_2 + S_1$$

$$= -\Delta S \tag{7.45}$$

Once again, using the definition of G, we observe that

$$-\Delta S = \frac{\Delta G - \Delta H}{T} \tag{7.46}$$

So

$$\left(\frac{\partial \Delta G}{\partial T}\right)_{p_1, p_2} = \frac{\Delta G - \Delta H}{T} \tag{7.47}$$

This is the second Gibbs-Helmholtz equation. The reader will observe once again that the initial and final pressures need not be equal; they are, however, constant so far as the derivative is concerned. If the isothermal change in state occurs at constant pressure, as it often does in chemical reactions, $p_2 = p_1$ and the derivative is written $(\partial \Delta G/\partial T)_p$.

EXERCISE:

Consider the following cycle, in which R and P denote reactants and products of a general change in state (V_1 and V_2 are dependent variables):

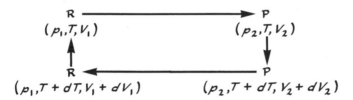

(A) Show that the reversible work attending the step at temperature T will equal $-\Delta G + p_2 V_2 - p_1 V_1$. Similarly, show that the work for the forward reaction at $T + dT$ will equal $-\Delta G - (\partial \Delta G/\partial T)_{p_1, p_2} dT + p_2(V_2 + dV_2) - p_1(V_1 + dV_1)$. Finally, observe that the work associated with the rise in temperature of the products by an amount dT will equal $p_2 dV_2$, whereas that associated with the fall in temperature of the reactants by dT will equal $-p_1 dV_1$. Show, as a consequence, that the total work for the cycle will equal $(\partial \Delta G/\partial T)_{p_1, p_2} dT$.

(B) If ΔS is the entropy change at T, deduce, by the use of equation 4.48 and the result of part A, the Gibbs-Helmholtz equation (7.47).

If the change in Gibbs energy for a certain process, such as a chemical reaction, is known at one temperature, and if the heat of reaction is also known, it is possible to calculate the change for the same reaction at another temperature. This is accomplished by integration of the appropriate Gibbs-Helmholtz equation. For convenience let us assume a single pressure and then rearrange equation 7.47 as follows:

$$\frac{1}{T}\left(\frac{\partial \Delta G}{\partial T}\right)_p - \frac{\Delta G}{T^2} = -\frac{\Delta H}{T^2} \tag{7.48}$$

Inspection of the left-hand side of this equation will disclose that it is precisely the derivative of $\Delta G/T$; we can therefore write that

$$\frac{\partial}{\partial T}\left(\frac{\Delta G}{T}\right) = -\frac{\Delta H}{T^2} \tag{7.49}$$

Hence

$$\frac{\Delta G}{T} = -\int \frac{\Delta H}{T^2}\, dT + I(p) \tag{7.50}$$

$I(p)$ being some function of pressure. If equation 7.49 is integrated between the limits T_1 and T_2, we obtain the equation

$$\frac{\Delta G_2}{T_2} - \frac{\Delta G_1}{T_1} = -\int_{T_1}^{T_2} \frac{\Delta H}{T^2}\, dT \tag{7.51}$$

This is the equation that enables one to calculate ΔG_2 from a knowledge of ΔG_1 and the heat of reaction. If the heat of reaction is known as a function of temperature, let us say in the form

$$\Delta H = \Delta H_0 + \alpha T + \beta T^2 + \ldots \tag{7.52}$$

then

$$\frac{\Delta G_2}{T_2} - \frac{\Delta G_1}{T_1} = \Delta H_0\left(\frac{1}{T_2} - \frac{1}{T_1}\right) - \alpha \ln \frac{T_2}{T_1} - \beta(T_2 - T_1) \ldots \tag{7.53}$$

Returning to equation 7.50 and utilizing equation 7.52, we can also write that

$$\Delta G = \Delta H_0 + I(p)T - \alpha T \ln T - \beta T^2 \ldots \tag{7.54}$$

It is evident from this equation that, to establish ΔG as a function of T, one must know not only ΔH but also the value of $I(p)$, which at a specified pressure will be a constant. It can be shown that, to gain a knowledge of $I(p)$, one must know the equivalent of ΔS or ΔG at some particular temperature.

The immediately preceding discussion has been carried out with the Gibbs energy form of the Gibbs-Helmholtz equation. A completely analogous treatment can be carried out for changes in Helmholtz free energy if one substitutes ΔA, ΔU, and V for ΔG, ΔH, and p, respectively. The formulation of the corresponding equations is left as an exercise for the student.

EXERCISE:

Assuming that $\Delta U = \Delta H - \Delta \nu RT$, where $\Delta \nu$ is the change in the number of moles of gases attending a chemical change in state, show that

$$\frac{\Delta A_2}{T_2} - \frac{\Delta A_1}{T_1} = \Delta H_0 \left(\frac{1}{T_2} - \frac{1}{T_1} \right) - (\alpha' - \Delta \nu R) \ln \frac{T_2}{T_1} - \beta(T_2 - T_1) \dots$$

and that

$$\Delta A = \Delta H_0 + I(V)T - (\alpha - \Delta \nu R)T \ln T - \beta T^2 \dots$$

if $I(V)$ is a function of integration.

Application of the Gibbs-Helmholtz Equation to Electrolytic Cells

Since the decrease in Gibbs free energy attending the operation of an electrolytic cell is equal to the reversible electrical work, it is clear that the Gibbs-Helmholtz equation for ΔG should give us valuable information concerning the temperature coefficient of an electromotive force. In particular, if we combine equation 7.14 with equation 7.47 and assume a single pressure, we obtain the following:

$$\left(\frac{\partial E}{\partial T} \right)_p = \frac{E}{T} + \frac{\Delta H}{zFT} \tag{7.55}$$

It is evident from this equation that the change in enthalpy associated with the change in state accompanying the cell reaction can be readily obtained from a knowledge of the electromotive force and its temperature coefficient. The change in entropy, which is directly proportional to the temperature coefficient of the electromotive force, can also be so determined. Equation 7.55, obviously, can be integrated in the same way as equation 7.47.

EXERCISE:

Show that equation 7.55 can be rearranged in the form

$$\Delta H = zFT^2 \frac{\partial}{\partial T}(E/T)$$

Heat Capacities at Temperatures Near Absolute Zero

Through use of the Gibbs-Helmholtz equation it is possible to deduce an interesting conclusion concerning heat capacities in the neighborhood of absolute zero. Although this discussion can be carried out without reference to the Third Law, it will nevertheless serve to amplify some of the Third Law arguments. To proceed, let us consider a chemical change in state at or near absolute zero, and let us assume that the corresponding change in entropy is a finite quantity. (It is not necessary, for our present purpose, to assume that ΔS approaches zero, even though we might be so inclined because of the Third Law.) Since $\Delta G = \Delta H - T\Delta S$, the change in Gibbs energy must approach the change in enthalpy as T goes to zero. Hence, from the Gibbs-Helmholtz equation,

$$\left(\frac{\partial \Delta G}{\partial T}\right)_p = \frac{\Delta G - \Delta H}{T} \tag{7.56}$$

we see that $(\partial \Delta G/\partial T)_p$ assumes an indeterminate form, $0/0$, as T approaches zero. Suppose that we should now attempt to evaluate the limit of $(\partial \Delta G/\partial T)_p$ as T approaches zero. To do so in accordance with the usual methods applicable to indeterminate forms, we shall differentiate numerator and denominator of the expression $(\Delta G - \Delta H)/T$ and then examine the limit of the new quotient. Following this procedure, we see that

$$\lim_{T \to 0} \left(\frac{\partial \Delta G}{\partial T}\right)_p = \lim_{T \to 0} \left\{ \left(\frac{\partial \Delta G}{\partial T}\right)_p - \left(\frac{\partial \Delta H}{\partial T}\right)_p \right\} \tag{7.57}$$

Subtracting the limit of $(\partial \Delta G/\partial T)_p$ from each side of equation 7.57 and recognizing that $(\partial \Delta H/\partial T)_p = \Delta C_p$, we conclude that

$$\lim_{T \to 0} \Delta C_p = \lim_{T \to 0} \left(\frac{\partial \Delta H}{\partial T}\right)_p = 0 \tag{7.58}$$

Thus we see that the change in heat capacity associated with a chemical reaction at absolute zero must be equal to zero.

This conclusion with respect to heat capacities might well have been anticipated in the light of our Third Law arguments concerning changes in entropy at absolute zero. According to the Third Law, the change in entropy at any temperature T should be

$$\Delta S = \int_0^T \frac{\Delta C_p}{T} \, dT \tag{7.59}$$

Now, if ΔC_p is a continuous function, it should go to zero as T approaches zero if the integral is to converge. If ΔC_p were different from zero for $T = 0$, the change in entropy at temperature T might be positively or negatively infinite because of the behavior of the integral in the neighborhood of its lower limit. We could resolve this difficulty, of course, by rejecting the Third Law and assigning at absolute zero an infinite change in entropy expressed in such asymptotic functional form that a finite change in entropy for temperatures other than absolute zero would result. Since this is a most unrealistic assumption to make, we are practically compelled to assume that ΔC_p goes to zero. By the use of the Gibbs-Helmholtz equation, expressed in terms of Helmholtz energy, we can also show that ΔC_v likewise approaches zero at absolute zero.

We can also write the counterpart of the Gibbs-Helmholtz equation for a pure substance:

$$\left(\frac{\partial G}{\partial T}\right)_p = \frac{G - H}{T} \tag{7.60}$$

From this we conclude, by an analysis similar to that employed above, that C_p (and C_v) must be zero at absolute zero. To deduce this, we must simply assume that the entropy of a substance is not infinite at absolute zero.

The conclusions of this section are not to be considered a part of the Third Law, but they are fully compatible with it and serve to fill out some of the discussion relating to it.

Effect of Gravitational Potential on Thermodynamic Functions

We have tacitly assumed, so far, that the location of a substance has no bearing on its thermodynamic properties; thus, for example, the properties of a mole of hydrogen at one atmosphere and 25 °C must be the same here as on the other side of the ocean. However, if we consider height above the earth's surface as a possible variable, we find that some of our thermodynamic functions depend upon that variable. Accordingly, we shall now inquire into the possible effect of a uniform gravitational field upon the thermodynamic functions with which we are already familiar.

If a mass, m, is raised to some height, h, above the earth's surface, it is clear that a certain amount of work must be done. If we neglect the change in gravitational field with change in distance from the earth's center, the work done in raising the mass will simply equal mgh if g is the acceleration of gravity. This work will also equal the potential energy of the object, relative to that at the earth's surface; it is, of course, readily convertible into other work in the thermodynamic sense.

From the foregoing considerations it is evident that we must introduce h, the height above the earth's surface, as an additional independent variable to characterize a substance in a gravitational field. We can take care of this by simply adding $-mgdh$ to our old expression for dW:

$$dW = pdV - mgdh \tag{7.61}$$

The negative sign appears in front of $mgdh$ because work must be done to raise the object. With this understanding, our differential expressions for energy, enthalpy, Helmholtz and Gibbs free energies will be modified as follows:

$$dU = TdS - pdV + mgdh \tag{7.62}$$

$$dH = TdS + Vdp + mgdh \tag{7.63}$$

$$dA = -SdT - pdV + mgdh \qquad (7.64)$$

$$dG = -SdT + Vdp + mgdh \qquad (7.65)$$

If we deal with one mole of material, the mass, m, will simply equal the molecular weight, M.

In connection with equations 7.62–7.65, we should observe that the entropy will not be directly dependent upon height, for the reversible heat attending the isothermal lifting of an object in a uniform gravitational field will be zero. Thus, if S is given as a function of T and p at the earth's surface, the identical function will be valid at height h in a uniform gravitational field; in other words, $(\partial S/\partial h)_{T,p} = 0$. On the other hand, the functions ordinarily representing U, H, A, and G must be augmented by mgh so that they take into account the contribution of potential energy to each of those quantities.

Thermodynamics of the Atmosphere

Utilizing the results of the preceding section, we are now in a position to describe, from a thermodynamic point of view, the nature of the earth's atmosphere. First of all, assuming the atmosphere to be in mechanical equilibrium, we can make a very general statement concerning the change in pressure with height. Since pressure is equal to force per unit area, the difference in pressure, dp, between height h and height $h + dh$, will be numerically equal to the weight of a column of air of unit cross-sectional area and of thickness dh (see Fig. 7-1). If ρ is the density of the air, that weight is simply equal to ρgdh, and, since the pressure diminishes with increasing height,

$$dp = -\rho gdh \qquad (7.66)$$

Upon multiplying both sides of this equation by the molar volume, V, we obtain the equation

$$Vdp = -V\rho gdh = -Mgdh \qquad (7.67)$$

in which we replace $V\rho$ by the molecular weight, M. Combining this result with equations 7.63 and 7.65, we deduce, for one mole of substance, the following very simple differential relationships:

$$dH - TdS = 0 \qquad (7.68)$$

$$dG + SdT = 0 \qquad (7.69)$$

It should be emphasized that equations 7.68 and 7.69 cannot be applied to any object, such as some solid material, that might be raised

Unit cross-section

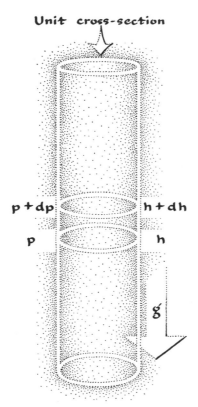

$p + dp$ $h + dh$

p h

g

FIGURE 7-1   Cylinder of unit cross-
section in the earth's gravitational
field.

in a gravitational field. (For that purpose one should employ equations
7.62–7.65.) Equations 7.68 and 7.69 are valid only for a fluid atmosphere
in a state of mechanical equilibrium in a uniform gravitational field.
One cannot, unfortunately, integrate them directly without making a
further assumption concerning the nature of the atmosphere, which is
equivalent to stipulating some kind of model.* There are, of course, an
infinite number of models that might be assumed, but we shall here
consider only two, so chosen because they are particularly simple and
instructive.

Our first model for the atmosphere will be an isothermal one, for
which it is supposed that the temperature is independent of height.

---

* Equations 7.68 and 7.69 are said to be nonintegrable differential equations in canonical
form. Such an equation can be solved only in conjunction with some other assumed relationship or in terms of what are known as pairs of integral equivalents. See Chapter II of *Ordinary
Differential Equations* by E. L. Ince (Longmans, Green & Co.).

Letting T be constant, we need only examine equation 7.69 to see that such an atmosphere must also be characterized by constant molar Gibbs energy. To reduce our expressions to practical form, let us now suppose that air behaves like an ideal gas of average molecular weight M. We have already seen, in accordance with equation 7.36, that the Gibbs free energy of an ideal gas (at the earth's surface) will be given by an equation of the type

$$G = \psi(T) + RT \ln p_0 \qquad (7.70)$$

in which p_0 is the pressure at the earth's surface and $\psi(T)$ is a function of temperature only. At any height, h, where the pressure is p, the Gibbs free energy will be given by the same kind of expression, except that the gravitational potential energy must be added as follows:

$$G = \psi(T) + RT \ln p + Mgh \qquad (7.71)$$

If we now stipulate that the Gibbs function is constant, we observe that

$$RT \ln p + Mgh = RT \ln p_0 \qquad (7.72)$$

This equation can be readily rearranged to give

$$p = p_0 e^{-Mgh/RT} \qquad (7.73)$$

We see that the pressure of an isothermal atmosphere in a uniform gravitational field will diminish exponentially with height. This also implies that there is no maximum height to the atmosphere, that it extends without end, becoming ever thinner but always present to some degree. One can also obtain equation 7.73 directly, without explicit reference to Gibbs free energy, by integrating equation 7.67 in conjunction with the equation of state for an ideal gas.

EXERCISE:

Derive equation 7.73 by direct integration of equation 7.67.

EXERCISE:

If air were an ideal gas of molecular weight 29 and the earth's atmosphere were isothermal at a temperature of 0 °C, at what height in meters would the air pressure equal 0.5 atm?

Though the isothermal model for the atmosphere is not very realistic, many of the considerations involved in its description are realized elsewhere—for example, in laboratory work with ultracentrifuges. During ultracentrifugation the equivalent of a very large gravitational field is imposed on a system that can be maintained at a uniform temperature, and at equilibrium the concentration of particles obeys an exponential law like that expressed by equation 7.73. The earth's atmosphere, on the other hand, is far from isothermal, it being common knowledge that the temperature usually drops with increasing altitude. The questions naturally arise: why does it drop, and what kind of model should be assumed to provide an adequate description of the actual state of affairs?

We obtain a more realistic model of the atmosphere by assuming that the air is not isothermal but adiabatic or isentropic. Since a rising mass of air will tend to expand, and since the reversible adiabatic expansion of a gas is accompanied by cooling, such a model will qualitatively agree with the nonisothermal character of the atmosphere. Accordingly, we shall now consider the situation that obtains when the entropy of a mass of air at the earth's surface is the same as its entropy at some height, h. This model turns out to be fairly good as a starting point for describing the atmosphere up to a limited height. For an atmosphere to be adiabatic, there must be little or no conduction or radiation of heat from one part to another, although there can be large-scale mixing without molecular diffusion. The thermodynamic description of an isentropic atmosphere is easily carried out in the light of equation 7.68; by setting $dS = 0$ we immediately recognize that the enthalpy must likewise be constant. Assuming that C_p is constant, we can assert that the enthalpy at the earth's surface will be given by the expression

$$H = C_p T_0 + K \qquad (7.74)$$

in which T_0 is the temperature at the earth's surface and K is a constant. At height h, where the temperature is T, the enthalpy will be given by the expression

$$H = C_p T + K + Mgh \qquad (7.75)$$

in which the ordinary expression for enthalpy has been increased by the gravitational potential energy. Stipulating constancy of enthalpy, we then see, upon combining equations 7.74 and 7.75, that

$$T = T_0 - \frac{Mgh}{C_p} \qquad (7.76)$$

Since the temperature can never be lower than absolute zero, we conclude from equation 7.76 that the atmosphere has a maximum height:

$$h_{\max} = \frac{C_p T_0}{Mg} \qquad (7.77)$$

The rate of fall of the temperature with respect to height is seen to be

$$-\frac{dT}{dh} = \frac{Mg}{C_p} \qquad (7.78)$$

which is a quantity known as the "normal lapse rate" of the atmosphere. The actual lapse rate may or may not be equal to the normal. If the lapse rate is greater than normal, the atmosphere is said to be superadiabatic, which is a condition of instability. This instability is a consequence of the fact that the upper gases, if we use reversible adiabatic movements for reference, will be relatively too dense compared with the lower gases. Under such circumstances the upper gases will tend to descend and displace the lower gases; thus, the atmosphere will be subject to considerable turbulence, which will ultimately lead to a rectification of the superadiabatic condition.

An atmosphere with less than the normal lapse rate will be mechanically stable, for the upper gases (if we again use reversible adiabatic mass movements for comparison) will be less dense than expected for an isentropic system. Under such circumstances the upper part of the atmosphere will float, so to speak, on the lower part—a condition that can persist undisturbed for some time. An isothermal atmosphere, which is said to be neutral, fulfills these conditions. If the temperature increases with altitude, the atmosphere is said to be in a state of inversion; this is a condition of considerable mechanical stability because the upper gases will be even lighter than under neutral conditions.

The foregoing discussion of the atmosphere is, of course, very much simplified. For one thing, we have omitted the effect of moisture in the air, which, if near saturation, can have a profound effect on the thermodynamics involved. Our cursory treatment does provide a starting point for the description of the atmosphere, however, and illustrates how thermodynamics can have an important bearing on meteorology.

Problems

1. By differentiating equation 7.52 with respect to T, obtain an expression for ΔC_p as a function of temperature. By indefinite integration of $(\Delta C_p/T)dT$, deduce a general expression for ΔS with an undetermined function of integration. By differentiating equation 7.54 partially with respect to T, obtain another expression, with appropriate sign, for ΔS. Compare the two forms, and explain.

2. (A) Derive an expression for the electromotive force of the cell

$$\text{Pt, } H_2(p_1), \text{ } H_2SO_4 \cdot x\text{Aq, } H_2(p_2), \text{ Pt}$$

assuming that hydrogen obeys the equation of state $pV = RT + \alpha p$. What is the total reversible work attending the passage of one faraday of electricity through the cell?

$$\text{Ans. } E = \frac{RT}{2F} \ln \frac{p_1}{p_2} + \frac{\alpha}{2F}(p_1 - p_2)$$

$$W_{\text{total}} = -\Delta A = \frac{RT}{2} \ln \frac{p_1}{p_2}$$

(B) If $\alpha = 0.0148$ liter mol^{-1} for hydrogen at 20 °C, what is the electromotive force in volts at that temperature if $p_1 = 10$ atm and $p_2 = 1$ atm? What is the percentage correction that can be attributed to gaseous nonideality?

3. Consider the electrical cell

$$\text{Pt, } Cl_2(p_1), \text{ NaCl} \cdot x\text{Aq, } Cl_2(p_2), \text{ Pt}$$

If chlorine obeys an equation of state of the form

$$pV = RT + \alpha p$$

in which α is a function of temperature, what is the electromotive force of the cell? What are ΔS and ΔH for the change of state accompanying the passage of one faraday of positive electricity from left to right within the cell? If α is given by the expression

$$\alpha = b - a/RT \text{ (from van der Waals' equation)}$$

at what temperature will $\Delta H = 0$? Explain why this is related to the Joule-Thomson inversion temperature. (See equation 6.46.)

4. Consider the change in Gibbs free energy attending the isothermal expansion or compression of some material:

$$\Delta G = \int_{p_1}^{p_2} V dp$$

Differentiating ΔG partially with respect to T, and noting that the differentiation will commute with the integration, show, in the light of the Gibbs-Helmholtz equation, that

$$\Delta H = \int_{p_1}^{p_2} \left[V - T \left(\frac{\partial V}{\partial T} \right)_p \right] dp$$

Observe that the last integrand must therefore be

$$\left(\frac{\partial H}{\partial p} \right)_T = V - T \left(\frac{\partial V}{\partial T} \right)_p$$

(This problem illustrates once more that the Gibbs-Helmholtz equation is simply another outgrowth of the Second Law.)

5. Consider the change in Helmholtz free energy attending the isothermal expansion of one mole of an ideal gas:

$$\Delta A = -RT \ln \frac{V_2}{V_1}$$

Substituting this expression for ΔA into the Gibbs-Helmholtz equation (7.44), show that the change in energy accompanying an isothermal expansion of an ideal gas is equal to zero.

6. Consider the change in Helmholtz free energy attending the isothermal expansion of a van der Waals fluid:

$$\Delta A = -RT \ln \frac{V_2 - b}{V_1 - b} - a \left(\frac{1}{V_2} - \frac{1}{V_1} \right)$$

Substituting this into the Gibbs-Helmholtz equation, show directly that

$$\Delta U = -a \left(\frac{1}{V_2} - \frac{1}{V_1} \right)$$

7. Consider the electrolytic cell

$$\text{Ag(c), AgCl(c), HCl} \cdot x\text{Aq, Cl}_2\text{(g), Pt-Ir}$$

(A) At 25 °C and 1 atm the electromotive force of the cell is 1.1362 volts. What is the Gibbs energy change in calories attending the passage of one faraday of electricity from left to right within the cell?

(B) The temperature coefficient of the electromotive force is -0.000595 volt per degree. What are ΔS in calories per degree and ΔH in calories for the change in state corresponding to the passage of one faraday? (Note that parts A and B can be worked out without any knowledge of the cell reaction.)

(c) Show that, when one faraday of electricity passes through the cell, the following change in state occurs:

$$\text{Ag(c)} + \tfrac{1}{2}\text{Cl}_2\text{(g)} \longrightarrow \text{AgCl(c)}$$

The heat of formation of AgCl(c) is reported as -30.37 kcal; compare this with ΔH of part B.

(D) Assuming chlorine to be an ideal gas, and neglecting the volumes of the solids compared with that of the gas, show that the total reversible work attending the operation of the cell will be given by the expression

$$W_{rev} = EF - \tfrac{1}{2}RT$$

From this and from the result of part A calculate ΔA for the cell reaction.

8. (A) By obtaining expressions for $\partial^2 S / \partial T \partial p$ in two different ways, prove that

$$\left(\frac{\partial C_p}{\partial p} \right)_T = -T \left(\frac{\partial^2 V}{\partial T^2} \right)_p$$

(B) Similarly, by considering expressions for $\partial^2 S/\partial T \partial V$, prove that

$$\left(\frac{\partial C_v}{\partial V}\right)_T = T\left(\frac{\partial^2 p}{\partial T^2}\right)_V$$

9. Assuming air to be an ideal gas with an average molecular weight of 29 and a constant heat capacity, C_p, equal to $\frac{7}{2}R$, calculate the normal lapse rate in degrees per meter of an adiabatic atmosphere. Assuming T_0, the temperature at the earth's surface, to be 0 °C, calculate the height of the atmosphere in kilometers.

10. Assuming the same conditions specified in Problem 9, calculate the height in meters at which the air pressure would be one-half of that at the earth's surface.

11. Suppose the earth's atmosphere consisted of a single nonideal gas obeying the equation of state $pV(1 - \beta p) = RT$, in which β is a function of T only. If the atmosphere were isothermal, how would the pressure depend upon height?

$$\text{Ans.} \quad p = \frac{p_0 e^{-Mgh/RT}}{1 - \beta p_0[1 - e^{-Mgh/RT}]}$$

12. Consider an atmosphere in mechanical equilibrium for which the total molar energy (thermal plus potential) is independent of height above the earth's surface.

   (A) Assuming the atmosphere to be an ideal gas, calculate the lapse rate, and show that the atmosphere would be superadiabatic and hence unstable.

   (B) What is the maximum height of such an atmosphere?

13. One-tenth mole of an ideal gas is subjected to the following sequence of steps:

   (A) Starting at 25 °C and 1 atm, the gas expands freely into a vacuum (Joule experiment) to double its volume.

   (B) The gas is next heated to 125 °C, the volume remaining constant.

   (C) The gas now expands reversibly and isothermally until its volume is again doubled.

   (D) The gas is finally cooled reversibly at constant pressure back to 25 °C.

   Calculate ΔU, ΔH, Q, W, ΔS, ΔA, and ΔG for the over-all change in state, A + B + C + D. (Express your answers in calories and in calories per degree.)

14. At 0 °C, liquid toluene obeys an equation of state of the form

$$V = V_1\left(0.8196 + \frac{409.5}{2,269 + p}\right)$$

in which V_1 is the volume under 1 atm pressure and p is the pressure in atmospheres. Evaluate in calories the changes in Gibbs and Helmholtz free energies attending the isothermal compression of 1 mole of toluene from 1 to 500 atm pressure at 0 °C. The density of toluene at 0 °C is 0.8841 g ml^{-1}.

15. According to the Tait equation, the volume of a liquid under pressure p is given by the formula

$$V = V_1 \left(1 - a \lg \left[\frac{b + p}{b + 1} \right] \right)$$

in which a and b are constants, p is measured in atmospheres, and V_1 is the volume under 1 atm pressure.

(A) Show that the change in Gibbs energy of a liquid compressed from 1 atm to pressure p is given by the expression

$$\Delta G = V_1(1 + a/2.303)(p - 1) - V_1 a(b + p) \lg \left(\frac{b + p}{b + 1} \right)$$

(B) For water at 25 °C, $a = 0.3150$ and $b = 2{,}996$ atm. Calculate ΔG and ΔA in calories for 1 mole of water compressed from 1 atm to 1,000 atm. The density of water at 25 °C is 0.997 g ml^{-1}.

16. A thermodynamic system consists of a rigid box containing materials that might undergo a change of state. The box is also kept in a thermostat so that its contents are maintained at a uniform temperature, T. If something should happen within the box, what if anything can be said for sure about the values of ΔU, ΔH, Q, W, ΔS, ΔA and ΔG?

           Ans. $W = 0$; $\Delta A < 0$; others uncertain

17. Consider the system described in Problem 16 with the following difference. Two wires lead out of the box and a potential difference corresponding to an electromotive force E is detected between them, provided no current is drawn.

(A) Assuming that one faraday of electricity is drawn out of the box at a finite, irreversible rate, what if anything can be said for sure about the values of ΔU, ΔH, Q, W, ΔS, ΔA and ΔG if only electrochemical reactions occur?

           Ans. $W < EF$; $\Delta A = -EF$; others uncertain

(B) The temperature coefficient of E is measured and found to have a value $(\partial E / \partial T)_V$. Now what else can be said about the quantities of which inquiry was made in part A?

$$\text{Ans.} \quad \Delta U = -EF + TF \left(\frac{\partial E}{\partial T} \right)_V ;$$

$$Q < TF \left(\frac{\partial E}{\partial T} \right)_V ; \quad \Delta S = F \left(\frac{\partial E}{\partial T} \right)_V$$

18. A cylinder containing materials that might undergo a change in state is provided with a frictionless piston that maintains a constant pressure, p, on the contents of the cylinder. The cylinder is also kept in a thermostat so that its contents are maintained at a uniform temperature, T. If something should happen within the cylinder giving rise to a change in volume equal to ΔV, what if anything can be said for sure about the values of ΔU, ΔH, Q, W, ΔS, ΔA and ΔG?

           Ans. $W = p\Delta V$; $\Delta A < -p\Delta V$; $\Delta G < 0$; others uncertain

19. The cylinder of Problem 18 has two wires leading out that exhibit a potential difference corresponding to an electromotive force E if no current is drawn.

 (A) Assuming that one faraday of electricity is drawn out at a finite, irreversible rate, what if anything can be said for sure about the values of ΔU, ΔH, Q, W, ΔS, ΔA and ΔG provided only electrochemical reactions occur.

Ans. $W < EF + p\Delta V$; $\Delta A = -EF - p\Delta V$; $\Delta G = -EF$; others uncertain

 (B) The temperature coefficient of E is found to have a value $(\partial E/\partial T)_p$. Now what else can be said about the quantities of which inquiry was made in part A?

$$\text{Ans.}\quad \Delta U = -EF + TF\left(\frac{\partial E}{\partial T}\right)_p - p\Delta V;$$

$$\Delta H = -EF + TF\left(\frac{\partial E}{\partial T}\right)_p; \quad Q < TF\left(\frac{\partial E}{\partial T}\right)_p; \quad \Delta S = F\left(\frac{\partial E}{\partial T}\right)_p$$

8

Equilibrium Between Phases;
Vapor Pressures

Introduction

The substances of which the world is made are found to exist in three different states of aggregation, gaseous, liquid, and solid. It is also observed that pure materials of definite chemical composition might exist in any of several solid modifications, each possessing a different crystal structure. Since substances exist in more than one form, it is appropriate, in this course of study, that we consider the thermodynamic factors affecting the conversion from one form to another. In this connection we shall be particularly concerned with the thermodynamics involved when two (or possibly three) phases of a substance co-exist in a state of equilibrium.

For illustrative purposes, let us consider the system water. Figure 8-1 schematically represents the phase diagram for water, a diagram in which pressures are plotted against temperature in such a way as to divide the area into three regions, representing the three common phases in which water occurs. The area below AOC is the region in which the vapor phase is stable, the area to the right of OB and above OC is the liquid-phase region, and that to the left of AOB represents the solid form, ice. The several lines that separate the regions are loci of points along which two phases can co-exist in equilibrium. The curve OC is commonly called the vapor-pressure curve for water, since it represents the pressures exhibited by liquid water in equilibrium with its vapor at the various temperatures. At the point O, and only at that point, the three phases—ice, liquid water, and water vapor—can all exist simultaneously in equilibrium. We shall now inquire into the nature of the lines separating the regions of the phase diagram; in other words, we shall find out what we can about the dependence of an equilibrium pressure on the temperature. To this end we shall specifically examine a liquid-vapor equilibrium, keeping the treatment sufficiently general, however, so that similar applications to other kinds of equilibria can be readily made when required.

Isothermals for Liquid-Vapor Systems

To acquire a better understanding of the factors involved in the conversion of a liquid to a vapor, we shall examine the isothermal curves that we obtain when we plot pressure versus volume for systems capable of existing in the two phases. Consider, for example, the isothermals of carbon dioxide, which are shown in Figure 8-2. At low temperatures and large enough volumes the system will exist in the gaseous or vapor form represented by the curves in the lower right-hand portion

FIGURE 8-1 Pressure-temperature phase diagram for water.

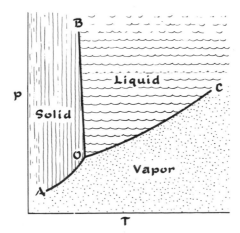

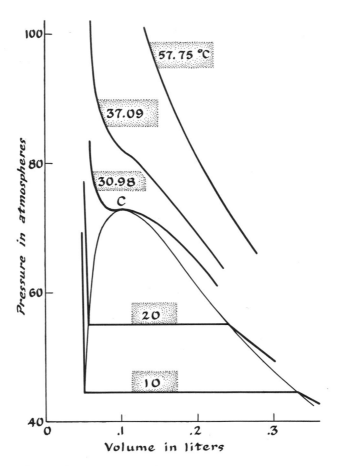

FIGURE 8-2   Isotherms for carbon dioxide with celsius temperatures indicated next
to curves. Point C is the critical point. Volume is given for one mole of material.

of the diagram. As the volume is diminished at constant temperature, the
pressure increases until a point of discontinuity is reached in the slope
of the curve; at this point a marked change in volume takes place without change in pressure. This phenomenon, which is associated with the
straight-line horizontal portion of a low-temperature isothermal, is the
process commonly known as liquefaction. The pressure corresponding
to such a flat portion of an isotherm is simply the equilibrium vapor
pressure. To the left of the flat portions, the isotherms represent the
liquid material; since the liquid is much less compressible than the gas,
the curves for the liquid are necessarily quite steep.

As the temperature rises, the change in volume attending an isothermal liquefaction will diminish, and the compressibility of the liquid

will increase. If the temperature is raised sufficiently, one finally reaches an isotherm for which that change in volume becomes zero and the discontinuity of slope vanishes. At this temperature one cannot detect any transition that could be considered a conversion from vapor to liquid. This temperature is called the critical temperature; the volume and pressure corresponding to the apex of the transition region are called the critical volume and the critical pressure (point C of Fig. 8-2). When the temperature is above the critical temperature, two distinct phases, such as liquid and vapor, cannot co-exist. It is said, accordingly, that a substance above its critical temperature cannot be liquefied, regardless of the pressure applied.

The isotherms plotted for carbon dioxide in Figure 8-2 depart markedly from those for ideal gases. The departures from ideality diminish, however, as the temperature rises, and at temperatures well above the critical temperature carbon dioxide can be regarded as an ideal gas. Moreover, at large enough volumes—that is to say, at very low pressures—carbon dioxide will behave quite ideally even if the temperature is below the critical temperature.

Let us now inquire about the change in Gibbs free energy accompanying the isothermal constant-pressure conversion of liquid to vapor. If the process is carried out reversibly (and this can be done at the equilibrium pressure), the entropy change of vaporization will be given by the expression

$$\Delta S_{vap} = \frac{\Delta H_{vap}}{T} \tag{8.1}$$

But, since $\Delta G = \Delta H - T\Delta S$ for any isothermal change in state, we conclude from equation 8.1 that

$$\Delta G_{vap} = \Delta H_{vap} - T\Delta S_{vap} = 0 \tag{8.2}$$

This can now be generalized for any kind of transition (such as solid-vapor or solid-liquid) between two phases of a pure substance; the Gibbs free-energy change attending a transition between phases at the equilibrium pressure is equal to zero. This is in keeping with our earlier conclusion, given by expression 7.9, that the Gibbs energy change for a constant-pressure isothermal process must be equal to or less than zero, the inequality holding for spontaneous processes and the equality indicating a state of equilibrium. An equilibrium between vapor and liquid, for example, can be most conveniently specified, from a thermodynamic point of view, if we simply stipulate the equality of the molar Gibbs free energies of the two phases.

The Clapeyron Equation

In view of the concepts discussed in the preceding section, we are now in a position to establish something about the dependence of vapor pressure, or other transition pressures, on temperature. If we recognize that the Gibbs energy change must equal zero at equilibrium, our only problem is to find how the pressure is constrained to change with temperature if ΔG is kept equal to zero. Using subscripts l and v to denote the liquid and vapor states respectively, we can write the two following equations for the total differentials of the Gibbs functions.

$$dG_v = -S_v dT + V_v dp \qquad (8.3)$$

$$dG_1 = -S_1 dT + V_1 dp \qquad (8.4)$$

In writing these equations, we assume that the temperature and pressure are the same for both phases but that the other thermodynamic properties may be different. By subtracting the second equation from the first, we get the total differential of the change in Gibbs energy for the evaporation process:

$$d\Delta G = -\Delta S dT + \Delta V dp \qquad (8.5)$$

Here $\Delta S = S_v - S_1$ and $\Delta V = V_v - V_1$. If we now stipulate that equilibrium is maintained both before and after the temperature and pressure have been changed, it follows that $d\Delta G$ must be equal to zero. Upon setting the right-hand side of equation 8.5 equal to zero, we conclude, after rearrangement, that

$$\frac{dp}{dT} = \frac{\Delta S}{\Delta V} \qquad (8.6)$$

Recognizing further that at equilibrium $\Delta S = \Delta H/T$, we see that

$$\frac{dp}{dT} = \frac{\Delta H}{T\Delta V} \qquad (8.7)$$

This result, known as the Clapeyron equation, is one of the more important equations of thermodynamics. It is applicable not merely to liquid-vapor equilibria but also to equilibria between solids and vapors, between solids and liquids, and between two solid modifications of a substance.

EXERCISE:

At 0 °C and 1 atm, the density of water is 1.000 g ml^{-1} and that of ice is 0.917 g ml^{-1}. If the specific heat of fusion is 79.7 cal, what is dp/dT, the slope of line OB in Fig. 8-1, expressed in atmospheres per degree?

The foregoing derivation was carried out on the assumption that the pressure applied to the liquid is equal to the pressure of the vapor in equilibrium with it. If, however, the applied pressure is different from the vapor pressure, we can readily modify the derivation to take this into account. If a liquid, for example, is confined to a vessel that also contains some inert gas that is insoluble in the liquid, the pressure applied to the liquid will be greater than the partial pressure of the vapor. Or we might have a system in which the liquid is held in position by a semipermeable membrane of such character that vapor but not liquid can pass through it. In either case the pressure applied to the liquid might be different from that of the vapor in equilibrium with it. Hence, if we let p_1 and p_v equal the applied pressure and the vapor pressure respectively, we can rewrite equations 8.3 and 8.4 as follows:

$$dG_v = -S_v dT + V_v dp_v \qquad (8.8)$$

$$dG_1 = -S_1 dT + V_1 dp_1 \qquad (8.9)$$

Once again we must stipulate that the Gibbs energy change for the evaporation process must remain equal to zero if equilibrium is to be maintained as we change the independent variables of the system. Hence we conclude that

$$d\Delta G = -\Delta S dT + V_v dp_v - V_1 dp_1 = 0 \qquad (8.10)$$

Upon solving this equation for dp_v, we conclude from the coefficient of dT that

$$\left(\frac{\partial p_v}{\partial T}\right)_{p_1} = \frac{\Delta S}{V_v} = \frac{\Delta H}{TV_v} \qquad (8.11)$$

and that

$$\left(\frac{\partial p_v}{\partial p_1}\right)_T = \frac{V_1}{V_v} \qquad (8.12)$$

The first of our new equations, (8.11), looks very much like the Clapeyron equation; in place of the change in volume associated with the transition, there appears the volume of the vapor; instead of a total derivative of pressure with respect to temperature, there is a partial derivative of vapor pressure with respect to temperature, the applied pressure being held constant. The Clapeyron equation is written with a *total* derivative because there is only one independent variable, T, and because there is only one pressure, which is given by a function of T alone, as represented by the appropriate curve in the phase diagram. Equation 8.11 involves a *partial* derivative because the two pressures are different, thus rendering the vapor pressure a function of two variables, T and p_1.

Practically speaking, there is little difference between the Clapeyron equation and its variant (8.11), for the volume of a liquid, when compared with that of its vapor, can generally be neglected, thus leaving ΔV substantially the same as V_v.

Equation 8.12 tells us how vapor pressures change with applied pressure at constant temperature. Since the volume of a vapor is much greater than that of a liquid, it is clear that moderate changes in applied pressure will produce little effect on the vapor pressure.

Finally, by combining equations 8.11 and 8.12, we conclude that

$$\left(\frac{\partial p_1}{\partial T}\right)_{p_v} = -\frac{\Delta S}{V_1} = -\frac{\Delta H}{TV_1} \tag{8.13}$$

This equation tells us how the applied pressure must change with temperature if the vapor pressure is to be kept constant. Equation 8.13 is not often used, but it is needed to complete our current discussion. Like that of Clapeyron, equations 8.11, 8.12, and 8.13 are valid for transitions other than evaporations as long as the various symbols denote actual pressures, volumes, etc.

EXERCISE:

A pan of water is carried isothermally and under constant applied pressure from the earth's surface to height h in an airplane with a pressurized cabin. At height h the molar Gibbs energy of the water is greater than it was at the earth's surface by amount Mgh. How does this increase in Gibbs energy affect the equilibrium vapor pressure of the water in the cabin of the airplane?

EXERCISE:

Under what applied pressure would water at 99.999 °C exhibit a vapor pressure of 1 atm? Assume the specific volume of liquid water to be 1 cm³ and the specific heat of vaporization to be 540 cal.

Applications of the Clapeyron Equation

The Clapeyron equation, one of the more important equations derived from Second Law considerations, can be applied in numerous ways to physical-chemical problems. One can use it, for example, in predicting the effect of pressure on melting points or on transition temperatures. One of the best-known examples of this application has to do with the equilibrium between ice and liquid water. Since ice is less dense than water, ΔV for the process of melting is negative. Moreover, since ΔH, the heat of fusion, is positive, we conclude from the Clapeyron equation that dp/dT must be a negative quantity. This tells us that the slope of the line OB appearing in Figure 8-1 must be negative even though it is exceedingly steep. This negative slope accounts for the fact that a piece of ice at or below its normal melting point can be melted by sufficient extra pressure. This is unusual behavior, exhibited by only a few substances, and, if water were not so common, the phenomenon would be regarded as one of the curiosities of nature.

One can also use the Clapeyron equation to calculate precisely a heat of transition from the change in volume and the derivative of pressure with respect to temperature. If the densities of the two phases are known, and if accurate experimental measurements of vapor pressures or transition pressures are available, ΔH can be determined quite accurately.

Perhaps the most widely used application of the Clapeyron equation, however, is in an approximate form to which we shall now turn.

The Clausius-Clapeyron Equation

If we apply the Clapeyron equation to a liquid-vapor equilibrium, we can make several approximations that permit us to arrive at some simple but useful conclusions. First of all, we can generally neglect the volume of a liquid, compared with that of the vapor, which is equivalent to saying that ΔV is simply equal to the volume of the vapor. (This amounts to the same thing as switching from equation 8.7 to equation 8.11 and then dismissing the effect of pressure applied to the liquid.) Next let us assume that the vapor behaves as an ideal gas, under which circumstance we can write that $V = RT/p$. Upon introducing these two approximations into the Clapeyron equation and rearranging the result, we obtain the following:

$$\frac{d \ln p}{dT} = \frac{\Delta H}{RT^2} \tag{8.14}$$

This equation, known as the Clausius-Clapeyron equation, is approximately valid for evaporation and sublimation processes but not, of course, valid for transitions between solids or for the melting of solids. In accordance with the preceding assumptions and approximations, ΔH must also be regarded as independent of the pressure, whereupon the variables can readily be separated and the equation integrated as follows:

$$\ln \frac{p_2}{p_1} = \int_{T_1}^{T_2} \frac{\Delta H}{RT^2} dT \qquad (8.15)$$

Finally, if we assume that ΔH is constant, we can perform the indicated integration and get

$$\ln \frac{p_2}{p_1} = \frac{\Delta H}{R}\left(\frac{1}{T_1} - \frac{1}{T_2}\right) \qquad (8.16)$$

which is a statement relating two vapor pressures (or sublimation pressures) to the corresponding temperatures. If the integration is carried out indefinitely, we can write the vapor-pressure equation in the following form:

$$\ln p = -\frac{\Delta H}{RT} + \text{constant} \qquad (8.17)$$

This equation suggests that, if the natural logarithm of the vapor pressure is plotted versus the reciprocal of the absolute temperature, one should obtain a straight line with slope equal to $-\Delta H/R$. It is observed experimentally that the vapor pressures of most substances agree quite well with an equation of this type. We can introduce further refinements, naturally, by taking into account the dependence of ΔH on temperature and by eliminating through various devices some of the other approximations that were made in the derivation.

EXERCISE:

Assuming that the heat of vaporization of a liquid is given by an equation of the form $\Delta H = \alpha + \beta T$, show upon integration of equation 8.14 that

$$\ln p = -\frac{\alpha}{RT} + \frac{\beta}{R} \ln T + \text{constant}$$

EXERCISE:

(A) According to Trouton's rule, the molar entropy of vaporization of a so-called "normal" liquid at its normal boiling point is equal to 21 calories per degree. ($\Delta S = \Delta H/T_0 = 21$, if T_0 is the normal boiling point.) Show, as a consequence, that the vapor pressure of any "normal" liquid should be given by an equation of the type

$$\ln p \ (\text{atm}) = 10.5(1 - T_0/T)$$

(B) Benzene normally boils at 80.1 °C. Estimate the vapor pressure of benzene in Torr at 85 °C.

Vapor Pressures and the Gibbs-Helmholtz Equation

The Clausius-Clapeyron equation, which is an approximation to that of Clapeyron, can also be derived from the Gibbs-Helmholtz relationship, provided the same approximations are made. This alternative derivation, which we shall now carry out, should prove to be quite instructive, for it will give us practice in the use of the Gibbs-Helmholtz equation and will suggest further applications along more general lines.

Consider, for example, the change in state

$$\begin{array}{ccc} H_2O(liq) & \longrightarrow & H_2O(g) \\ T,p & & T,p \end{array}$$

in which both T and p are independent variables whose values are unchanged for the over-all change in state. Now, if T and p are both independent, p is not necessarily equal to the equilibrium vapor pressure, but might have any value within limits. (Actually, if p is less than p_{eq}, the vapor pressure, we should expect the change in state to occur spontaneously; conversely, if p is greater than p_{eq}, the reverse change in state will tend to occur.)

Let us now evaluate the change in Gibbs free energy attending the change in state described above. To do so, we shall formulate a reversible means of bringing about the change isothermally and shall then evaluate the integral of Vdp along that reversible path. The most convenient reversible path for accomplishing this consists of three stages, the second of which involves the reversible evaporation at the equilibrium pressure, p_{eq}. These three steps are as follows:

$$\begin{array}{llll} \text{I} & H_2O(liq) & \longrightarrow & H_2O(liq) \\ & T,p & & T,p_{eq} \end{array}$$

$$\begin{array}{llll} \text{II} & H_2O(liq) & \longrightarrow & H_2O(g) \\ & T,p_{eq} & & T,p_{eq} \end{array}$$

$$\begin{array}{llll} \text{III} & H_2O(g) & \longrightarrow & H_2O(g) \\ & T,p_{eq} & & T,p \end{array}$$

The Gibbs energy change for process I will simply equal

$$\Delta G_I = \int_p^{p_{eq}} V_1 dp \qquad (8.18)$$

That for process II, on the other hand, will be

$$\Delta G_{II} = 0 \qquad (8.19)$$

since the Gibbs free-energy change for an equilibrium reversible evaporation must be zero. Finally, for process III, we can write

$$\Delta G_{\mathrm{III}} = \int_{p_{\mathrm{eq}}}^{p} V_v dp \tag{8.20}$$

The total Gibbs energy change for the specified net change in state will then be given by the sum of the three:

$$\Delta G = \int_{p_{\mathrm{eq}}}^{p} (V_v - V_1) dp = \int_{p_{\mathrm{eq}}}^{p} \Delta V dp \tag{8.21}$$

The reader is reminded that equation 8.21 gives the actual Gibbs energy change for the specified change in state at pressure p and temperature T. The fact that we followed a three-step reversible route for making the calculation does not restrict the validity of the equation to such indirect paths. Our indirect method was simply used as an expedient for getting at the general answer.

Let us now modify equation 8.21 by introducing the assumptions that we made in going from the Clapeyron to the Clausius-Clapeyron equation. First, let us neglect V_1, compared with V_v, and then let us assume the vapor to be an ideal gas. Upon making these approximations and performing the indicated integration, we obtain the equation

$$\Delta G = RT \ln \frac{p}{p_{\mathrm{eq}}} \tag{8.22}$$

from which it is evident that ΔG will equal zero only if p equals p_{eq}. If p is less than p_{eq}, then ΔG will be negative, and the evaporation process can be expected to occur spontaneously. On the other hand, if p is greater than p_{eq}, then ΔG will be positive for the change in state as written; this indicates that evaporation will not occur and that the reverse process, condensation, will tend to take place spontaneously.

We shall now subject equation 8.22 to further manipulation. Neglecting the effect of applied pressure on the vapor pressure, which is an excellent approximation compatible with the assumption that V_1 is negligible compared with V_v, we can assert that p_{eq} is a function of temperature only. Hence, dividing equation 8.22 by T and differentiating partially with respect to T (keeping p constant), we obtain, in accordance with equation 7.49, the result

$$\frac{\partial \left(\dfrac{\Delta G}{T} \right)}{\partial T} = -R \frac{d \ln p_{\mathrm{eq}}}{dT} = -\frac{\Delta H}{T^2} \tag{8.23}$$

Upon rearrangement of this result, we see that

$$\frac{d \ln p_{eq}}{dT} = \frac{\Delta H}{RT^2} \qquad (8.24)$$

which is the same as equation 8.14.

Application to Reactions Involving Single Gases and Solids or Liquids

The foregoing treatment of vapor pressure through the Gibbs-Helmholtz equation can now be extended to processes other than evaporations (or sublimations), such as chemical reactions involving a single gaseous species reacting with any number of separate solid or liquid phases. The following are examples of the kinds of reactions that can be so treated:

$$CaCO_3(c) \longrightarrow CaO(c) + CO_2(g)$$
$$CuSO_4 \cdot 3H_2O(c) \longrightarrow CuSO_4 \cdot H_2O(c) + 2H_2O(g)$$
$$Ag_2O(c) \longrightarrow 2Ag(c) + \tfrac{1}{2}O_2(g)$$

In each example only one gaseous substance is generated, such gas formation being accomplished by a change in composition of the solids. This class of reactions can now be generalized in the form

$$\begin{array}{cc} A(c) \longrightarrow & B(c) + \nu G(g) \\ T,p & T,p \end{array}$$

in which the solid A at T and p decomposes into solid B and ν moles of gas G, also at T and p. At any given temperature there will be an equilibrium pressure, p_{eq}, under which the three forms can co-exist without experiencing any tendency to react. A reversible means of bringing about the indicated change in state can then be formulated as follows:

$$\begin{array}{lll} \text{I} & A(c) \longrightarrow & A(c) \\ & T,p & T,p_{eq} \end{array}$$

$$\begin{array}{lll} \text{II} & A(c) \longrightarrow & B(c) + \nu G(g) \\ & T,p_{eq} & T,p_{eq} \end{array}$$

$$\begin{array}{lll} \text{III} & B(c) + \nu G(g) \longrightarrow & B(c) + \nu G(g) \\ & T,p_{eq} & T,p \end{array}$$

The Gibbs free-energy changes can be evaluated as for a simple evaporation if we remember that $\Delta G_{II} = 0$ since the second stage is a reversible equilibrium process carried out at constant pressure. The total

Gibbs energy change will then be given by the expression

$$\Delta G = \int_{p_{eq}}^{p} \Delta V dp \qquad (8.25)$$

in which $\Delta V = \nu V_G + V_B - V_A$, the V's denoting molar volumes. If we neglect the volumes of the solids, compared with that of the gas, and assume that the gas is ideal, we obtain after integration the result

$$\Delta G = \nu RT \ln \frac{p}{p_{eq}} \qquad (8.26)$$

Here again the relative magnitudes of p and p_{eq} will determine the sign of ΔG and the direction in which the reaction will go. Applying the Gibbs-Helmholtz equation, we finally conclude that

$$\frac{d \ln p_{eq}}{dT} = \frac{\Delta H}{\nu RT^2} \qquad (8.27)$$

which is the same as equation 8.24 except for the factor ν. Since $\Delta H/\nu$ obviously equals the heat of reaction per mole of gas generated, the presence of this factor is not unexpected. Equation 8.27 can be integrated, naturally, and treated in much the same way as equation 8.14.

EXERCISE:

Copper sulfate forms three different hydrates according to the following chemical equations:

$$CuSO_4(c) + H_2O(g) \longrightarrow CuSO_4 \cdot H_2O(c)$$

$$CuSO_4 \cdot H_2O(c) + 2H_2O(g) \longrightarrow CuSO_4 \cdot 3H_2O(c)$$

$$CuSO_4 \cdot 3H_2O(c) + 2H_2O(g) \longrightarrow CuSO_4 \cdot 5H_2O(g)$$

The equilibrium vapor pressures associated with these reactions are respectively 0.8, 5.6, and 7.8 Torr at 25 °C. A sample of pure $CuSO_4 \cdot 5H_2O(c)$ is put into an evacuated container; after equilibrium has been established, the observed pressure is 7.0 Torr. Describe the final system.

Isothermals of a van der Waals Fluid

Although van der Waals' modification of the ideal-gas law equation is by no means exact, it nevertheless implies a number of striking qualitative features that correspond remarkably well to reality. When properly interpreted, this equation,

$$\left(p + \frac{a}{V^2}\right)(V - b) = RT \qquad (8.28)$$

can be used for predicting the phenomenon of liquefaction as well as the existence of a critical point. One can, in fact, derive an equation for the vapor pressure of a van der Waals fluid in terms of a, b, and the temperature.

Typical isotherms calculated from van der Waals' equation appear in Figure 8-3. If the temperature is below the critical temperature (the critical isotherm is the one that passes through the critical point, C), the pressure exhibits a maximum and a minimum, a behavior not actually observed in nature. If the temperature is sufficiently low, the calculated pressures may even be negative for a certain range of volumes, as shown by the lowest curve.

If we assume the temperature to be below the critical temperature, the left-hand side (l) of a van der Waals isotherm will correspond to the liquid state and the right-hand side (v) will represent the vapor. The middle portion provides a hypothetical continuous path by which liquid

FIGURE 8-3   Isotherms for a van der Waals fluid. The curve passing through point C is the critical isotherm.

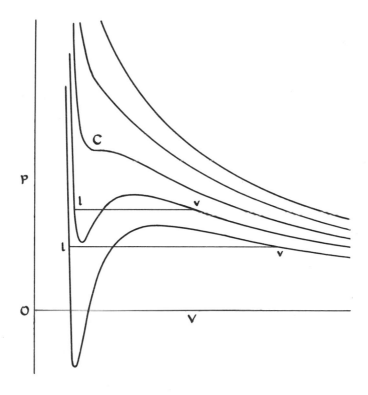

and vapor can be converted into each other. To make van der Waals' equation correspond more closely to reality (in other words, to make the transition from liquid to vapor a constant-pressure process), we employ the following modification, suggested by Maxwell. Horizontal straight lines (1 — v) are drawn in the p, V diagrams to connect the liquid and vapor portions of an isotherm in such a way that the area under the horizontal line is equal to the area under the original curve. Equating areas in this way will render the Gibbs energy change for the evaporation process equal to zero if we assume that the original equation provides a thermodynamically valid and continuous link between the two phases. The pressure corresponding to such a horizontal line is then the vapor pressure of the liquid when it is under an applied pressure equal to its own vapor pressure.

The validity of the equal area criterion for establishing the horizontal part of an isotherm can easily be proved. Let p equal the pressure corresponding to the line to be established, and let V_1 and V_v equal the volumes of liquid and vapor. Then the Gibbs free-energy change in a transition from liquid to vapor along a continuous van der Waals path will be given by the integral of $V dp$, which, after integration by parts, yields the expression

$$\Delta G = p(V_v - V_1) - \int_{V_1}^{V_v} p \, dV \tag{8.29}$$

Since, for equilibrium, $\Delta G = 0$, we conclude that

$$p(V_v - V_1) = \int_{V_1}^{V_v} p \, dV \tag{8.30}$$

But the left-hand side of this equation is precisely the area under the horizontal line, whereas the integral is the area under the curve; hence the proposition has been proved.

When we modify van der Waals isotherms by replacing the intermediate portions with such horizontal lines, we obtain approximate representations of actual isotherms such as those for carbon dioxide (Fig. 8-2). We should not infer, however, that the entire intermediate portion of a van der Waals isotherm is meaningless. The portion above the horizontal vapor-pressure line and to the right of the maximum can be regarded as representing supersaturated vapor—that is, vapor in a physically realizable condition that will spontaneously tend to undergo condensation. Similarly, the portion below the horizontal line and to the left of the minimum represents a metastable liquid of such character that a moderate disturbance will cause it to flash into vapor. Such metastable liquids may even exist under negative pressures, as predicted by the low-temperature van der Waals isotherms and verified by careful experiments. The portion of a van der Waals isotherm between the maximum and the minimum is, of course, physically meaningless, for it implies that an increase in pressure would produce an increase in volume, an impossibility.

Critical Point of a van der Waals Fluid

The coordinates of the critical point (p_c, V_c, T_c) for a van der Waals fluid can readily be found. Let us first rewrite van der Waals' equation in the form of a cubic in V:

$$V^3 - \left(\frac{RT}{p} + b\right) V^2 + \frac{a}{p} V - \frac{ab}{p} = 0 \qquad (8.31)$$

It is clear, from the nature of the equation, that for any given pair of values of T and p there will be three roots for V, of which at least one must be real. If T is less than T_c, then van der Waals' equation will have three real roots, as can be seen from in Figure 8-3. As the temperature and pressure approach their critical values, however, the three roots for V will tend to coalesce, and at the critical point the three roots become equal to the critical volume, V_c. Under these circumstances it follows that the cubic equation reduces to the form

$$(V - V_c)^3 = 0 \qquad (8.32)$$

which must be identical with equation 8.31 if p equals p_c and T equals T_c. Equating coefficients of the different powers of V appearing in equations 8.31 and 8.32, we conclude that

$$p_c = a/27b^2$$
$$V_c = 3b$$

and

$$T_c = \frac{8a}{27Rb} \qquad (8.33)$$

We can obtain the coordinates of the critical point equally well by recognizing that the critical point is a horizontal point of inflection, a point at which

$$\frac{\partial p}{\partial V} = 0 \quad \text{and} \quad \frac{\partial^2 p}{\partial V^2} = 0 \qquad (8.34)$$

These two conditions, in combination with the equation of state, are sufficient to establish equations 8.33.

EXERCISE:

Derive equations 8.33 by means of equations 8.34 in conjunction with van der Waals' equation of state.

Van der Waals' Reduced Equation of State; Corresponding States

Instead of describing a fluid in terms of its actual pressure, volume, and temperature, we may find it convenient, for some purposes, to utilize a new set of variables, called reduced variables, which are defined as the ratios of the actual variables to their critical values. The reduced pressure, π, the reduced volume, ϕ, and the reduced temperature, θ, are defined as follows:

$$\pi = p/p_c$$
$$\phi = V/V_c$$
$$\theta = T/T_c \qquad (8.35)$$

Introducing the reduced variables into van der Waals' equation of state and making use of equations 8.33, we obtain the equation

$$\left(\pi + \frac{3}{\phi^2}\right)\left(\phi - \frac{1}{3}\right) = \frac{8}{3}\theta \qquad (8.36)$$

which is known as van der Waals' reduced equation of state. This equation, containing nothing that can be regarded as characteristic of any particular substance, appears to be of a universal character and should be valid for all fluids. It states, in particular, that at specified reduced pressures and reduced temperatures all fluids should exhibit the same reduced volumes. Now, even though van der Waals' equation is not exact, the reduced form does suggest an important principle, which is found to have fairly broad validity—namely, the law of corresponding states. This law says that all fluids at the same reduced temperature and pressure have the same reduced volume (the precise analytical relationship is not that of van der Waals). In general terms the law of corresponding states enables us to assert that the reduced pressure is a universal function of ϕ and θ of the form

$$\pi = \pi(\phi,\theta) \qquad (8.37)$$

Vapor Pressure of a van der Waals Liquid

An equation for the vapor pressure of a van der Waals liquid can now be obtained by recourse to equation 8.30. Utilizing the equation of state and carrying out the indicated integration, we obtain the equation

$$p(V_v - V_1) = RT \ln \frac{V_v - b}{V_1 - b} + a \left(\frac{1}{V_v} - \frac{1}{V_1} \right) \tag{8.38}$$

Let us now make the following reasonable approximations. First, let us assume that the volume of the vapor is much larger than that of the liquid and also much larger than b. Then we can neglect V_1 and b compared with V_v; we can likewise neglect $1/V_v$ compared with $1/V_1$. Making these approximations, we obtain the expression

$$pV_v = RT \ln V_v - RT \ln (V_1 - b) - a/V_1 \tag{8.39}$$

Next, let us suppose that the vapor is an ideal gas—that is, that pV_v equals RT. Introducing this assumption and rearranging equation 8.39, we get the expression

$$\ln p = \ln \left(\frac{RT}{V_1 - b} \right) - \frac{a}{V_1 RT} - 1 \tag{8.40}$$

It should now be clear that we should have an equation for the vapor pressure if we knew V_1, the volume of the liquid. An adequate approximation to the volume of a van der Waals liquid can be obtained as follows. Let us rewrite equation 8.31 in the following form, which simply involves isolation of the linear term in V:

$$V = b + \left(\frac{bp + RT - pV}{a} \right) V^2 \tag{8.41}$$

If one neglects the cubic and quadratic terms in this equation, the solution of the remaining linear equation will be a crude approximation to the lowest root of the original equation, a root that must correspond to the volume of the liquid. This crude approximation, which is the lowest possible value that V could approach, is simply $V = b$. A substitution of this approximation into the right-hand side of equation 8.41 then yields a better approximation,

$$V_1 = b \left(1 + \frac{bRT}{a} \right) \tag{8.42}$$

which is the expression that we shall use. Since, as we shall see later, bRT/a is small compared with unity, the reciprocal of V_1 can be written approximately as

$$\frac{1}{V_1} = \frac{1}{b} \left(1 - \frac{bRT}{a} \right) = \frac{1}{b} - \frac{RT}{a} \tag{8.43}$$

According to the approximation expressed by equation 8.42, the volume of a liquid is independent of the pressure although it is a function of the temperature; this implies an incompressible liquid with a coefficient of thermal expansion different from zero. Substituting equations 8.42 and 8.43 into equation 8.40, we now obtain the equation

$$\ln p = \ln \left(\frac{a}{b^2}\right) - \frac{a}{bRT} \qquad (8.44)$$

The most striking thing about this equation is that it has the same functional form as equation 8.17, obtained by integration of the Clausius-Clapeyron equation. Further comparison discloses, moreover, that a/b should equal the heat of vaporization, ΔH. Since this is generally large compared with RT, bRT/a should be small compared with unity, as previously assumed.

We can also obtain a reduced vapor-pressure equation by combining equations 8.44, 8.33, and 8.35:

$$\ln \pi = \ln 27 - \frac{27}{8\theta} \qquad (8.45)$$

The correctness of this functional form has been established for a large number of so-called "normal" liquids, although the precise numbers appearing in equation 8.45 are not correct.

Even though the vapor-pressure equations 8.44 and 8.45 are not quantitatively exact, the fact that they possess the correct functional characteristics is evidence of the ingenious nature of van der Waals' equation. Other equations of state either do not lend themselves to the preceding treatment or may give rise to incorrect functional forms. It appears, therefore, that van der Waals' equation contains numerous hidden implications, many of which appear to go beyond its author's original goal.

Vapor Pressures of Small Liquid Drops

In discussing equilibria between phases, we have up to now disregarded the curvature or shape of the interfacial surfaces and have tacitly assumed that the interfaces were planar in nature. Actually, if a surface has a curvature, the phases on each side of the surface will be subject to a pressure difference attributable to surface tension forces. Thus if a small drop of liquid is suspended in its own vapor, the pressure within the liquid will be greater than its vapor pressure. This in turn accounts for a small increase in the vapor pressure of the liquid in accordance with equation 8.12.

To calculate the excess pressure within a drop of liquid, let us proceed as follows. Consider a box of fixed volume V containing vapor and one spherical drop of liquid, of radius r, all kept at constant temperature, T. Such a system will be at equilibrium when its Helmholtz free energy is at a minimum. Let p_v and V_v be the pressure and volume of the vapor, with p_1 and V_1 the corresponding quantities for the liquid drop. Evidently $V = V_v + V_1$, and, of course, $V_1 = (4/3)\pi r^3$. Now the surface tension, γ, measures the work that must be done to increase the surface area by unit amount. Since the surface area is $4\pi r^2$, it follows that the surface tension work required to increase the radius of the drop by an amount dr will be given by $8\pi r\gamma dr$; moreover, since $dV_1 = 4\pi r^2 dr$, it follows that the work to be done upon increasing the surface area will be simply equal to $(2\gamma/r)dV_1$.

Since the temperature is constant, it follows that dA, the differential of the Helmholtz free energy associated with an evaporation or condensation producing a change in drop volume, will be given by

$$dA = -p_1 dV_1 - p_v dV_v + \frac{2\gamma}{r} dV_1 \qquad (8.46)$$

But since the total volume $V_1 + V_v$ is constant, it follows further that $dV_v = - dV_1$. Hence

$$dA = \left\{-(p_1 - p_v) + \frac{2\gamma}{r}\right\} dV_1 \qquad (8.47)$$

But at equilibrium, $dA = 0$; therefore

$$p_1 - p_v = \frac{2\gamma}{r} \qquad (8.48)$$

The quantity $p_1 - p_v$ is precisely the excess pressure on the liquid over and above its vapor pressure. If, instead of having a liquid drop suspended in its vapor, we had a bubble of vapor surrounded by liquid, the treatment could be carried out the same way. The final result will resemble that expressed by equation 8.48 except that the sign is changed so that the vapor will be at a higher pressure than the liquid.

EXERCISE:

Show that, if a liquid surrounds a bubble of its vapor, $p_v - p_1 = 2\gamma/r$.

The vapor pressure of a liquid drop of radius r can now be calculated by use of equations 8.12 and 8.48. Assuming the liquid to be incompressible and its vapor to be an ideal gas, we obtain upon integration of equation 8.12

$$\ln p_v = \frac{p_1 V_1}{RT} + \text{constant} \qquad (8.49)$$

If p_0 is the vapor pressure of the liquid when the applied pressure is equal to its own vapor pressure, then

$$\ln \frac{p_v}{p_0} = \frac{V_1 \Delta p}{RT} \qquad (8.50)$$

where Δp is the excess pressure on the drop. Substituting equation 8.48 into equation 8.50, bearing in mind that $\Delta p = p_1 - p_v$, we then obtain

$$\ln \frac{p_v}{p_0} = \frac{2\gamma V_1}{rRT} \qquad (8.51)$$

EXERCISE:

Show that if $(p_v - p_0)/p_0$ is small compared with unity, then equation 8.51 becomes

$$\frac{p_v - p_0}{p_0} \approx \frac{2\gamma V_1}{rRT}$$

The increased vapor pressure of small drops gives rise to a matter of practical concern. A fog consisting of many very small uniform drops of water can be in equilibrium with supersaturated water vapor. This accounts for the stability of supersaturated fogs, which we might expect would spontaneously turn into rain.

Problems

1. The surface tension of water at 25 °C is 72 dyn cm^{-1}, and the vapor pressure of water at the same temperature is 23.7 mm. Calculate the excess vapor pressure of water drops of radius 0.1 mm.

2. Consider the following reversible cycle for substance A, which has vapor pressure p at temperature T and vapor pressure $p + dp$ at $T + dT$ (V_v and V_1 are to be considered dependent variables):

$$
\begin{array}{ccc}
A\,(\text{liq}) & \longrightarrow & A(v) \\
(T, p, V_1) & & (T, p, V_v) \\
\uparrow & & \downarrow \\
A\,(\text{liq}) & \longleftarrow & A(v) \\
(T + dT, p + dp, V_1 + dV_1) & & (T + dT, p + dp, V_v + dV_v)
\end{array}
$$

Neglecting infinitesimals of second or higher order, show that the reversible work attending this cycle is equal to $-(V_v - V_1)dp$. Hence, in view of equation 4.48, derive the Clapeyron equation.

3. Consider the chemical reaction

$$Ag(c) + \tfrac{1}{2}Cl_2(g) \longrightarrow AgCl(c)$$

From the Gibbs energy change for this reaction at 1 atm pressure computed from data given in Problem 7, Chapter Seven, calculate the equilibrium chlorine pressure by use of equation 8.26. (Watch the signs carefully.)

4. Consider any electrolytic cell all parts of which are at the same pressure, p. (See, for example, the cell of Prob. 7, Chap. Seven.)

   (A) Show that the pressure coefficient of the electromotive force of the cell is given by the expression

$$\left(\frac{\partial E}{\partial p}\right)_T = -\frac{\Delta V}{zF}$$

   in which ΔV is the change in volume attending the passage of z faradays through the cell.

   (B) Neglecting the volumes of liquids and solids compared with those of the gases, and assuming the gases to be ideal, show that

$$\left(\frac{\partial E}{\partial p}\right)_T = -\frac{\Delta \nu RT}{zFp}$$

   where $\Delta \nu$ is the net increase in the number of moles of gases attending the passage of z faradays through the cell.

   (C) Integrating the result of part B, show that

$$E_2 - E_1 = -\frac{\Delta \nu RT}{zE} \ln \frac{p_2}{p_1}$$

   where E_1 and E_2 are the electromotive forces corresponding to p_1 and p_2.

   (D) For the cell described in Problem 7, Chapter Seven, $p_1 = 1$ atm and $E_1 = 1.1362$ volts. Calculate the pressure, p_2, at which $E_2 = 0$, and thus obtain the equilibrium dissociation pressure of AgCl at 25 °C. (Compare with Prob. 3 above.)

5. Consider the following three chemical changes in state:

$$CuSO_4(c) + H_2O(g) \longrightarrow CuSO_4 \cdot H_2O(c)$$

$$CuSO_4 \cdot H_2O(c) + 2H_2O(g) \longrightarrow CuSO_4 3H_2O(g)$$

$$CuSO_4 \cdot 3H_2O(c) + 2H_2O(g) \longrightarrow CuSO_4 \cdot 5H_2O(c)$$

At 25 °C the equilibrium water-vapor pressures for the three reactions are respectively $p_I = 0.8$, $p_{II} = 5.6$, and $p_{III} = 7.8$ Torr.

   (A) Derive an expression for the change in Gibbs free energy attending the reaction

$$CuSO_4(c) + 5H_2O(g) \longrightarrow CuSO_4 \cdot 5H_2O(c)$$

   when the total pressure, p, is greater than p_{III}.

   (B) If $p = 23.8$ Torr (the vapor pressure of water at 25 °C), what is ΔG in calories for the over-all reaction given in A above?

(c) If $p < p_{III}$, would the formula derived in part A be meaningful? Explain the situation corresponding to $p = \sqrt[5]{p_{I}p_{II}^2p_{III}^2}$.

$$\text{Ans.} \quad \text{(A)} \quad \Delta G = -RT \ln \frac{p^5}{p_{I}p_{II}^2p_{III}^2}$$

6. The equilibrium dissociation pressure of calcium carbonate is given as a function of temperature by the equation

$$\lg p_{Torr} = -\frac{11,355}{T} - 5.388 \lg T + 29.119$$

Derive an expression for ΔH in calories as a function of T.

7. The vapor pressures of water at 0, 50, and 100 °C are 4.579, 92.51, and 760.0 Torr. Evaluate the constants a, b, and c of an equation of the type

$$\lg p_{Torr} = a + b/T + c \lg T$$

Hence deduce an expression for ΔH_{vap} and ΔC_p for the change in state

$$H_2O(liq) \longrightarrow H_2O(g)$$

8. Show from equation 8.42 that the coefficient of cubical expansion of a van der Waals liquid is given approximately by the expression

$$\alpha = \frac{bR}{a}$$

Hence, in view of the statement that $\Delta H_{vap} = a/b$, show that $\alpha\Delta H_{vap} = R$ (a constant). (Actually $\alpha\Delta H_{vap}$ is quite constant for normal liquids, but the value of the product is nearer 10 cal deg^{-1} than 2 cal deg^{-1}.)

9. (A) Recognizing that the Gibbs free energy of fusion of ice at 0 °C and 1 atm is equal to zero, and assuming the heat capacities of ice and super-cooled water to be constant, derive by integration of the Gibbs-Helmholtz equation an expression for the Gibbs free energy of fusion of ice as a function of temperature. Also, derive expressions for ΔH_{fus} and ΔS_{fus} as functions of temperature

(B) If the specific heat of fusion of ice at 0 °C and 1 atm is 79.7 cal, and the specific heat capacities of ice and water are 0.48 and 1.00 cal deg^{-1}, what are the molar values for ΔH_{fus}, ΔG_{fus}, and ΔS_{fus} at -3 °C?

10. (A) Combining the Clapeyron equation with the general equation for heat capacity (6.15), derive an expression for the heat capacity of a saturated vapor—that is, a vapor heated along such a path that its pressure is always equal to the equilibrium vapor pressure of the liquid.

(B) If you neglect the volume of liquid compared with that of vapor, and assume the vapor to be an ideal gas, to what does the expression reduce?

(C) What is the molar heat capacity in cal deg^{-1} of saturated water vapor at 100 °C and 1 atm? C_p for the vapor equals 9 cal deg^{-1} mol^{-1} and $\Delta H_{vap} = 9,720$ cal mol^{-1}. Explain the negative value.

$$\text{Ans.} \quad \text{(B)} \quad C = C_p - \Delta H_{vap}/T$$

11. The Dieterici equation of state is the equation

$$p = \frac{RT}{V-b} e^{-a/VRT}$$

in which a and b have the same qualitative significance as in van der Waals' equation.

(A) Show by the use of equations 8.34 that the critical constants, according to Dieterici's equation, are

$$p_c = \frac{a}{4b^2e^2} \qquad V_c = 2b \qquad T_c = \frac{a}{4Rb}$$

(B) Show that the Dieterici reduced equation of state is

$$\pi = \frac{\theta}{2\phi - 1} e^{2-2/\theta\phi}$$

12. Consider the chemical reaction

$$Ag_2O(c) \longrightarrow 2Ag(c) + \tfrac{1}{2}O_2(g)$$

The dissociation pressure of silver oxide is 422 Torr at 173 °C and 605 Torr at 183 °C. Making the assumptions implicit in equation 8.27, calculate the heat of the reaction.

13. The vapor pressure of water at several temperatures is given in the following brief table:

t °C	p Torr
0.0	4.579
1.0	4.926
99.0	733.24
100.0	760.00

(A) Using the pair of values corresponding to 0 and 1 °C, calculate a value for ΔH_{vap}.

(B) Similarly calculate a value for ΔH_{vap} in the range from 99 to 100 °C.

(C) Using the values for 0 and 100 °C, calculate an average heat of vaporization over the hundred-degree range, and compare it with those obtained in A and B.

14. One mole of liquid water is originally contained in a capsule that is enclosed in an evacuated tank of fixed volume that is kept at 100 °C. The capsule is broken, and the water is allowed to evaporate into the previously evacuated tank. After equilibrium has been established (at 100 °C), the pressure of the water vapor in the tank is found to be equal to 0.5 atm. Assuming water vapor to be an ideal gas, and neglecting the volume of liquid water compared with that of the vapor, calculate ΔU, ΔH, Q, W, ΔS, ΔA, and ΔG for the process. The "heat of vaporization" of water at 100 °C and 1 atm is 9,720 cal mol^{-1}.

15. One mole of liquid water is evaporated at 100 °C and 1 atm pressure in a cylinder provided with a piston that is subject to friction. To make the

piston move against the friction, one must establish a pressure difference between the two sides of the piston equal to 0.1 atm.

(A) If the molar heat of vaporization (ΔH) of water at 100 °C and 1 atm is 9,720 cal, what are ΔU, Q, W, ΔS, ΔA, and ΔG for the evaporation? Assume water vapor to be an ideal gas, and neglect the volume of the liquid. (See Prob. 9, Chap. Two.)

(B) Compare ΔS with Q/T, $-\Delta A$ with W, and $-\Delta G$ with $W - p\Delta V$.

16. Consider the change in state

$$H_2O(\text{liq}) \longrightarrow H_2O(\text{g})$$
$$T,p \qquad\qquad\quad T,p$$

in which p is not necessarily the equilibrium vapor pressure.

(A) Neglecting the volume of the liquid, and assuming the vapor to be an ideal gas, derive an equation for the change in ΔA for the change in state.

(B) Applying the Gibbs-Helmholtz equation to the expression derived in part A, show that

$$
\begin{aligned}
\frac{d \ln p_{\text{eq}}}{dT} &= \frac{\Delta U}{RT^2} + \left(\frac{\partial \ln p}{\partial T}\right)_v \\
&= \frac{\Delta U}{RT^2} + \frac{1}{T} \\
&= \frac{\Delta H}{RT^2}
\end{aligned}
$$

17. A vessel of fixed volume, V, contains n_1 moles of liquid water in equilibrium with n_v moles of water vapor at temperature T. Suppose the system is heated reversibly, at constant total volume, so that some of the liquid evaporates to form more vapor at a greater pressure.

(A) Neglecting the molar volume of liquid water compared with that of vapor, and assuming the vapor to be an ideal gas, show that the apparent heat capacity of the contents of the vessel will be given by the expression

$$C_{v(\text{apparent})} = n_1 C_{v1} + n_v C_{vv} + \frac{n_v \Delta U_{\text{vap}}^2}{RT^2}$$

in which C_{v1} and C_{vv} are the molar heat capacities of liquid and vapor and ΔU_{vap} is the energy of vaporization. Note that, as n_v approaches zero, the heat capacity approaches that of the liquid, which, under such limiting conditions, fills the entire volume. However, as n_1 approaches zero, the heat capacity does not approach that of the vapor present. Why? (n_1 cannot equal zero; if it did, the problem would lose its meaning so far as heating is concerned. The problem would be meaningful with n_1 equal to zero if the system were cooled.)

(B) Calculate the apparent heat capacity of 0.5 mole of liquid water in equilibrium with 0.5 mole of water vapor at 100 °C when the system is heated reversibly in a vessel of fixed volume.

9

Partial Molar Quantities and Systems of Variable Composition

Intensive and Extensive Properties

In describing a single-phase thermodynamic system consisting of a pure substance, we find it convenient to employ certain variables, both independent and dependent, such as pressure, temperature, volume, mass, energy, and entropy. When we deal with systems made up of more than one substance, particularly systems in which the chemical make-up might change, the description also calls for a knowledge of the composition. We could express the composition of a mixture by specifying the masses or numbers of moles of each of the chemical species present, but in practice we usually give the weight percentages or the mole fractions of the different substances.

A general examination of the properties of a system, and of the variables that might be used to characterize those properties, discloses that they fall into two distinct classes, to be designated *intensive* and *extensive*.

An *intensive* variable of a homogeneous system is one whose magnitude is simultaneously characteristic both of the whole system and of any arbitrarily chosen macroscopic portion of it. An intensive variable measures an inherent property of a system, a property whose value does not depend on the size of the system. Pressure and temperature, for example, are intensive properties, for the same values can be used to describe either the entire system or any part of it.* No reference need be made to the total amount of material present, and the magnitudes of such variables are unaffected by an arbitrary division of the system into parts. Another intensive property is the density. It should be clear that the density, being the mass per unit volume, does not depend on the total volume or the total mass, but is only the ratio of the two or that ratio for any arbitrary part of the system. Other intensive properties are the mole fraction, the weight percentage, the index of refraction, and a specific extinction coefficient for radiation. For heterogeneous systems, some intensive variables will assume different values in different phases, but the variables are still considered intensive if they fulfill the criteria laid down for a homogeneous system.

The variables that have magnitudes proportional to the total amount of material under consideration, and other properties associated with them, are said to be *extensive*; examples are mass, volume, energy, heat capacity, and entropy. Defined precisely, an extensive property is a homogeneous function of first degree in the masses of materials present, such homogeneity being fulfilled while all intensive variables are held constant. Before this definition of an extensive property can be fully appreciated, it is necessary that we define a homogeneous function. For this purpose consider any function such as $f(x, y, z)$. This function is said to be homogeneous of degree n in the variables x, y, and z if

$$f(\lambda x, \lambda y, \lambda z) = \lambda^n f(x, y, z) \tag{9.1}$$

for all values of λ. (This can, of course, be generalized for any number of variables.) If expression 9.1 is not an identity, the function is not homogeneous. If the degree of homogeneity is unity, the operation of multiplying each of the independent variables by a parameter, λ, produces the same effect as multiplying the over-all function by the same factor. It is precisely this condition that identifies an extensive property.

We can now readily see why the volume, for example, is a homogeneous function of first degree in the masses of materials that make up a system. For illustrative purposes, consider an aqueous solution of sodium chloride consisting, let us say, of 10 grams of salt dissolved in

---

* This assumes, of course, a state of mechanical and thermal equilibrium and the absence of any external fields. A fluid transmitting an acoustic wave will not have a uniform pressure, and a heat conductor in a temperature gradient will not have a uniform temperature. Also, because of gravitation, the pressure of the earth's atmosphere will depend upon the distance above the earth's surface.

100 grams of water at 25 °C and one atmosphere. Whatever the volume of that solution might be, it is obvious that, if we took a solution of 20 grams of sodium chloride in 200 grams of water at the same temperature and pressure, we should have double the original volume. Alternatively, if we took exactly half by volume of the original solution, we should simultaneously take half of each of the masses.

Instead of identifying an extensive property as one that is homogeneous with respect to masses, we can equally well say that it is homogeneous with respect to the numbers of moles of materials present, for the mass of any substance is directly proportional to its number of moles, the proportionality factor being the molecular weight, a constant. In this connection we recognize that masses and numbers of moles are themselves extensive properties. Bearing in mind that one set of independent variables might be replaced by another set, we can now assert, as a conclusion, that an extensive property will exhibit homogeneity of first degree with respect to any and all extensive variables that are actually used for specifying the function. This permits us to select our independent variables to suit our convenience, without jeopardizing application of the homogeneity condition.

To illustrate how different properties can depend explicitly on different numbers of extensive variables, let us consider the examples afforded by volume, Helmholtz free energy, and entropy. To specify the volume of a system, we naturally choose as independent variables the temperature, the pressure, and the numbers of moles, so

$$V = V(T, p, n_1, n_2, \ldots) \tag{9.2}$$

With this choice of variables, the homogeneity of V applies only with respect to the n_i's; that is to say

$$\lambda V = V(T, p, \lambda n_1, \lambda n_2, \ldots) \tag{9.3}$$

In other words, when we multiply each number of moles of material by λ to indicate the homogeneity, the intensive variables such as T and p must be kept unchanged.

Should we express the Helmholtz free energy as a function of $T, p, n_1, n_2, \ldots$, the homogeneity condition for that function would be specified in the same way as for volume. However, the natural variables for A might logically be taken as T, V, and the numbers of moles, so that

$$A = A(T, V, n_1, n_2, \ldots) \tag{9.4}$$

Under these circumstances the criterion for homogeneity is fulfilled if we stipulate that

$$\lambda A = A(T, \lambda V, \lambda n_1, \lambda n_2, \ldots) \tag{9.5}$$

Since V is itself an extensive variable, it must, in accordance with equation 9.3, also be multiplied by λ when the n_i's are so multiplied. The homogeneity condition is thus applied to all of the extensive variables upon which the function under consideration is explicitly dependent. Finally, if we consider the entropy, which might naturally be considered a function of energy, volume, and the numbers of moles of materials present, we can write

$$S = S(U, V, n_1, n_2, \ldots) \tag{9.6}$$

in which no intensive variables appear explicitly in the formulation. The homogeneity condition now requires that

$$\lambda S = S(\lambda U, \lambda V, \lambda n_1, \lambda n_2, \ldots) \tag{9.7}$$

The entropy could be written, of course, as a function of T, p, and n_i's, in which case its homogeneity would be expressed in the form of equation 9.3.

Euler's Theorem on Homogeneous Functions

Homogeneous functions possess a remarkable property discovered by Euler and described by a theorem under his name. Referring to equation 9.1 for a general example of homogeneous functions of nth degree, we see, upon differentiating partially with respect to λ, that

$$x \frac{\partial f}{\partial(\lambda x)} + y \frac{\partial f}{\partial(\lambda y)} + z \frac{\partial f}{\partial(\lambda z)} = n\lambda^{n-1} f \tag{9.8}$$

Since this equation is valid for all values of λ, it must, in particular, be valid for λ equal to unity. Hence, if we set λ equal to unity, we find that

$$x \frac{\partial f}{\partial x} + y \frac{\partial f}{\partial y} + z \frac{\partial f}{\partial z} = nf \tag{9.9}$$

This result, known as Euler's theorem on homogeneous functions, can, of course, be generalized for any number of independent variables.

If the theorem is applied to an extensive property of a thermodynamic system, such as the volume, it follows, in accordance with equations 9.3, 9.8, and 9.9, that

$$V = n_1 \left(\frac{\partial V}{\partial n_1}\right)_{T,p,n_2,\ldots} + n_2 \left(\frac{\partial V}{\partial n_2}\right)_{T,p,n_1,\ldots} + \cdots$$

$$= \sum_i n_i \left(\frac{\partial V}{\partial n_i}\right)_{T,p,n_1,n_2,\ldots} \tag{9.10}$$

This equation shows that the total volume of a system can be written as the sum of a number of terms, each associated more or less with a particular component of the system. If we apply equation 9.10 to a single pure substance, the partial derivative appearing in that equation will be precisely equal to the volume of one mole of the substance. Similarly, if we deal with a mixture of such character that its total volume, regardless of composition, is exactly equal to the sum of the volumes of its separate components, the partial derivatives will each equal the corresponding molar volumes of the pure substances.

The total volume of a mixture, in general, will not be exactly equal to the sum of the separate volumes. Under such circumstances, the partial derivatives appearing in equation 9.10 can be regarded as the molar contributions toward the total volume, corrected, in a precise mathematical way, for shrinkage or expansion. In this way the departures from additivity of volumes are apportioned among the several components of the system. The afore-mentioned partial derivatives are called *partial molar volumes;* they constitute one example of a more general concept to which we shall now turn.

EXERCISE:

Test the following functions for homogeneity, and show by direct test the validity of equation 9.9:

$$f(x,y) = ax^2y + by^{7/2}/x^{1/2} + cy^3 \tan^{-1}\left(\frac{y}{x}\right)$$

$$f(x,y) = ax^2/y^2 + b(x^{3/2} + y^{3/2})/xy^{1/2}$$

EXERCISE:

Consider a homogeneous function of first degree expressed in terms of masses m_1, m_2, m_3, . . . :

$$Z = Z(m_1, m_2, m_3, \ldots)$$

If each mass, m_i, is now replaced by a homogeneous function of first degree in another set of variables, ζ_1, ζ_2, ζ_3, . . . , so that

$$m_i = m_i(\zeta_1, \zeta_2, \zeta_3, \ldots)$$

show that Z becomes a homogeneous function of first degree in ζ_1, ζ_2, ζ_3, ... Show also, in view of equation 9.9, that

$$Z = \zeta_1 \left(\frac{\partial m_1}{\partial \zeta_1}\frac{\partial Z}{\partial m_1} + \frac{\partial m_2}{\partial \zeta_1}\frac{\partial Z}{\partial m_2} + \ldots\right)$$

$$+ \zeta_2 \left(\frac{\partial m_1}{\partial \zeta_2}\frac{\partial Z}{\partial m_1} + \frac{\partial m_2}{\partial \zeta_2}\frac{\partial Z}{\partial m_2} + \ldots\right)$$

$$+ \ldots$$

EXERCISE:

(A) Show that, if the molar volume of a substance is a homogeneous function of zeroth degree in pressure and temperature, the energy of the substance must be a function of temperature only. Suggestion: Use Euler's theorem on homogeneous functions to show that $(\partial p/\partial T)_V = -p/T$, or use the result of Problem 16, Chapter Six.

(B) Similarly, show that, if the pressure is a homogeneous function of zeroth degree with respect to volume and temperature, the enthalpy must be a function of temperature only. (Note: This example is designed to illustrate the homogeneity concept and has nothing to do with extensive or intensive properties *per se*.)

Partial Molar Quantities

Consider any extensive property, Z, which can be expressed as a function of T, p, n_1, n_2, ... We shall now introduce a new function, to be called the *partial molar Z*, with the definition *

$$Z_i = \left(\frac{\partial Z}{\partial n_i}\right)_{T,p,n_1,n_2, \ldots} \tag{9.11}$$

As pointed out in the preceding section, if Z represents the volume, then the derived quantity, which will now be denoted by V_i, is the partial molar volume of the ith component. Similarly, if Z is the enthalpy, we obtain from equation 9.11 the partial molar enthalpy, H_i. As another example, if Z is the total mass, the derived quantity, or partial molar mass, will simply equal the molecular weight.

Applying Euler's theorem to our general example, we now observe that

$$Z = n_1 Z_1 + n_2 Z_2 + \ldots = \sum_i n_i Z_i \tag{9.12}$$

We have already seen, in the case of volume, how a partial molar quantity might be interpreted, but one additional comment would appear to be appropriate at this time. Z_i, since it represents the rate of change of Z with respect to addition of substance i, will just equal the increase in Z attending the addition of one mole of substance i to an infinite amount of the system. For example, the partial molar volume, V_i, will be precisely equal to the increase in volume of a solution when one mole of the ith component is added to an infinite volume of the solution.

---

* Partial molar functions were generally denoted with a bar above the symbol, such as $\bar{Z}_i$. In accordance with a recommendation of the International Union of Pure and Applied Chemistry, the bars will be omitted in this book. See reference 7 in Appendix F-IV.

This formulation of partial molar quantities presupposes that the independent variables are chosen as $T, p, n_1, n_2, \ldots$ If we express a function like the Helmholtz free energy in terms of $T, V, n_1, n_2, \ldots$, Euler's theorem, by reason of equation 9.5, yields the result

$$A = V\left(\frac{\partial A}{\partial V}\right)_{T,n_1,n_2,\ldots} + n_1\left(\frac{\partial A}{\partial n_1}\right)_{T,V,n_2,\ldots} + n_2\left(\frac{\partial A}{\partial n_2}\right)_{T,V,n_1,\ldots} + \ldots \quad (9.13)$$

When this choice of independent variables is made, the partial derivatives of A with respect to numbers of moles are not called partial molar Helmholtz energies, but will be identified with something else, the significance of which will appear later. The presence of the term $\partial A/\partial V$ in equation 9.13 is a consequence of the fact that V itself is an extensive variable appearing among the independent variables used for expressing A.

Some General Properties of Partial Molar Quantities

From equation 9.12, which gives the value of an extensive property in terms of its partial molar quantities, we find, upon differentiating, that

$$dZ = n_1 dZ_1 + n_2 dZ_2 + \ldots + Z_1 dn_1 + Z_2 dn_2 + \ldots \quad (9.14)$$

But, since Z was presumed to be a function of T, p, and the number of moles, we can equally well write that

$$dZ = \left(\frac{\partial Z}{\partial T}\right)_{p,n_1,n_2,\ldots} dT + \left(\frac{\partial Z}{\partial p}\right)_{T,n_1,n_2,\ldots} dp + Z_1 dn_1 + Z_2 dn_2 + \ldots \quad (9.15)$$

Combining the two equations, we conclude that

$$n_1 dZ_1 + n_2 dZ_2 + \ldots = \left(\frac{\partial Z}{\partial T}\right)_{p,n_1,n_2,\ldots} dT + \left(\frac{\partial Z}{\partial p}\right)_{T,n_1,n_2,\ldots} dp \quad (9.16)$$

This equation tells us that changes in partial molar quantities are not mutually independent. If T and p are held constant, for example, we can use equation 9.16 to determine the change in one partial molar quantity from the changes in the others.

Partial molar quantities are homogeneous functions of degree zero in the numbers of moles of substances that make up the system. This proposition can be readily proved in the following way. Differentiating equation 9.1 with respect to x, we observe, after dividing through by λ, that

$$\frac{\partial f(\lambda x, \lambda y, \lambda z)}{\partial(\lambda x)} = \lambda^{n-1}\frac{\partial f}{\partial x} \quad (9.17)$$

But this equation expresses precisely the condition that must be fulfilled if $\partial f/\partial x$ is to be a homogeneous function of degree $n - 1$. Since our general extensive property, Z, is homogeneous of degree unity, Z_i must be homogeneous of degree zero. Hence Z_i must be an intensive property, depending on the ratios of the n_i's and not on their absolute values. Accordingly, Z_i can be expressed as a function of temperature, pressure, and the mole fractions; one of those fractions will be redundant since their sum must equal unity.

Applying Euler's theorem to a partial molar quantity such as Z_i, and taking cognizance of the degree of homogeneity, we obtain the result

$$n_1 \left(\frac{\partial Z_1}{\partial n_1} \right)_{T,p,n_2, \dots} + n_2 \left(\frac{\partial Z_1}{\partial n_2} \right)_{T,p,n_1, \dots} + \dots = 0 \qquad (9.18)$$

But we also know, since the order of partial differentiation is immaterial, that

$$\frac{\partial Z_1}{\partial n_2} = \frac{\partial^2 Z}{\partial n_1 \partial n_2} = \frac{\partial Z_2}{\partial n_1}, \text{ etc.} \qquad (9.19)$$

Hence we conclude that

$$n_1 \left(\frac{\partial Z_1}{\partial n_1} \right)_{T,p,n_2, \dots} + n_2 \left(\frac{\partial Z_2}{\partial n_1} \right)_{T,p,n_2, \dots} + \dots = 0 \qquad (9.20)$$

This is an important result, and we shall have frequent occasion to use it. It is generally used in a slightly different form, which we obtain by transforming the independent variables from numbers of moles to mole fractions. The net result is

$$x_1 \left(\frac{\partial Z_1}{\partial x_1} \right)_{T,p,x_2, \dots} + x_2 \left(\frac{\partial Z_2}{\partial x_1} \right)_{T,p,x_2, \dots} + \dots = 0 \qquad (9.21)$$

The reader is reminded, however, that the number of mole fractions that can be used for independent variables is one less than the number of substances present. In a three-component system we can assume x_1 and x_2 (but not x_3) to be independent; in a two-component system only one mole fraction (say x_1) can be taken as independent. In writing equation 9.21 for a two-component system, we must omit the subscript x_2 from the indicated partial derivatives, for x_2 obviously cannot be kept constant as x_1 is changed. Under such circumstances the equation becomes

$$x_1 \left(\frac{\partial Z_1}{\partial x_1} \right)_{T,p} + x_2 \left(\frac{\partial Z_2}{\partial x_1} \right)_{T,p} = 0 \qquad (9.22)$$

in which $x_2 = 1 - x_1$. When applied to the Gibbs free energy, equation 9.22 is generally known as the Gibbs-Duhem equation.

EXERCISE:

Prove, for a two-component system, the result expressed by equation 9.22.

EXERCISE:

Since Z_i can be regarded as a function of T, p, x_1, x_2 . . . , the total differential of Z_i will be

$$dZ_i = \left(\frac{\partial Z_i}{\partial T}\right)_{p,x_1,x_2,\,\ldots} dT + \left(\frac{\partial Z_i}{\partial p}\right)_{T,x_1,x_2,\,\ldots} dp$$

$$+ \sum_j \left(\frac{\partial Z_i}{\partial x_j}\right)_{T,p,x_1,\,\ldots} dx_j$$

Moreover, since $\sum_i n_i \left(\dfrac{\partial Z_i}{\partial T}\right)_{p,x_1,x_2,\,\ldots} = \dfrac{\partial}{\partial T}\sum_i n_i Z_i = \dfrac{\partial Z}{\partial T}$, and, simi-

larly, since $\sum_i n_i \left(\dfrac{\partial Z_i}{\partial p}\right)_{T,x_1,x_2,\,\ldots} = \dfrac{\partial Z}{\partial p}$, it follows from equation 9.16 that

$$\left[n_1 \left(\frac{\partial Z_1}{\partial x_1}\right) + n_2 \left(\frac{\partial Z_2}{\partial x_1}\right) + \ldots \right] dx_1$$

$$+ \left[n_1 \left(\frac{\partial Z_1}{\partial x_2}\right) + n_2 \left(\frac{\partial Z_2}{\partial x_2}\right) + \ldots \right] dx_2$$

$$+ \ldots = 0$$

Hence observe the correctness of equation 9.21 in view of the fact that the differentials dx_1, dx_2 . . . are mutually independent.

Partial Molar Volumes

Although partial molar volumes are by no means the most important partial molar quantities to be dealt with in chemical thermodynamics, a considerable amount of attention is given to them because they provide convenient examples for illustrating partial molar concepts. To determine partial molar volumes experimentally, we must measure volumes or densities of mixtures over a range of concentrations. We can then analyze the data so obtained graphically or otherwise to establish values for the appropriate partial derivatives. If empirical formulas are available for relating densities to concentrations, we can readily compute partial molar volumes by analytical methods.

An obvious way of getting partial molar volumes from experimental data would be to take slopes from graphs obtained by plotting the total volume of a mixture versus the number of moles of one of the substances present while keeping all other independent variables constant. Since this method, although direct, often proves to be numerically inaccurate, other procedures have been devised.

One such method is to analyze the dependence of the average molar volume of a mixture upon composition. To illustrate this procedure, let us consider a two-component system with mole numbers n_1 and n_2 and mole fractions x_1 and x_2. The average molar volume, $\langle V_m \rangle$, can then be written in the form

$$\langle V_m \rangle = \frac{V}{n_1 + n_2} = x_1 V_1 + x_2 V_2 \tag{9.23}$$

Differentiating $\langle V_m \rangle$ partially with respect to x_1 and remembering that x_2 equals $1 - x_1$, we obtain the result

$$\frac{\partial \langle V_m \rangle}{\partial x_1} = V_1 - V_2 + x_1 \frac{\partial V_1}{\partial x_1} + x_2 \frac{\partial V_2}{\partial x_1} \tag{9.24}$$

But, in view of equation 9.22, which is applicable to all partial molar quantities, we see that the last two terms in equation 9.24 add up to zero. Combining the remaining equation with the definition of the average molar volume, we conclude that

$$V_1 = \langle V_m \rangle + x_2 \frac{\partial \langle V_m \rangle}{\partial x_1}$$

and that

$$V_2 = \langle V_m \rangle - x_1 \frac{\partial \langle V_m \rangle}{\partial x_1} \tag{9.25}$$

It is clear from these equations that partial molar volumes can be obtained from the average molar volume and its dependence on composition. Numerous other methods have also been devised for getting at partial molar volumes, but they will not be discussed here.

Let us now consider the isopiestic and isothermal change in volume attending the formation of a solution from $n_1, n_2, \ldots$ moles of pure materials. If we represent the molar volumes of the pure materials by the symbols $V_{m1}, V_{m2}, \ldots$, the change in volume, ΔV_{mix}, will be given by the expression

$$\Delta V_{\text{mix}} = V - n_1 V_{m1} - n_2 V_{m2} \ldots \tag{9.26}$$

in which V is the total volume of the solution. Differentiating equation 9.26 partially with respect to n_i, we obtain

$$\left(\frac{\partial \Delta V_{\text{mix}}}{\partial n_i} \right)_{T, p, n_1, \ldots} = V_i - V_{mi} \tag{9.27}$$

It is evident, therefore, that the partial molar volume change of mixing is precisely the difference between the partial molar volume of a substance in solution and the molar volume of the pure substance.

In an infinitely dilute solution, the partial molar volume of solvent (which we shall consider to be the material present in great excess) is generally the same as the molar volume of the pure solvent. The partial molar volume of a solute, however, in the same infinitely dilute solution, will usually be different from that of the pure solute, for solute molecules experience a marked change in environment in going from pure solute to an infinitely dilute solution, whereas solvent molecules experience no perceptible change when undergoing the corresponding process.

Partial Molar Enthalpy

The enthalpy function also measures an extensive property of a thermodynamic system and accordingly is amenable to partial molar analysis. Unlike volumes, for which absolute values can be directly determined, the enthalpies dealt with in ordinary thermodynamics are all uncertain to the extent of an additive constant. We might anticipate, therefore, that partial molar enthalpies will likewise be indeterminate in an absolute sense, a conclusion that will be readily substantiated. However, this does not preclude the determination of changes in partial molar enthalpies; and, since changes are what we actually use, we shall experience no logical difficulty or hardship from the limitation.

Consider several pure substances of mole numbers $n_1, n_2, \ldots$, and let them be isothermally mixed to form a homogeneous solution. This process of mixing will be accompanied by an enthalpy change, ΔH_{mix}, which will, in general, be a function of the composition. If we let $H_{m1}, H_{m2}, \ldots$ denote the molar enthalpies of the pure materials, the total enthalpy of the mixture will be

$$H = n_1 H_{m1} + n_2 H_{m2} + \ldots + \Delta H_{\mathrm{mix}} \qquad (9.28)$$

We can now obtain the partial molar enthalpies of the various substances in the solution by differentiating equation 9.28 partially with respect to the number of moles of the material in question:

$$H_i = \left(\frac{\partial H}{\partial n_i}\right)_{T,p,n_1,\ldots} = H_{mi} + \left(\frac{\partial \Delta H_{\mathrm{mix}}}{\partial n_i}\right)_{T,p,n_1,\ldots} \qquad (9.29)$$

The change in partial molar enthalpy upon dissolving a pure substance in a solution will then be given by the expression

$$\Delta H_i = H_i - H_{mi} = \left(\frac{\partial \Delta H_{\mathrm{mix}}}{\partial n_i}\right)_{T,p,n_1,\ldots} \qquad (9.30)$$

It follows that a partial molar analysis of heats of mixing will yield differences between the partial molar enthalpies in solution and the molar enthalpies of the pure substances.

Numerous empirical formulas have been employed to describe the heats of mixing for various solutions. Since such formulas generally give ΔH_{mix} as a function of concentration, appropriate manipulation in accordance with equation 9.30 will enable one to get at changes in partial molar enthalpies.

When two nonelectrolytes are mixed, the heat involved is often given by a formula of the type

$$\Delta H_{\text{mix}} = V\phi_1\phi_2 B \tag{9.31}$$

in which ϕ_1 and ϕ_2 are the volume fractions of the two substances, V the total volume, and B a constant. Assuming that no shrinkage or expansion attends the mixing process, we can then write that

$$\Delta H_{\text{mix}} = \frac{n_1 n_2 V_{m1} V_{m2}}{n_1 V_{m1} + n_2 V_{m2}} B \tag{9.32}$$

Substituting into equation 9.30, we obtain

$$\Delta H_1 = H_1 - H_{m1} = V_{m1}\phi_2^2 B$$
$$\Delta H_2 = H_2 - H_{m2} = V_{m2}\phi_1^2 B \tag{9.33}$$

It is evident from these equations that, if $\phi_2 = 0$, then $H_1 = H_{m1}$ and $H_2 = H_{m2} + V_{m2}B$. This is in direct correspondence with the situation pertaining to volumes; as a solution becomes infinitely dilute, the partial molar enthalpy of solvent approaches that of pure solvent, while the partial molar enthalpy of solute generally approaches something different from that of pure solute. These conclusions are not a consequence of the specific form of equation 9.31 but can be regarded as valid for the same reasons advanced in connection with partial molar volumes in infinitely dilute solutions.

In Chapter Three an infinitely dilute solution was defined as one so dilute that further addition of solvent produced no heat effect. This is in keeping with our present conclusions. It will also be recalled that we referred to heats of solution of various substances in infinitely dilute solution. Those heats of solution can now be identified as the partial molar heats of solution in infinitely dilute solution. Moreover, the heats of formation of materials in infinitely dilute solutions can also be identified as the partial molar heats of formation of the solutes, the actual numerical values being subject to the conventions that were adopted with respect to the enthalpies of the elements.

Partial Molar Energy

Partial molar energies are seldom used, although they can be obtained by combination of partial molar enthalpies and partial molar volumes. Since $U = H - pV$, it is readily seen, if we differentiate partially with respect to n_i, that

$$U_i = H_i - pV_i \qquad (9.34)$$

Naturally, partial molar energies will be uncertain to the extent of an additive constant in the same way as partial molar enthalpies.

Partial Molar Gibbs Free Energy; Chemical Potential

The most important partial molar quantity with which we shall be concerned in chemical thermodynamics is the partial molar Gibbs free energy, a quantity also known as the *chemical potential*. The chemical potential will be represented* by the symbol μ_i, with the definition

$$\mu_i = \left(\frac{\partial G}{\partial n_i}\right)_{T,p,n_1,n_2,\,\ldots} \qquad (9.35)$$

Since it is natural to choose T, p, n_1, n_2, . . . as the independent variables for the Gibbs energy function, we can write that

$$dG = \left(\frac{\partial G}{\partial T}\right)_{p,n_1,n_2,\,\ldots} dT + \left(\frac{\partial G}{\partial p}\right)_{T,n_1,n_2,\,\ldots} dp + \left(\frac{\partial G}{\partial n_1}\right)_{T,p,n_2,\,\ldots} dn_1$$

$$+ \left(\frac{\partial G}{\partial n_2}\right)_{T,p,n_1,\,\ldots} dn_2 + \ldots \qquad (9.36)$$

But, in the light of equations 7.25 and 9.35, we can rewrite equation 9.36 in the form

$$dG = -SdT + Vdp + \mu_1 dn_1 + \mu_2 dn_2 + \ldots \qquad (9.37)$$

Now, in view of the defined relationship between the Helmholtz and Gibbs free energies (namely, $A = G - pV$,) we see that $dA = dG - pdV - Vdp$; so

$$dA = -SdT - pdV + \mu_1 dn_1 + \mu_2 dn_2 + \ldots \qquad (9.38)$$

Moreover, since $H = G + TS$, we can also write that

$$dH = TdS + Vdp + \mu_1 dn_1 + \mu_2 dn_2 + \ldots \qquad (9.39)$$

---

* The chemical potential could, of course, also be represented by G_i, but because of its special significance, it is given a distinctive symbol.

Furthermore, since $U = H - pV$, we deduce that

$$dU = TdS - pdV + \mu_1 dn_1 + \mu_2 dn_2 + \cdots \tag{9.40}$$

Finally, we can rewrite equations 9.39 and 9.40 by solving for dS, getting

$$dS = \frac{1}{T} dH - \frac{V}{T} dp - \frac{\mu_1}{T} dn_1 - \frac{\mu_2}{T} dn_2 - \cdots \tag{9.41}$$

$$dS = \frac{1}{T} dU + \frac{p}{T} dV - \frac{\mu_1}{T} dn_1 - \frac{\mu_2}{T} dn_2 - \cdots \tag{9.42}$$

It should be clear from the six preceding differential equations that the chemical potential has a significance transcending that of a simple partial molar quantity. Inspection of those equations will disclose that the chemical potential can be defined equally well in any of six different ways:

$$
\begin{aligned}
\mu_i &= \left(\frac{\partial G}{\partial n_i}\right)_{T,p,n_1,n_2,\,\ldots} \qquad = \left(\frac{\partial A}{\partial n_i}\right)_{T,V,n_1,n_2,\,\ldots} \\[2mm]
&= \left(\frac{\partial H}{\partial n_i}\right)_{S,p,n_1,n_2,\,\ldots} \qquad = \left(\frac{\partial U}{\partial n_i}\right)_{S,V,n_1,n_2,\,\ldots} \\[2mm]
&= -T \left(\frac{\partial S}{\partial n_i}\right)_{H,p,n_1,n_2,\,\ldots} \quad = -T \left(\frac{\partial S}{\partial n_i}\right)_{U,V,n_1,n_2,\,\ldots}
\end{aligned}
\tag{9.43}
$$

It is the universality of the chemical potential, demonstrated by equations 9.43, that justifies giving it a special name, thereby emphasizing that it is not derived solely from one particular thermodynamic variable such as the Gibbs free energy. It is evident that one can use changes in chemical potentials to calculate changes not only for the Gibbs function but also for the Helmholtz function, energy, enthalpy, and entropy, depending upon the constraints imposed on the processes.

Other Partial Molar Quantities in Relation to Chemical Potential

Just as the partial molar energy can be obtained from the partial molar enthalpy and the partial molar volume, so, too, other partial molar quantities can be obtained by similar means. Because of the defined relationship between A and G (namely, $A = G - pV$), we see, upon differentiating partially with respect to n_i at constant T and p, that

$$A_i = G_i - pV_i = \mu_i - pV_i \tag{9.44}$$

The partial molar Helmholtz energy, A_i, is obviously not the same as the chemical potential, even though the latter can be represented as an appropriate partial derivative of A.

In the light of the definitions of G and A, we can also assert that

$$\mu_i = H_i - TS_i \tag{9.45}$$

and that

$$A_i = U_i - TS_i \tag{9.46}$$

The partial molar quantities that we shall use most frequently are μ_i (the partial molar Gibbs energy), H_i, and V_i; the others can be conveniently expressed in terms of these three.

Let us now investigate further the connection among the several definitions of chemical potential given by equations 9.43, using the definition involving Helmholtz free energy as our first example. If we assume that A is given as a function of T, V, n_1, n_2, . . . , we can write, by appropriate transformation of variables, that

$$A_i = \left(\frac{\partial A}{\partial n_i}\right)_{T,p,n_1,n_2, \ldots}$$

$$= \left(\frac{\partial A}{\partial n_i}\right)_{T,V,n_1,n_2, \ldots} + \left(\frac{\partial A}{\partial V}\right)_{T,n_1,n_2, \ldots} \left(\frac{\partial V}{\partial n_i}\right)_{T,p,n_1,n_2, \ldots} \tag{9.47}$$

But $(\partial A/\partial n_i)_{T,V,n_1,n_2, \ldots} = \mu_i$ and $(\partial A/\partial V)_{T,n_1,n_2, \ldots} = -p$, and $(\partial V/\partial n_i)_{T,p,n_1,n_2, \ldots} = V_i$. Substituting these statements into equation 9.47, we obtain the result $A_i = \mu_i - pV_i$, which is the same as equation 9.44. Making the same substitutions into equation 9.13, which expresses Euler's theorem applied to the Helmholtz energy, we find that

$$A = -pV + n_1\mu_1 + n_2\mu_2 + \ldots = -pV + G \tag{9.48}$$

This conclusion is nothing new, of course, but it does serve to check our manipulations and to strengthen our understanding of Euler's theorem as applied to extensive properties expressed in terms of variables other than T, p, n_1, n_2, ...

A similar treatment can be given to the entropy, which might, for example, be regarded as a function of U, V, n_1, n_2, ... The partial molar entropy must therefore be given by the expression

$$S_i = \left(\frac{\partial S}{\partial n_i}\right)_{T,p,n_1,n_2, \ldots}$$

$$= \left(\frac{\partial S}{\partial n_i}\right)_{U,V,n_1,n_2, \ldots} + \left(\frac{\partial S}{\partial U}\right)_{V,n_1,n_2, \ldots} \left(\frac{\partial U}{\partial n_i}\right)_{T,p,n_1,n_2, \ldots}$$

$$+ \left(\frac{\partial S}{\partial V}\right)_{U,n_1,n_2, \ldots} \left(\frac{\partial V}{\partial n_i}\right)_{T,p,n_1,n_2, \ldots} \tag{9.49}$$

But identification of the several partial derivatives on the right-hand side of equation 9.49 enables us to rewrite that equation in the form

$$S_i = -\frac{\mu_i}{T} + \frac{U_i}{T} + \frac{pV_i}{T} \tag{9.50}$$

This conclusion is also obvious in view of the definition of Gibbs free energy. Finally, if we apply Euler's theorem to the entropy, assumed to be a function of U, V, n_1, n_2, . . . , we write

$$S = U\left(\frac{\partial S}{\partial U}\right)_{V,n_1,n_2,\,...} + V\left(\frac{\partial S}{\partial V}\right)_{T,n_1,n_2,\,...} + n_1\left(\frac{\partial S}{\partial n_1}\right)_{U,V,n_2,\,...}$$

$$+ n_2\left(\frac{\partial S}{\partial n_2}\right)_{U,V,n_1,\,...} + \cdots \tag{9.51}$$

This directly reduces to

$$S = \frac{U}{T} + \frac{pV}{T} - \frac{G}{T} \tag{9.52}$$

which serves to check our treatment once again.

Partial Derivatives of the Chemical Potential

Since dG is an exact differential, it follows from equation 9.37 that

$$\frac{\partial^2 G}{\partial T \partial n_i} = -\left(\frac{\partial S}{\partial n_i}\right)_{T,p,n_1,n_2,\,...} = \left(\frac{\partial \mu_i}{\partial T}\right)_{p,n_1,n_2,\,...} \tag{9.53}$$

and that

$$\frac{\partial^2 G}{\partial p \partial n_i} = \left(\frac{\partial V}{\partial n_i}\right)_{T,p,n_1,n_2,\,...} = \left(\frac{\partial \mu_i}{\partial p}\right)_{T,n_1,n_2,\,...} \tag{9.54}$$

These equations can now be rewritten as

$$\left(\frac{\partial \mu_i}{\partial T}\right)_{p,n_1,n_2,\,...} = -S_i \tag{9.55}$$

and

$$\left(\frac{\partial \mu_i}{\partial p}\right)_{T,n_1,n_2,\,...} = V_i \tag{9.56}$$

These two equations closely resemble equations 7.25 and, indeed, could be derived directly from those equations by partial differentiation with respect to n_i. Since dA is also an exact differential, we see from equation 9.38 that

$$\left(\frac{\partial \mu_i}{\partial T}\right)_{V,n_1,n_2,\,...} = -\left(\frac{\partial S}{\partial n_i}\right)_{T,V,n_1,n_2,\,...} \tag{9.57}$$

and that

$$\left(\frac{\partial \mu_i}{\partial V}\right)_{T,n_1,n_2,\,...} = -\left(\frac{\partial p}{\partial n_i}\right)_{T,V,n_1,n_2,\,...} \tag{9.58}$$

These equations, and many more like them, are not as frequently used as equations 9.55 and 9.56. It should be emphasized that the quantities on the right-hand sides of equations 9.57 and 9.58 are not partial molar quantities. At this point we should note that $(\partial p/\partial n_i)_{T,V,n_1,n_2,\ldots}$ cannot even be regarded as a partial pressure, a proposition we can readily demonstrate by considering a mixture of ideal gases.

EXERCISE:

Prove that for a mixture of ideal gases $(\partial \mu_i/\partial V)_{T,n_1,n_2,\ldots} = -RT/V$.

EXERCISE:

Show in general that

$$\left(\frac{\partial \mu_i}{\partial V}\right)_{T,n_1,n_2} = V_i \left(\frac{\partial p}{\partial V}\right)_{T,n_1,n_2\ldots}$$

Show also that

$$\left(\frac{\partial U}{\partial n_i}\right)_{T,V} = H_i - TV_i \left(\frac{\partial p}{\partial T}\right)_{V,n_i}$$

and that

$$\left(\frac{\partial S}{\partial n_i}\right)_{T,V} = S_i - V_i \left(\frac{\partial p}{\partial T}\right)_{V,n_i}$$

Partial Molar Properties of Ideal Gases

Before ending this chapter on partial molar quantities, we shall find it worth our while, in anticipation of subsequent developments, to inquire about the partial molar properties of ideal gaseous mixtures. The simplest of these quantities, the partial molar volume, is readily obtained from the equation of state,

$$V = \frac{(n_1 + n_2 + \ldots)RT}{p} \tag{9.59}$$

Obviously,

$$V_i = \left(\frac{\partial V}{\partial n_i}\right)_{T,p,n_1,n_2,\ldots} = \frac{RT}{p} \tag{9.60}$$

The partial molar volume of a gas in an ideal gaseous mixture is therefore independent of the composition and is the same as the molar volume of any pure ideal gas measured at the same temperature and pressure.*

---

* The reader will note that the partial pressure does not appear in the expression for partial molar volume.

Since the energy of a mixture of ideal gases is equal to the sum of the separate energies, a partial molar energy is simply equal to the molar energy of the corresponding pure gas, a function of temperature only. An equivalent statement can be made with respect to enthalpy.

The partial molar entropy of a particular component of an ideal gaseous mixture can be obtained by partial differentiation of equations of the type of 6.55 and 6.57. By this method we find that

$$S_i = C_{pi} \ln T - R \ln p - R \ln x_i + k_i'$$
$$= C_{pi} \ln T - R \ln p_i + k_i' \tag{9.61}$$

and that
$$S_i = C_{vi} \ln T + R \ln V + k_i \tag{9.62}$$

These equations tell us that the partial molar entropy of a substance in an ideal gaseous mixture is the same as the molar entropy of the pure gas at the same temperature occupying the total volume all by itself. The explicit dependence on temperature indicated by equations 9.61 and 9.62 rests upon the assumption that C_v and C_p are constants. If these heat capacities are not constant, the equations will assume somewhat different forms, but the general statement above will still be valid.

Since the Helmholtz and Gibbs free energies depend on linear combinations of U, H, and TS, the partial molar free energies must be the same for ideal gaseous mixtures as the molar functions for the pure gases under like conditions of temperature and volume. Hence we find from equation 7.39 that

$$A_i = C_{vi}T + K_i - C_{vi}T \ln T - RT \ln V - k_iT$$
$$= \phi_i(T) - RT \ln V \tag{9.63}$$

$\phi_i(T)$ being a function of T only but characteristic of substance i. Similarly, from equation 7.40 we obtain the expression

$$\mu_i = C_{pi}T + K_i - C_{pi}T \ln T + RT \ln p_i - k_i'T$$
$$= \psi_i(T) + RT \ln p_i$$
$$= \psi_i(T) + RT \ln p + RT \ln x_i \tag{9.64}$$

The function $\psi_i(T)$, which equals the value of the chemical potential when the partial pressure of the gas is equal to unity, is called the *standard chemical potential* or *standard Gibbs free energy* and is denoted by the symbol μ_i°. It is customary to measure the pressure in atmospheres; when the partial pressure of an ideal gas is equal to one atmosphere, we say that the gas is in its *standard state* for the temperature. Incorporating this new notation into equation 9.64, we now write

$$\mu_i = \mu_i^\circ + RT \ln p_i \tag{9.65}$$

The form of this equation is so convenient that it will be retained, with an appropriate interpretation, for systems other than ideal gases.

EXERCISE:

Deduce equation 9.60 by the use of equation 9.56 in conjunction with equation 9.64.

Problems

1. Suppose that the density of a solution, in grams per milliliter, is known as a function of the concentration, c, of solute expressed in moles per liter— that is, that $\rho = \rho(c)$ if $c = n_1/V$ and n_1 is the number of moles of solute and V is the volume of solution in liters.

   (A) Show that the partial molar volume of solute will be given in liters by the expression

   $$V_1 = \frac{M_1 - 1{,}000\, \dfrac{d\rho}{dc}}{1{,}000 \left(\rho - c\, \dfrac{d\rho}{dc} \right)}$$

   Show also that the partial molar volume of solvent will be given by the expression

   $$V_2 = \frac{M_2}{1{,}000 \left(\rho - c\, \dfrac{d\rho}{dc} \right)}$$

   (M_1 and M_2 denote molecular weights.)

   (B) If the density of the solution is a linear function of the molar concentration of solute, show that the partial molar volume of solvent is the same as the molar volume of pure solvent and that the partial molar volume of solute is independent of concentration.

2. The density of aqueous ammonium nitrate solutions at 25 °C is given by the following function of the molar concentration:

   $$\rho = 0.99708 + 3.263 \times 10^{-2}c - 9.63 \times 10^{-4}c^{3/2} - 4.73 \times 10^{-5}c^2 \text{ g ml}^{-1}$$

   Calculate V_1 and V_2 for $c = 0$, 0.1, 0.5, and 1.0 molar.

3. Consider a solution consisting of two or more components, and let us define the apparent molar volume, $\mathcal{V}_1$, of component 1 by the equation

   $$\mathcal{V}_1 = \frac{V - n_2 V_{m2} - n_3 V_{m3} - \cdots}{n_1}$$

   in which V_{mi} is the molar volume of pure substance i. (The concept "apparent molar volume" can be considered an outgrowth of attributing departures from additivity of volumes solely to a particular component of interest.) Assuming that $\mathcal{V}_1$ is known as a function of composition, prove that

$$V_1 = \mathcal{V}_1 + n_1 \frac{\partial \mathcal{V}_1}{\partial n_1}$$

and that

$$V_i = V_{mi} + n_1 \frac{\partial \mathcal{V}_1}{\partial n_i} \text{ for } i \neq 1$$

4. Suppose that the density of a solution, ρ, in grams per milliliter is known as a function of the weight percentage of solute, w. Show that the partial molar volumes of solute, V_1, and solvent, V_2, are given in liters by the expressions

$$V_1 = \frac{M_1}{1,000\rho} \left\{ 1 - \frac{(100 - w)}{\rho} \frac{\partial \rho}{\partial w} \right\}$$

$$V_2 = \frac{M_2}{1,000\rho} \left\{ 1 + \frac{w}{\rho} \frac{\partial \rho}{\partial w} \right\}$$

If ρ is a linear function of w, and if $w \frac{\partial \rho}{\partial w}$ is small compared with ρ, show to a good approximation that V_2 is constant.

5. At 25 °C the density of succinic acid in water is given by the expression (valid up to 5.5% acid)

$$\rho = 0.9970 + 0.00304w \text{ g ml}^{-1}$$

Calculate V_1 and V_2 for $w = 0, 2,$ and 4%.

At 20 °C the density of an aqueous solution of urea in water is given by the expression (valid up to 35% urea)

$$\rho = 0.9982 + 2.702 \times 10^{-3}w + 3.712 \times 10^{-6}w^2 + 22.85 \times 10^{-9}w^3 \text{ g ml}^{-1}$$

Calculate V_1 and V_2 for $w = 0, 15,$ and 30%.

6. Consider a binary solution whose pressure is expressed as a function of the concentrations, c_1 and c_2, in moles per unit volume, and temperature, T. Prove that

$$V_1 = \frac{\dfrac{\partial p}{\partial c_1}}{c_1 \dfrac{\partial p}{\partial c_1} + c_2 \dfrac{\partial p}{\partial c_2}}$$

7. (A) Consider a series of aqueous solutions containing one mole of sulfuric acid in n_2 moles of water. For such a solution the apparent molar enthalpy of the acid, $\mathcal{H}_1$, is defined as

$$\mathcal{H}_1 = H - n_2 H_{m2}$$

where H is the total enthalpy of the solution and H_{m2} the enthalpy of pure water. Remembering that n_1 is kept constant and equal to unity, show that

$$H_1 = \mathfrak{K}_1 - n_2 \frac{\partial \mathfrak{K}_1}{\partial n_2}$$

and that

$$H_2 = H_{m2} + \frac{\partial \mathfrak{K}_1}{\partial n_2}$$

How can you reconcile these formulas with those of Problem 3?

(B) The following table gives apparent molar enthalpies of formation of sulfuric acid in solutions containing 1 mole of acid in n_2 moles of water (Aq) at 25 °C:

n_2	$\mathfrak{K}_1$	n_2	$\mathfrak{K}_1$
0	−194.548 kcal	20	−211.500 kcal
1	−201.193	100	−212.150
1.5	−203.128	500	−212.833
2	−204.455	1000	−213.275
2.5	−205.452	2000	−213.740
3	−206.241	3000	−214.015
4	−207.428	4000	−214.220
6	−208.944	5000	−214.390
8	−209.865	10000	−215.060
10	−210.451	100000	−216.855
12	−210.835	∞	−217.32

By graphical means, or otherwise, evaluate $\partial \mathfrak{K}/\partial n_2$ for $n_2 = 2$, 8, and 3000, and then calculate values for the partial molar heat of formation of sulfuric acid for those compositions.

8. By differentiating the Gibbs-Helmholtz equation for Gibbs free energy partially with respect to n_i, show that

$$\left(\frac{\partial \mu_i}{\partial T}\right)_{p,n_1,n_2,\ldots} = \frac{\mu_i - H_i}{T}$$

By operating on the Gibbs-Helmholtz equation expressed for the Helmholtz function, show that

$$\left(\frac{\partial \mu_i}{\partial T}\right)_{V,n_1,n_2,\ldots} = \frac{\mu_i - H_i}{T} + \left(\frac{\partial V}{\partial T}\right)_{p,n_1,n_2,\ldots}\left(\frac{\partial p}{\partial n_i}\right)_{T,V,n_1,n_2,\ldots}$$

$$= \frac{\mu_i - H_i}{T} + V_i\left(\frac{\partial p}{\partial T}\right)_{V,n_1,n_2,\ldots}$$

Show, as a consequence, that for a mixture of ideal gases

$$\left(\frac{\partial \mu_i}{\partial T}\right)_{V,n_1,n_2,\ldots} = \left(\frac{\partial \mu_i}{\partial T}\right)_{p,n_1,n_2,\ldots} + R$$

9. When a pair of liquids with molar volumes V_{m1} and V_{m2} are mixed together, the entropy of mixing is given approximately by the expression

$$\Delta S_{\text{mix}} = -n_1 R \ln \phi_1 - n_2 R \ln \phi_2$$

in which ϕ_1 and ϕ_2 are volume fractions and n_1 and n_2 the numbers of moles. Assuming the volume change on mixing to be zero, derive expressions for the entropies of mixing, ΔS_1 and ΔS_2.
If the heat of mixing is given by an equation like equation 9.31, what are $\Delta \mu_1$ and $\Delta \mu_2$?

10. (A) Prove for a homogeneous solution that

$$n_1 \left(\frac{\partial p}{\partial n_i}\right)_{T,V,n_2,\,\ldots} + n_2 \left(\frac{\partial p}{\partial n_2}\right)_{T,V,n_1,\,\ldots} + \ldots = \frac{1}{\kappa}$$

if κ is the coefficient of compressibility of the solution.

(B) Show for an ideal gaseous mixture that

$$n_i \left(\frac{\partial p}{\partial n_i}\right)_{T,V,n_1,n_2,\,\ldots} = p_i$$

and that

$$1/\kappa = p$$

11. At moderate pressures, one mole of a gaseous mixture (as well as a mole of pure gas) can be assumed to obey the equation of state, $pV = RT + \alpha p$, in which α is a function of temperature and composition.

(A) Prove that, if x_A moles of substance A are mixed with x_B moles of substance B, and if x_A and x_B equal mole fractions in the final mixture, the heat of mixing at pressure p will be given by the expression

$$\Delta H_{\text{mix}} = \{\alpha_M - x_A \alpha_A - x_B \alpha_B\} p - Tp \frac{\partial}{\partial T} \{\alpha_M - x_A \alpha_A - x_B \alpha_B\}$$

in which α_M refers to the mixture.

(B) The values of α for helium, carbon dioxide, and an equimolar mixture of the two are given in the following table:*

t	α for He	α for CO_2	α for equimolar He-CO_2
30 °C	0.0117 liter mol^{-1}	-0.1193 liter mol^{-1}	-0.0161 liter mol^{-1}
60 °C	0.0115	-0.0955	-0.0087

Calculate approximately the heat of mixing in calories for the range 30–60 °C when 1/2 mole of helium is mixed with 1/2 mole of carbon dioxide at 1 atm pressure.

(c) Suppose that the value of α_M is expressed by an equation of the type

$$\alpha_M = x_A^2 \alpha_A + 2 x_A x_B \alpha_{AB} + x_B^2 \alpha_B$$

---

* T. L. Cottrell and R. A. Hamilton, *Trans. Faraday Soc.*, 52, 156 (1956).

in which α_{AB} is a parameter characteristic of the pair of substances. Prove that the heat of mixing will be given by an equation of the type

$$\Delta H_{\text{mix}} = x_A x_B p \beta$$

provided β is given by the following function of temperature:

$$\beta = T^2 \frac{d}{dT} \left\{ \frac{\alpha_A - 2\alpha_{AB} + \alpha_B}{T} \right\}$$

(D) From the results of parts B and C evaluate the quantity β for the mixture of helium and carbon dioxide.

10

Chemical Reactions; the Phase Rule

Reaction Potential (Affinity of a Reaction)

We introduced the concept of partial molar quantities to help us describe systems of variable composition, and for illustrative purposes we considered the physical addition of one or more substances to a system. From the point of view of the chemist, however, the most important way of producing a change in composition is through a chemical reaction, a process in which certain reacting species vanish as reaction products appear. Accordingly, we shall now investigate the thermodynamics of chemical reactions by utilizing the concept of chemical potential.

Consider a perfectly general chemical reaction, occurring in a homogeneous system (gaseous or liquid), that can be represented by the following balanced chemical equation:

$$aA + bB + \ldots \longrightarrow eE + fF + \ldots$$

At any instant we can specify the system by stipulating $T, p, n_A, n_B, \ldots$ $n_E, n_F \ldots$, the n's representing the numbers of moles of materials present. If the only way in which the composition can change is through the chemical reaction, certain definite constraints can be imposed on the relative magnitudes of the changes in the amounts of the various substances. Now the changes in the amounts of reactants and products are related to one another through the coefficients of the corresponding substances in the balanced chemical equation as follows:

$$-\frac{dn_A}{a} = -\frac{dn_B}{b} = \ldots = \frac{dn_E}{e} = \frac{dn_F}{f} = \ldots \quad (10.1)$$

The changes in numbers of moles of reactants and of products will, of course, have opposite signs, since the former class of substances will disappear as the other appears. The proportions expressed by equations 10.1 can now be replaced by several separate equations through introduction of a proportionality factor, $d\xi$, which we shall call the differential extent of reaction. Hence

$$dn_A = -ad\xi$$
$$dn_B = -bd\xi$$
$$dn_E = ed\xi$$
$$dn_F = fd\xi \quad (10.2)$$
$$\text{etc.}$$

ξ is simply defined as the number of times the reaction as written has occurred, so $d\xi$ will measure an increment of reaction.

In accordance with equation 9.37 the differential of the total Gibbs energy of the system will be

$$dG = -SdT + Vdp + \mu_A \, dn_A + \mu_B \, dn_B + \ldots \mu_E \, dn_E + \mu_F \, dn_F + \ldots \quad (10.3)$$

But this equation, in the light of equations 10.2, can now be rewritten as

$$dG = -SdT + Vdp + (e\mu_E + f\mu_F + \ldots - a\mu_A - b\mu_B - \ldots)d\xi \quad (10.4)$$

The coefficient of $d\xi$ (namely, $e\mu_E + f\mu_F + \ldots - a\mu_A - b\mu_B \ldots$) is precisely the rate of change of Gibbs free energy (at constant T and p) with respect to the extent of reaction. We shall denote this quantity by the symbol $\Delta\tilde{\mu}$, and we shall call it the *reaction potential*.* Accordingly, we write the definition:

$$\Delta\tilde{\mu} = (e\mu_E + f\mu_F + \ldots) - (a\mu_A + b\mu_B + \ldots) \quad (10.5)$$

It is clear that the reaction potential is simply measured by the sum of the chemical potentials of the products of the reaction less the sum for

---

* The reaction potential is often called the *reaction free energy* (meaning Gibbs free energy, $\Delta\tilde{G}$) and its negative is also called the *affinity* of the reaction (with the symbol **A**).

the reactants, each chemical potential being multiplied by the coefficient of the corresponding substance in the balanced chemical equation.

It is evident that, if changes in composition can occur only by a chemical reaction, the independent variables of the total Gibbs energy can be taken as T, p, and ξ. (The values of the n's for a particular value of ξ, say $\xi = 0$, must also be known as initial conditions for defining the system, but the n's need no longer be regarded as mutually independent variables, for their changes are determined by ξ alone.) We can therefore write that

$$\left(\frac{\partial G}{\partial \xi}\right)_{T,p} = \Delta \tilde{\mu} \tag{10.6}$$

Moreover, since dG is an exact differential, we can also assert that

$$\left(\frac{\partial \Delta \tilde{\mu}}{\partial T}\right)_{p,\xi} = -\left(\frac{\partial S}{\partial \xi}\right)_{p,T} = -\Delta \tilde{S} \tag{10.7}$$

in which expression

$$\Delta \tilde{S} = (eS_E + fS_F + \ldots) - (aS_A + bS_B + \ldots) \tag{10.8}$$

$\Delta \tilde{S}$ can be called the partial molar entropy of reaction. Similarly, we can write

$$\left(\frac{\partial \Delta \tilde{\mu}}{\partial p}\right)_{T,\xi} = \left(\frac{\partial V}{\partial \xi}\right)_{T,p} = \Delta \tilde{V} \tag{10.9}$$

in which expression

$$\Delta \tilde{V} = (eV_E + fV_F + \ldots) - (aV_A + bV_B + \ldots) \tag{10.10}$$

$\Delta \tilde{V}$ can be called the partial molar volume of reaction; it is precisely the rate of change of total volume per unit of reaction for reactions carried out at constant temperature and pressure.

EXERCISE:

Prove that for an ideal gaseous reaction the quantity $\Delta \tilde{V}$ appearing in equation 10.10 must be

$$\Delta \tilde{V} = \frac{\Delta \nu RT}{p}$$

if p is the total pressure and $\Delta \nu = (e + f + \ldots) - (a + b + \ldots)$. Show also that

$$\left(\frac{\partial n}{\partial \xi}\right)_{T,p} = \Delta \nu$$

provided n equals the total amount of material present expressed in moles.

The foregoing treatment can now be extended to any extensive property, Z, that can be considered a function of T, p, and ξ. The partial molar change of Z per unit of reaction, $\Delta \tilde{Z}$, is defined as

$$\left(\frac{\partial Z}{\partial \xi}\right)_{T,p} = \Delta \tilde{Z} = (eZ_E + fZ_F + \ldots) - (aZ_A + bZ_B + \ldots) \quad (10.11)$$

Returning to the Gibbs energy, $G = H - TS$, we find, upon differentiating partially with respect to ξ, that

$$\Delta \tilde{\mu} = \Delta \tilde{H} - T\Delta \tilde{S} \quad (10.12)$$

Combining equations 10.12 and 10.7, we can deduce a variant of the Gibbs-Helmholtz equation:

$$\left(\frac{\partial \Delta \tilde{\mu}}{\partial T}\right)_{p,\xi} = \frac{\Delta \tilde{\mu} - \Delta \tilde{H}}{T} \quad (10.13)$$

This equation can be manipulated and integrated by the methods that were used earlier (p. 144).

Just as the chemical potential could be defined in any of several ways (eqs. 9.43), so, too, the reaction potential is amenable to several alternative definitions. Replacing G by $A + pV$, we find from equations 10.4 and 10.5 that

$$dA = -SdT - pdV + \Delta \tilde{\mu}d\xi \quad (10.14)$$

Similarly, four other equations can be written:

$$dH = TdS + Vdp + \Delta \tilde{\mu}d\xi \quad (10.15)$$

$$dU = TdS - pdV + \Delta \tilde{\mu}d\xi \quad (10.16)$$

$$dS = \frac{1}{T}dH - \frac{V}{T}dp - \frac{\Delta \tilde{\mu}}{T}d\xi \quad (10.17)$$

$$dS = \frac{1}{T}dU + \frac{p}{T}dV - \frac{\Delta \tilde{\mu}}{T}d\xi \quad (10.18)$$

It is evident, therefore, that the reaction potential can be written in at least six different ways, as follows:

$$\Delta \tilde{\mu} = \left(\frac{\partial G}{\partial \xi}\right)_{T,p} = \left(\frac{\partial A}{\partial \xi}\right)_{T,V}$$

$$= \left(\frac{\partial H}{\partial \xi}\right)_{S,p} = \left(\frac{\partial U}{\partial \xi}\right)_{S,V}$$

$$= -T\left(\frac{\partial S}{\partial \xi}\right)_{H,p} = -T\left(\frac{\partial S}{\partial \xi}\right)_{U,V} \quad (10.19)$$

The great significance of the reaction potential should now be apparent. Consider an infinitesimal change in state occurring at constant T and p; in accordance with expression 7.9 we can assert that $dG \leqslant 0$. But, in the light of equations 10.4 and 10.5, this means that

$$\Delta \tilde{\mu}d\xi \leqslant 0 \quad (10.20)$$

Similarly, if an infinitesimal change in state occurs at constant T and V, then, according to expression 7.4, we write that $dA \leqslant 0$. But this statement, combined with equation 10.14, leads to the same inequality, 10.20, that we have above. Moreover, if a system is adiabatically enclosed, a condition that we can guarantee, for example, by holding the pairs U-V or H-p constant, then $dS \geqslant 0$ because of statement 4.37. But this, in combination with equation 10.17 or 10.18, gives the result

$$-\frac{\Delta\bar{\mu}d\xi}{T} \gtrless 0 \tag{10.21}$$

which, since T must always be greater than zero, is likewise equivalent to statement 10.20. Since inequalities 10.20 and 10.21 hold for spontaneous processes, and the equalities indicate states of equilibrium, we can now state a general conclusion of prime importance. *A chemical reaction will tend to take place spontaneously (thus rendering* $d\xi > 0$) *as long as the reaction potential* ($\Delta\bar{\mu}$) *is negative. If the reaction potential is positive, the reverse reaction will tend to occur (thereby making* $d\xi < 0$). *If the reaction potential is equal to zero, the system is in a state of equilibrium.*

It is important to observe that the same condition for spontaneity applies regardless of the constraints imposed, such as constant T and p, constant T and V, or constant U and V. To be sure, the ultimate course of the reaction will depend upon the constraints, since different final states will be attained under different circumstances. At any instant, however, the tendency for a reaction to occur will depend upon the reaction potential under the existing internal circumstances, and not upon the external constraints. This is most reasonable, for the molecules within a given system will tend to behave in a manner determined by their properties at the particular time under consideration and not by their properties after reaction occurs.

Simultaneous Reactions

The foregoing treatment can be generalized to cover the situation when two or more chemical reactions occur simultaneously. To illustrate, let us consider a pair of reactions like the following:

$$(I) \quad CO + \tfrac{1}{2}O_2 \longrightarrow CO_2$$

$$(II) \quad CO_2 + H_2 \longrightarrow CO + H_2O$$

If $d\xi_I$ and $d\xi_{II}$ represent the differential extents of the reactions, then

$$dn_{CO} = -d\xi_I + d\xi_{II}$$

$$dn_{O_2} = -\tfrac{1}{2}d\xi_I$$

$$dn_{CO_2} = d\xi_I - d\xi_{II}$$

$$dn_{H_2} = -d\xi_{II}$$

$$dn_{H_2O} = d\xi_{II} \tag{10.22}$$

Substituting equations 10.22 into the general expression for dG, equation 9.37, we obtain, upon simplification,

$$dG = -SdT + Vdp + \Delta\tilde{\mu}_I d\xi_I + \Delta\tilde{\mu}_{II} d\xi_{II} \tag{10.23}$$

in which expression

$$\Delta\tilde{\mu}_I = \mu_{CO_2} - \mu_{CO} - \tfrac{1}{2}\mu_{O_2}$$

$$\Delta\tilde{\mu}_{II} = \mu_{CO} + \mu_{H_2O} - \mu_{CO_2} - \mu_{H_2} \tag{10.24}$$

Equation 10.23 can, of course, be written for any pair of simultaneous chemical reactions and can also be generalized for any number of reactions. It is clear from that equation that the Gibbs energy of a system capable of undergoing a pair of reactions can be regarded as a function of T, p, ξ_I, and ξ_{II}. Evidently

$$\left(\frac{\partial G}{\partial \xi_I}\right)_{T,p,\xi_{II}} = \Delta\tilde{\mu}_I$$

$$\left(\frac{\partial G}{\partial \xi_{II}}\right)_{T,p,\xi_I} = \Delta\tilde{\mu}_{II} \tag{10.25}$$

We can also write for each of the separate reactions the equivalents of equations 10.7, 10.9, etc., plus an additional one of the form

$$\frac{\partial \Delta\tilde{\mu}_I}{\partial \xi_{II}} = \frac{\partial \Delta\tilde{\mu}_{II}}{\partial \xi_I} \tag{10.26}$$

In this equation the independent variables, other than ξ_I and ξ_{II}, can be chosen in several different ways. Equation 10.26 takes cognizance of the dependence of one reaction potential upon the extent of another reaction; the effect can be important when the two reactions involve one or more chemical substances in common, such as CO and CO_2 in the reactions cited above.

When two or more reactions occur simultaneously, the most general condition for spontaneity can be written as

$$\Delta\tilde{\mu}_I d\xi_I + \Delta\tilde{\mu}_{II} d\xi_{II} < 0 \tag{10.27}$$

For complete equilibrium, both reaction potentials must be equal to zero. Partial equilibrium can exist if only one of the reaction potentials is zero, but that condition will be disturbed by the occurrence of the other reaction unless the partial derivatives of equation 10.26 are equal to zero.

If two simultaneous reactions perturb each other, it is conceivable that one of them, if its reaction potential is favorable, can carry along the other, even if it is unfavorable by itself, provided the over-all inequality (10.27) is fulfilled. This possible behavior is analogous to that of a mechanical system consisting of two unequal masses in a gravitational field joined together by a string passing over a pulley. Although the separate masses each tend to fall, the smaller one will actually rise under the conditions specified because of the constraint imposed by the string. When certain violent chemical reactions occur, small quantities of unexpected by-products often appear through reactions that might by themselves be thermodynamically unfavorable. However, the over-all thermodynamics must be consistent with inequality 10.27, although specific examples are usually explained in terms of kinetic reaction mechanisms.

A simple example of this kind of behavior is given by the reaction of wet zinc with oxygen:

$$Zn + H_2O + O_2 \longrightarrow ZnO + H_2O_2$$

From a purely thermodynamic point of view, this reaction can be regarded as a composite of

$$Zn + \tfrac{1}{2}O_2 \longrightarrow ZnO$$

and

$$H_2O + \tfrac{1}{2}O_2 \longrightarrow H_2O_2$$

Although the oxidation of zinc takes place readily, the oxidation of water to hydrogen peroxide is thermodynamically most unfavorable; nevertheless, the combination can occur. It is presumed, mechanistically, that zinc first forms a peroxide that is subsequently hydrolyzed, both steps being spontaneous. Other, less understood combinations of reactions that illustrate this principle can occur in flames.

Reactions of Ideal Gases; Equilibrium Constants

Since the chemical potential of an ideal gas can be written as

$$\mu_i = \mu_i^{\circ} + RT \ln p_i \tag{10.28}$$

the reaction potential (defined by eq. 10.5) of a general reaction involving only ideal gases will assume the form

$$\Delta\tilde{\mu} = \Delta\tilde{\mu}^{\circ} + RT \ln \left\{ \frac{p_E^e p_F^f \cdots}{p_A^a p_B^b \cdots} \right\} \tag{10.29}$$

In this expression, $\Delta\tilde{\mu}^{\circ}$ is called the standard reaction potential (or standard Gibbs free energy of reaction) and is defined as

$$\Delta\tilde{\mu}^{\circ} = (e\mu_E^{\circ} + f\mu_F^{\circ} + \ldots) - (a\mu_A^{\circ} + b\mu_B^{\circ} + \ldots) \tag{10.30}$$

It is evident that the reaction potential, $\Delta\tilde{\mu}$, will depend on the partial pressures of the gases in the system as well as on the temperature. The condition for equilibrium (namely, $\Delta\tilde{\mu} = 0$) can now be realized if, and only if,

$$\frac{p_E^e p_F^f \cdots}{p_A^a p_B^b \cdots} = \exp\left(-\Delta\tilde{\mu}^\circ/RT\right) \qquad (10.31)$$

Since this condition for equilibrium is of paramount importance, the quantity on the right-hand side of equation 10.31 is given a special name and symbol. It is called the equilibrium constant for the reaction and is denoted by K. Hence, by definition,

$$K = \exp\left(-\Delta\tilde{\mu}^\circ/RT\right) \qquad (10.32)$$

or

$$-RT \ln K = \Delta\tilde{\mu}^\circ \qquad (10.33)$$

Since $\Delta\tilde{\mu}^\circ$ is a function of temperature only, it is evident that K likewise will depend only on the temperature. Combining equations 10.33 and 10.29, we see that

$$\Delta\tilde{\mu} = -RT \ln K + RT \ln\left[\frac{p_E^e p_F^f \cdots}{p_A^a p_B^b \cdots}\right] \qquad (10.34)$$

Examination of this equation discloses that $\Delta\tilde{\mu}$ will be negative if the partial-pressure expression in brackets is less than K; under such circumstances the reaction will go forward, thus tending to increase p_E, p_F, $\ldots$ while decreasing p_A, p_B, $\ldots$. Similarly, $\Delta\tilde{\mu}$ will be positive if the aforementioned pressure expression is greater than K; in that event the reverse reaction will tend to occur, thus diminishing the bracketed expression until equilibrium is reached.

EXERCISE:

Consider the following reaction, assuming the gases to be ideal:

$$PCl_3(g) + Cl_2(g) \rightleftharpoons PCl_5(g)$$

Set up the expression for $\Delta\tilde{\mu}$, using mole fractions and the total pressure. Differentiating partially with respect to p, show that

$$\Delta\tilde{V} = -\frac{RT}{p}$$

and compare with the direct calculation using equation 10.10.

Temperature Coefficient of Equilibrium Constant

If equation 10.34 is substituted into the Gibbs-Helmholtz equation (10.13), one obtains, upon rearrangement, the result

$$\frac{d \ln K}{dT} = \frac{\Delta\tilde{H}}{RT^2} \qquad (10.35)$$

Since $\Delta \tilde{H}$ is a function of temperature only for ideal gaseous systems, it is evident that equation 10.35 can be directly integrated to give

$$\ln \frac{K_2}{K_1} = \int_{T_1}^{T_2} \frac{\Delta \tilde{H}}{RT^2} \, dT \tag{10.36}$$

Assuming further that $\Delta \tilde{H}$ is constant, we see that

$$\ln \frac{K_2}{K_1} = \frac{\Delta \tilde{H}}{R} \left(\frac{1}{T_1} - \frac{1}{T_2} \right) \tag{10.37}$$

In accordance with this assumption, it is clear from the above that a plot of $\ln K$ against $1/T$ should yield a straight line of slope $-\Delta \tilde{H}/R$. On the other hand, if $\Delta \tilde{H}$ is given by a power-series expression like equation 3.34, then

$$\ln \frac{K_2}{K_1} = \frac{\Delta \tilde{H}_0}{R} \left(\frac{1}{T_1} - \frac{1}{T_2} \right) + \frac{\Delta a}{R} \ln \frac{T_2}{T_1} + \frac{\Delta b}{2R} (T_2 - T_1)$$

$$+ \frac{\Delta c}{6R} (T_2^2 - T_1^2) + \ldots \tag{10.38}$$

Dropping one of the subscripts and combining all terms involving the other subscript into a constant, A, we obtain

$$\ln K = A - \frac{\Delta \tilde{H}_0}{RT} + \frac{\Delta a}{R} \ln T + \frac{\Delta b}{2R} T + \frac{\Delta c}{6R} T^2 + \ldots \tag{10.39}$$

Equation 10.39 provides a very general representation of an equilibrium constant as a function of temperature.

EXERCISE:

Consider the ideal gaseous reaction

$$aA + bB + \ldots \rightleftarrows eE + fF + \ldots$$

(A) Suppose that each partial pressure (p_i) appearing in the expression for the equilibrium constant is replaced by $c_i RT$, where c_i is the concentration in moles per unit volume. Show that at equilibrium

$$\frac{c_E^e c_F^f \cdots}{c_A^a c_B^b \cdots} = K(RT)^{-\Delta \nu} = K'$$

provided

$$\Delta \nu = (e + f + \ldots) - (a + b + \ldots)$$

and K' is a new kind of equilibrium constant (also a function of T only).

(B) Show also that

$$\frac{d \ln K'}{dT} = \frac{\Delta \tilde{U}}{RT^2}$$

Extension to Reactions Involving Pure Solids and Liquids

Equation 10.35 is identical in form with the Clausius-Clapeyron vapor-pressure equation, so the results obtained by integration will, of course, exhibit corresponding similarities. This resemblance is not accidental, for we saw in Chapter Eight that the Clausius-Clapeyron could also be derived from the Gibbs-Helmholtz equation, provided one neglected the volume of liquid (or solid) and assumed the vapor to be an ideal gas. Subsequently an identical procedure was carried out for chemical reactions involving pure solids or liquids in conjunction with one ideal gaseous species. Now the developments of Chapter Eight can be reconciled with those of the immediately preceding section by the simple expedient of identifying the equilibrium pressure with the equilibrium constant for the reaction. Thus let us reconsider the three reactions indicated on page 169:

(I) $CaCO_3(c) \longrightarrow CaO(c) + CO_2(g)$

(II) $CuSO_4 \cdot 3H_2O(c) \longrightarrow CuSO_4 \cdot H_2O(c) + 2H_2O(g)$

(III) $Ag_2O(c) \longrightarrow 2Ag(c) + \frac{1}{2}O_2(g)$

In each instance, equilibrium will be realized only if the partial pressure of the gaseous material has a definite value at a given temperature. Accordingly, the reactions will have equilibrium constant expressions of the form

$$K_I = p_{CO_2}$$
$$K_{II} = p_{H_2O}^2$$
$$K_{III} = p_{O_2}^{1/2} \tag{10.40}$$

The foregoing arguments can now be applied to more general chemical reactions, such as those involving pure solids and liquids (not solutions) reacting with two or more gases. Such a reaction can be represented as follows:

$$aA(c) + bB(g) + \ldots \longrightarrow eE(c) + fF(g) + \ldots$$

Assuming the gases to be ideal and neglecting the volumes of solids, one obtains the expression

$$\Delta \tilde{\mu} = -RT \ln K + RT \ln \frac{p_F^f \ldots}{p_B^b \ldots} \tag{10.41}$$

from which the partial pressures of $A(c)$ and $E(c)$ are omitted. In this treatment it is tacitly assumed that the pure solids (or liquids) have constant chemical potentials, at a given temperature, and that the reaction potential can be appreciably affected only by a change in the partial pressures of the gases. (This is also equivalent to assuming that the vapor pressure of a pure solid or liquid is constant at a given temperature.) Hence, in writing an equilibrium-constant expression for a reaction involving pure solids (or liquids) as well as ideal gases, we proceed just as we would with an all-gaseous reaction except that the partial pressures of the condensed phases are omitted. (As constants, they can be considered absorbed in the equilibrium constant.) Such omission, however, does not extend to the calculation of reaction enthalpies, for $\Delta \tilde{H}$, like a heat of vaporization or sublimation, represents the change in enthalpy corresponding to a balanced chemical equation.

EXERCISE:

The standard Gibbs energy of $I_2(g)$ relative to $I_2(c)$ is equal to 4.627 kcal mol^{-1} at 25 °C. The heat of sublimation of iodine is equal to 14.923 kcal mol^{-1}. Calculate the vapor pressure of $I_2(c)$ at 25 °C and at 100 °C.

Calculation of Equilibrium Constants from Gibbs Free-energy Data

One of the important objectives of chemical thermodynamics is to determine whether a chemical reaction might be expected to go. We can do this by calculating reaction potentials and equilibrium constants. Since the direct determination of an equilibrium constant for a particular reaction is very often not feasible, it becomes necessary to devise indirect methods for getting at the desired quantities.

We can realize the ultimate objective by establishing tables of Gibbs energies for the various compounds of interest. We can obtain such information in any of several ways. The Gibbs energies of some compounds can be determined by direct equilibrium measurements for certain simple reactions. In other instances, studies of electromotive force may yield the desired results. In any event, we can use the results so obtained to calculate reaction potentials for other conceivable chemical reactions.

A very important method of calculating entropies from purely thermal data involves use of the Third Law. Such results, in combination with enthalpies of formation, will provide information about reaction potentials of molecular' formation. Still another method, which has become increasingly important, involves the calculation of thermodynamic properties from data on molecular structure. This way of arriving at thermodynamic properties will be dealt with later, when we turn to the subject of statistical mechanics.

TABLE 10-1

Standard Enthalpies and Gibbs Energies of Formation at 25 °C

(Values given in kcal mol^{-1} relative to pure elements in standard states)

SUBSTANCE	ΔH_f°	ΔG_f°	SUBSTANCE	ΔH_f°	ΔG_f°
$O_3(g)$	34.1	39.0	$CH_2O(g)$	−28.	−27.
$H_2O(liq)$	−68.315	−56.687	$HCOOH(liq)$	−101.51	−86.38
$H_2O(g)$	−57.796	−54.634	$CH_3OH(liq)$	−57.04	−39.76
$H_2O_2(g)$	−32.58	−25.24	$CF_4(g)$	−221.	−210.
$HF(g)$	−64.8	−65.3	$CCl_4(liq)$	−32.37	−15.60
$HCl(g)$	−22.062	−22.777	$CCl_4(g)$	−24.6	−14.49
$Br_2(g)$	7.387	0.751	$CHCl_3(g)$	−24.65	−16.82
$HBr(g)$	−8.70	−12.77	$CBr_4(g)$	19.	16.
$I_2(g)$	14.923	4.627	$CS_2(g)$	28.05	16.05
$HI(g)$	6.33	0.41	$COS(g)$	−33.96	−40.47
$SO_2(g)$	−70.944	−71.748	$HCN(g)$	32.3	29.8
$SO_3(g)$	−94.58	−88.69	$C_2H_2(g)$	54.19	50.00
$H_2S(g)$	−4.93	−8.02	$C_2H_4(g)$	12.49	16.28
$H_2SO_4(liq)$	−194.548	−164.938	$C_2H_6(g)$	−20.24	−7.86
$NO(g)$	21.57	20.69	$CH_3CHO(g)$	−39.72	−30.81
$NO_2(g)$	7.93	12.26	$CH_3COOH(liq)$	−115.8	−93.2
$N_2O(g)$	19.61	24.90	$CH_3COOH(g)$	−103.31	−89.4
$N_2O_4(g)$	2.19	23.38	$CH_3CN(g)$	20.9	25.0
$NH_3(g)$	−11.02	−3.94	$SiO_2(c, quartz)$	−217.72	−204.75
$N_2H_4(g)$	22.80	38.07	$SiH_4(g)$	8.2	13.6
$HNO_3(liq)$	−41.61	−19.31	$SnCl_4(g)$	−112.7	−103.3
$NH_4Cl(c)$	−75.15	−48.51	$ZnO(c)$	−83.24	−76.08
$P_4(g)$	14.08	5.85	$ZnCl_2(c)$	−99.20	−88.296
$P_4O_{10}(c)$	−713.2	−644.8	$Ag_2O(c)$	−7.42	−2.68
$PH_3(g)$	1.3	3.2	$AgCl(c)$	−30.370	−26.244
$H_3PO_4(c)$	−305.7	−267.5	$Fe_2O_3(c)$	−197.0	−177.4
$PF_3(g)$	−219.6	−214.5	$MgO(c)$	−143.81	−136.10
$PCl_3(g)$	−68.6	−64.0	$MgCl_2(c)$	−153.28	−141.45
$PCl_5(g)$	−89.6	−72.9	$CaO(c)$	−151.79	−144.37
$C(c, diamond)$ *	0.453 3	0.693 0	$CaCl_2(c)$	−190.2	−178.8
$CO(g)$	−26.416	−32.780	$CaCO_3(c, calcite)$	−288.46	−269.80
$CO_2(g)$	−94.051	−94.254	$SrO(c)$	−141.5	−134.3
$CH_4(g)$	−17.88	−12.13	$BaO(c)$	−132.3	−125.5

Additional data will be found in Appendix D.

* Relative to graphite.

Without concerning ourselves now with the detailed methods for establishing such data, we shall simply observe that by various means one can arrive at Gibbs energies of formation of chemical compounds. In this connection we shall adopt for the Gibbs function the same convention that we followed with respect to enthalpies; that is, the Gibbs energies of elements in their standard states will be set equal to zero. For ideal gases the standard state will be taken as unit pressure, generally one atmosphere. For nongaseous materials, the standard state is usually chosen as the pure solid or pure liquid. A fuller appreciation of the reason for this choice will be deferred to Chapter Nineteen; meanwhile we see that this is fully consistent with the arguments of the preceding section, since the chemical potentials of solids and liquids can be regarded as substantially constant.

Standard chemical potentials or Gibbs energies of formation of a number of common substances, together with the corresponding enthalpies, are shown in Table 10-1.

Illustration

Let us consider, for example, the reaction for the synthesis of ammonia from its elements:

$$\tfrac{1}{2}N_2(g) + \tfrac{3}{2}H_2(g) \longrightarrow NH_3(g)$$

The standard reaction potential will be

$$\Delta\tilde{\mu}^\circ = \Delta\mu^\circ_{NH_3(g)} - \tfrac{1}{2}\Delta\mu^\circ_{N_2(g)} - \tfrac{3}{2}\Delta\mu^\circ_{H_2(g)} \tag{10.42}$$

Referring to the table and observing our convention with respect to elements, we see that, at 25 °C, $\Delta\tilde{\mu}^\circ = \Delta\mu_{NH_3} = -3.94$ kcal. Hence the equilibrium constant of the reaction will be $K = \exp\left(-\Delta\tilde{\mu}^\circ/RT\right) = 772.8$ per atm. We also see from the table that $\Delta H = -11.02$ kcal. From this we can compute the temperature coefficient of $\ln K$ and hence show by integration and appropriate manipulation that

$$\ln K = \frac{5546}{T} - 11.95 \qquad \text{or} \qquad \lg K = \frac{2408}{T} - 5.190 \tag{10.43}$$

Actually, these equations are not valid for temperatures far removed from 25 °C. To obtain a better expression, one should also observe the variability of ΔH with temperature.

EXERCISE:

Using data from Table 10-1, calculate the equilibrium constant and its temperature coefficient at 25 °C for the reaction

$$CO_2(g) + CS_2(g) \rightleftarrows 2COS(g)$$

The Phase Rule

In Chapter Eight we considered in some detail the equilibrium existing between different phases of a single substance. We shall now turn to the more general question of the factors affecting equilibrium between phases in systems made up of more than one substance. Our immediate object will be to derive a very important and general relationship known as the *Phase Rule* of J. Willard Gibbs.

In discussing the general subject of phase equilibria, we shall find it necessary to speak of the following three quantities: (1) the number of *phases*, P; (2) the number of *components*, C; (3) the number of *degrees of freedom*, or *variance*, F. Of these quantities, the first and the third are relatively easy to define and understand. The number of components, on the other hand, is sometimes difficult to ascertain, so we shall be obliged to discuss in some detail just what is meant by "component."

When we speak of the number of phases, P, in a system, we shall mean the number of distinct kinds of homogeneous regions characterized by definite intensive properties and separated from one another by boundaries determined by discontinuities in one or more of those intensive properties. In this connection it is important to note that we do not count the total number of regions, but rather the number of *kinds* of regions; a system consisting of crushed ice and water is a two-phase system even though the ice may be broken up into many pieces separated by water.

By "number of components," C, we shall mean the minimum number of pure chemical substances that are required for the preparation of arbitrary amounts of all the phases of the system. The stipulation concerning arbitrary amounts is important, for it means, as will be illustrated later, that there must be no stoichiometric constraints on the relative masses of the different phases that can be prepared from the components.

If no chemical equilibria are involved, the number of components will simply equal the number of chemical species that can be identified. However, the number of components may often be less than the total number of distinct kinds of molecules that exist in a system, for many extra varieties may arise from chemical dissociations, associations, or other equilibrium reactions. An aqueous solution of sodium chloride is a two-component system, for it can be prepared from two pure substances, sodium chloride and water. The fact that sodium chloride is ionized in such a solution does not increase the number of pure materials necessary for making the solution.

Consider now an aqueous solution containing Na^+, K^+, Cl^-, and NO_3^- with concentrations subject to the electroneutrality constraint: $[Na^+] + [K^+] = [Cl^-] + [NO_3^-]$. Since the positive and negative ions can be paired in four different ways, it might be supposed that the solution is made up of five components—four salts plus water. Actually this is not so, for it is always possible to select three salts plus water to make up a solution of the type specified. Therefore the solution described above will, in general, be a four-component system. However, if one of the positive ions has a concentration precisely equal to that of one of the negative ions—for example, if $[Na^+] = [Cl^-]$—then the system reduces to one of three components, since only two pure salts plus water will be required.

To illustrate further the calculation of the number of components, consider a gaseous system containing PCl_5, PCl_3, and Cl_2, subject to the following chemical equilibrium:

$$PCl_3 + Cl_2 \rightleftarrows PCl_5$$

In general, this will be a two-component system, for it can be prepared from two pure materials, PCl_3 and Cl_2. However, if the phosphorus trichloride and the chlorine are present in precisely equimolar amounts, the system is a one-component system, for it can be made from the pentachloride alone.

That a system made up of PCl_5, PCl_3, and Cl_2 might reduce to a one-component system under certain circumstances is possible only because phosphorus trichloride and the chlorine are in the same phase. Under some conditions, an equivalence in the numbers of moles of materials derivable from a third substance may be of no consequence so far as reducing the number of components is concerned. To illustrate, consider a three-phase system made up of $CaCO_3(c)$, $CaO(c)$, and $CO_2(g)$ and subject to the equilibrium

$$CaCO_3(c) \rightleftarrows CaO(c) + CO_2(g)$$

This system is always to be regarded as comprised of two components, even if the carbon dioxide and the calcium oxide are present in equimolar amounts, for only the intensive characteristics are of significance in connection with phase equilibria. In other words, at a given temperature and pressure, a small piece of calcium oxide is just as good as a large piece for participating in the specified equilibrium reaction. Hence a stoichiometric equivalence of amounts of materials in different phases is irrelevant, and such equivalence cannot be used as a basis for reducing the apparent number of components. (See the definition of components on p. 219.) This is to be contrasted with the case of PCl_3 and Cl_2, where an equimolar equivalence is significant because it fixes an intensive composition.

A somewhat more complicated system is that made up of solid silver chloride in equilibrium with an aqueous solution containing Na^+, Cl^-, NO_3^-, and a trace of Ag^+. This will generally be a four-component system (three salts plus water) even though we might make it by simply adding sodium chloride to an aqueous solution of silver nitrate. Such a method of preparation, however, would give rise to a definite amount of precipitated silver chloride in relation to the amount of aqueous solution. Since such a stoichiometric constraint is inadmissible as a basis for reducing the number of components, an additional component (say silver chloride) must be introduced, thus bringing the total up to four.

By "number of degrees of freedom," F, of a system at equilibrium, we shall mean the number of intensive variables that we can arbitrarily change, within limits, without causing any phase to disappear or any new phase to appear. To illustrate what is meant by "degree of freedom," let us consider once more the system water, whose phase diagram appears in Figure 8-1. If water is in a condition represented by any point not lying on a line in the diagram, two intensive variables—temperature and pressure—can be independently varied without change of phase. If one has liquid water in the neighborhood of one atmosphere pressure and 25 °C, one can change both the temperature and the pressure by reasonable amounts without causing the water either to evaporate or to freeze. Under those circumstances we say that the system has two degrees of freedom.

Consider now the situation that exists when two phases are in equilibrium, as, for example, liquid water with its vapor. Under those circumstances the system will be represented by a point on line OC in Figure 8-1. Since that line represents a functional relationship between T and p, it appears that the number of independent variables has been reduced by one and that we can arbitrarily change only one of the variables without causing a phase to disappear. We can change the temperature without causing disappearance of a phase, but only if we change the pressure simultaneously by a predetermined amount. When two phases of a single substance are in equilibrium, the system has, therefore, only one degree of freedom, which can be chosen either as temperature or as pressure.

Let us now consider the triple point, O, of Figure 8-1, the point at which solid, liquid, and vapor might co-exist in a state of equilibrium. It is clear from the diagram that the three phases can co-exist at that point only and that changing any variable will necessarily remove the system from that point and thus cause the disappearance of at least one phase. Therefore, when the three phases are in a state of equilibrium, the system is said to have zero degrees of freedom or to be invariant.

We shall now examine the situation that obtains when more than one substance is involved. Suppose that we have a single-phase, two-component system—an aqueous solution, let us say, of sodium chloride. To characterize the intensive variables of such a system, one must specify not only the temperature and pressure but also the composition, which can be expressed by a mole fraction or a weight percentage. The system is accordingly said to possess three degrees of freedom, since we can change all three of those variables, within limits, without producing a new phase. If we have a single-phase solution made up of C components, we can specify the intensive state of the solution by giving values for the temperature, the pressure, and $C - 1$ mole fractions or weight percentages. Under those circumstances there is a total of $C + 1$ degrees of freedom.

Having discussed what we mean when we speak of phases, components, and degrees of freedom, we are now in a position to derive Gibbs's Phase Rule, which is a statement relating F to C and P. For our derivation we shall provisionally assume that all of the C components are present to some extent in each of the P phases. To facilitate the discussion, let us number the phases 1, 2, 3, ... P, and the components 1, 2, 3, ..., C; T_j and p_j shall denote the temperature and the pressure of the jth phase, and μ_{ij} shall equal the chemical potential of the ith component in the jth phase. Now, if all the phases were completely separated from one another, and if there were no equilibria between phases, each phase would have $C + 1$ degrees of freedom, representing temperature, pressure, and $C - 1$ mole fractions. Hence a system of P separate phases, each consisting of C components, would have $P(C + 1)$ potential degrees of freedom. If the phases are in equilibrium, however, we can write certain equations that will impose constraints on some of the intensive variables, thereby reducing the number of degrees of freedom. Thus, if P phases are in equilibrium, it is clear that the P temperatures and the P pressures must all be equal, else heat transfer or mass transfer would occur spontaneously. To assert that P quantities are equal requires $P - 1$ equations; the statement of uniform temperature and uniform pressure therefore implies $2(P - 1)$ equations of constraint.

Let us now inquire into the conditions existing with respect to chemical potentials. If the several phases are all in equilibrium, the chemical potential of a given substance must be the same in each of the phases. If this were not true, the substance would tend to move from a phase in which its chemical potential is high to one in which it is lower. Such movement of material from a region of high chemical potential to one of lower potential would occur spontaneously and would imply a state of nonequilibrium in accordance with the Second Law of Thermodynamics applied to systems at constant temperature and pressure. The transfer of material from one phase to another would continue until each substance had a uniform chemical potential throughout the system; only then would there be complete equilibrium. Now to specify that the chemical potential of component i is the same in each of the P phases will require $P - 1$ additional equations. For C components this will require $C(P - 1)$ equations, which, when added to those obtained for temperature and pressure, will give a total of $(C + 2)(P - 1)$ constraints. These can be represented by the equality signs in the following array:

$$\begin{array}{lcccccc}
\text{Phase:} & 1 & 2 & 3 & \cdot \ \cdot \ \cdot & P \\
\text{Temperature:} & T_1 = & T_2 = & T_3 & \cdot \ \cdot \ \cdot & = T_P \\
\text{Pressure:} & p_1 = & p_2 = & p_3 & \cdot \ \cdot \ \cdot & = p_P \\
\text{Chemical potentials:} & \mu_{11} = & \mu_{12} = & \mu_{13} & \cdot \ \cdot \ \cdot & = \mu_{1P} \\
& \mu_{21} = & \mu_{22} = & \mu_{23} & \cdot \ \cdot \ \cdot & = \mu_{2P} \\
& \cdot & \cdot & \cdot & & \cdot \\
& \cdot & \cdot & \cdot & & \cdot \\
& \cdot & \cdot & \cdot & & \cdot \\
& \mu_{C1} = & \mu_{C2} = & \mu_{C3} & \cdot \ \cdot \ \cdot & = \mu_{CP} \quad (10.44)
\end{array}$$

But we have already seen that the potential number of degrees of freedom in a system of P separate phases is $P(C + 1)$. The net number of degrees of freedom in such a system at equilibrium will therefore be $P(C + 1)$ less the number of constraints, $(C + 2)(P - 1)$, as follows:

$$\begin{aligned}
F &= P(C + 1) - (C + 2)(P - 1) \\
&= C - P + 2 \qquad\qquad\qquad\qquad (10.45)
\end{aligned}$$

This equation is a statement of the famous rule of J. Willard Gibbs.

The preceding derivation of the Phase Rule was carried out on the assumption that every component is present in each of the phases. Suppose, however, that one of the components is completely absent from a particular phase and incapable of existing in that phase. What effect would this have on the Phase Rule? First we recognize that, if a given component cannot be present in a particular phase, the potential number of degrees of freedom for that phase will be reduced by one, since there will be one less composition variable. At the same time the chemical potential of that component will have no value for the particular phase of interest and therefore will not appear in the array represented by equations 10.44. Removal of that particular chemical potential from the array will then necessarily remove one equality sign, which would otherwise appear as an equilibrium constraint. It is evident, therefore, that the absence of any components from a phase will simultaneously diminish by equal amounts the potential number of degrees of freedom and the number of constraints. As a result, the net degrees of freedom, F, will remain unchanged at $C - P + 2$ in accordance with equation 10.45.

Applications of the Phase Rule

We shall now illustrate the use of the Phase Rule by applying it to a few simple systems. Actually, the rule is important, not because it is of great practical use, but because it is an elegant formulation of a proposition of fundamental thermodynamic significance. Although the applications that we shall now consider will help us understand the full implications of the rule, they will really tell us nothing new in a descriptive sense.

Consider first any one-component system like pure water. In accordance with the Phase Rule, the number of degrees of freedom, F, must equal $3 - P$. Since there must be at least one phase present, the maximum number of degrees of freedom in a one-component system will be two; these degrees of freedom can be chosen as temperature and pressure. If, however, two phases are in equilibrium, the number of degrees of freedom is reduced to one, which can be chosen either as temperature or as pressure, the other variable becoming dependent. Finally, if three phases are present, as at a triple point, a one-component system will become invariant.

For a two-component system the number of degrees of freedom will be $4 - P$. If such a system exists in only one phase, the three degrees of freedom can be chosen as temperature, pressure, and mole fraction of one of the components. If a two-component system exists in two phases, the number of degrees of freedom is reduced to two. These can be chosen in any of several ways, but practically they might be taken as temperature and the composition of one of the phases. Having specified those two variables, the pressure is determined and the composition of the other phase likewise becomes fixed. An example of such a system would be ammonia and water existing in the liquid and vapor phases. At a certain temperature and composition of the aqueous phase, a vapor phase will exist in equilibrium at a definite total pressure and a definite composition. Should our system consist of sodium chloride and water in equilibrium with vapor, there will still be two degrees of freedom even though we can neglect the partial vapor pressure of the sodium chloride.

A system consisting of PCl_5, PCl_3, and Cl_2 at equilibrium in a gaseous state will, in general, have three degrees of freedom since it is a one-phase system made up of two components. These degrees of freedom can be taken as temperature, pressure, and one mole fraction (or one partial pressure). Because of the chemical equilibrium, the other mole fractions (or partial pressures) will be determined by the equilibrium.

Suppose that we now consider a two-phase system consisting of solid silver chloride in equilibrium with an aqueous solution containing Na^+, Cl^-, NO_3^-, and Ag^+. We have seen that this will generally be a four-component system and, in accordance with the Phase Rule, will exhibit four degrees of freedom. These four degrees of freedom can be chosen in any of several different ways, but most conveniently they might be taken as temperature, pressure, concentration of sodium ions, and concentration of chloride ions. Having fixed those quantities, we have determined everything else; specifically, the other two ion concentrations (those for Ag^+ and NO_3^-) will be fixed because the silver and chloride ions are in equilibrium with solid silver chloride and the over-all solution must be electrically neutral.

EXERCISE:

One liter of 0.1 N $BaCl_2$ is mixed with two liters of 0.2 N K_2SO_4. Discuss the resulting system in relation to the Phase Rule, assuming only two phases (aqueous and solid). How many degrees of freedom are there, and how can they be chosen?

We have seen that, if chemical equilibria are involved, the number of components may be less than the number of identifiable chemical species. Sometimes, however, certain chemical equilibria that are conceivable are not realized in practice. Under such circumstances one can apply the Phase Rule, more or less precisely, by disregarding the chemical reaction and assuming the presence of a greater number of components. To illustrate, let us consider a gaseous system made up of hydrogen, oxygen, and water vapor. At elevated temperatures the following equilibrium might exist:

$$H_2(g) + \tfrac{1}{2}O_2(g) \rightleftarrows H_2O(g)$$

At room temperature, however, the attainment of the indicated equilibrium would be so slow that for all practical purposes the system could be regarded as one of three components. We see, therefore, that proper application of the Phase Rule requires an understanding of the equilibria that are actually realized; this is necessary if one is to obtain a correct value for the number of components in the system. In this connection we observe that the presence or absence of catalysts can affect the number of components by affecting the possibility of attainment of chemical equilibrium.

Problems

1. Consider an ideal gaseous reaction mixture capable of undergoing the reaction

$$aA + bB + \ldots \rightleftharpoons eE + fF + \ldots$$

Originally the mixture contains n_A moles of A at partial pressure p_A, $\ldots$ n_E moles of E at partial pressure p_E, $\ldots$ etc. Suppose that the reaction now takes place to an extent, measured by ξ, such that the final partial pressures become $p'_A, \ldots p'_E, \ldots$ etc. Show that the total Gibbs energy change attending this extent of reaction is given by the expression

$$\Delta G = n_A RT \ln \frac{p'_A}{p_A} + \ldots + n_E RT \ln \frac{p'_E}{p_E} + \ldots + \xi \Delta \tilde{\mu}'$$

in which $\Delta \tilde{\mu}'$ is the reaction potential corresponding to the final state. [You can work this problem most readily by setting up expressions for the initial and final chemical potentials ($\mu_A = \mu_A^o + RT \ln p_A$ and $\mu'_A = \mu_A^o + RT \ln p'_A$), etc., and then summing over all chemical species to obtain total Gibbs energies of initial and final states.] From the result above show that, if the final state is one of equilibrium, the total Gibbs energy change is simply that required to take all substances initially present to their equilibrium pressures. Show also that

$$\Delta G = n'_A RT \ln \frac{p'_A}{p_A} + \ldots n'_E RT \ln \frac{p'_E}{p_E} + \ldots + \xi \Delta \tilde{\mu}$$

in which expression $\Delta \tilde{\mu}$ refers to the reaction potential at the *start* of the reaction.

2. Suppose that we start with a moles of A, b moles of B, etc., without any E, F, etc. (See the reaction in Prob. 1.) In view of the result of Problem 1, show that, if the reaction is carried to completion (that is, if $\xi = 1$),

$$\Delta G = -RT \ln K + RT \ln \frac{p'^e_E p'^f_F \ldots}{p^a_A p^b_B \ldots}$$

(Note that, although this result has the same form as that exhibited by $\Delta\tilde{\mu}$, it has a quite different significance. This result would apply to a reaction carried out with the aid of a "reaction box," a vessel containing an equilibrium mixture and provided with semi-permeable membranes through which reactants can be introduced and products removed.)

3. Consider the following chemical reaction carried out at temperature T and constant total pressure p:

$$N_2O_4 \longrightarrow 2NO_2$$

Suppose that we start with $n_{N_2O_4}$ and n_{NO_2} moles of the two gases and the reaction then proceeds to an extent, ξ, such that the final numbers of moles are $n'_{N_2O_4} = n_{N_2O_4} - \xi$ and $n_{NO_2} = n'_{NO_2} + 2\xi$. Show that the reaction potential, $\Delta\tilde{\mu}$, will be given as the following function of ξ:

$$\Delta\tilde{\mu} = -RT \ln K + RT \ln \frac{(n_{NO_2} + 2\xi)^2 p}{(n_{N_2O_4} - \xi)(n_{NO_2} + n_{N_2O_4} + \xi)}$$

By direct integration of $\Delta\tilde{\mu}$ show, in agreement with the result of Problem 1, that

$$\Delta G = \int_0^\xi \Delta\tilde{\mu}\, d\xi = n_{NO_2} RT \ln \frac{p'_{NO_2}}{p_{NO_2}} + n_{N_2O_4} RT \ln \frac{p'_{N_2O_4}}{p_{N_2O_4}} + \xi \Delta\tilde{\mu}'$$

in which expression the primes denote final states. (Observe how involved the method of this problem is compared with that of Problem 1. The general result of Problem 1 can also be obtained by direct integration, but the derivation is quite tedious.)

4. Consider a mixture of two ideal gaseous substances, A and B, subject to the equilibrium

$$A \rightleftharpoons B$$

with an equilibrium constant K. A tube containing a total of n moles of gases A and B is slowly heated at constant pressure in the presence of a catalyst so that the system is always in equilibrium. Show that the apparent heat capacity of the system is

$$C_p \text{ (apparent)} = \frac{nC_{pA}}{K+1} + \frac{nKC_{pB}}{K+1} + \frac{\Delta\tilde{H}^2 nK}{RT^2(K+1)^2}$$

Notice that the third term is always positive, regardless of the sign of $\Delta\tilde{H}$. (Compare with Prob. 17, Chap. Eight; this problem will also help you understand the unusual heat capacity of an equilibrium mixture of ortho and para hydrogen discussed in Chap. Thirteen.)

5. Prove in general that the apparent heat capacity of a mixture of gases heated under equilibrium conditions can never be less than the sum of the heat capacities of the substances present at any instant.

6. The standard Gibbs energies of formation of $NO_2(g)$ and $N_2O_4(g)$ at 25 °C are 12.26 and 23.38 kcal mol^{-1}, and their corresponding heats of formation are 7.93 and 2.19 kcal mol^{-1}.

   (A) Assuming ΔC_p to be equal to zero, set up an expression for ln K as a function of temperature valid in the neighborhood of 25 °C for the reaction

$$N_2O_4 \rightleftharpoons 2NO_2$$

   (B) The molar heat capacities of $NO_2(g)$ and $N_2O_4(g)$ at 25 °C are 8.89 and 18.47 cal deg^{-1}. Assuming the heat capacities to be constant, derive an improved expression for ln K as a function of T.

   (C) How much difference is there between the values of ln K calculated from the expressions of parts A and B at 100 °C? at 200 °C?

7. Consider the dissociation of PCl_5 into PCl_3 and Cl_2.

   (A) Show that, if one starts with pure PCl_5 and a fraction, α, dissociates at equilibrium, the total pressure being equal to p, the equilibrium constant will be

$$K_p = \frac{\alpha^2 p}{1 - \alpha^2}$$

   (B) At 250 °C the equilibrium constant equals 1.78 atm. Calculate α for $p = 0.01$, 0.10, and 1.00 atm.

   (C) Calculate $\Delta\tilde{\mu}$ for an equimolar mixture at a pressure of 1 atm at 250 °C.

   (D) Calculate $\Delta\tilde{\mu}$ for a 10-percent dissociation of PCl_5 at 250 °C and 1 atm.

8. At 25 °C, the molar heat capacities of $NH_3(g)$, $N_2(g)$, and $H_2(g)$ are 8.38, 4.968, and 6.889 cal deg^{-1}, respectively. Regarding the synthesis of ammonia treated on page 218, derive improved expressions for ln K and lg K, assuming the heat capacities to be constant.

9. Consider a vessel containing some solid ammonium hydrogen sulfide in equilibrium with gaseous hydrogen sulfide and ammonia:

$$NH_4HS(c) \rightleftharpoons H_2S(g) + NH_3(g)$$

If the ammonia and hydrogen sulfide are present in equimolar amounts, what are C, P, and F for the system? If the two gases are not present in equimolar amounts, what are C, P, and F?

10. In deriving the Phase Rule, we suppose that the pressures of all the phases must be equal at equilibrium. How can this be reconciled with equilibrium in a gravitational field, under which circumstances the pressure is a function of height? Consider, for example, a tank of water at 0 °C under a pressure of 1 atm at the surface. Would it be possible to hold ice at 0 °C in equilibrium with water at the bottom of the tank, where the pressure is greater than 1 atm? What is the Gibbs energy change in calories attending the transfer of 1 mole of ice at 0 °C from the bottom to the surface of a water tank 10 meters deep? At 0 °C the density of water is 1.000 g cm^{-3}, and that of ice is 0.917 g cm^{-3}. The acceleration of gravity can be taken as 980 cm sec^{-2}.

11. (A) Suppose that we carry out studies of equilibrium systems at one fixed total pressure, say 1 atm. Show that, if pressure is thus eliminated as a variable, the Phase Rule can be written as $F = C - P + 1$. (This formulation of the Phase Rule is often adopted under appropriate circumstances. When this is done, however, the reader is warned against inadvertently reintroducing pressure as a variable in implicit form, thereby rendering invalid the justification for the modified equation.)

(B) Suppose that, in addition to pressure and temperature, there were a third intensive variable, other than a chemical potential, that must be the same for all phases in equilibrium. Show that the Phase Rule would become $F = C - P + 3$.

12. Consider the Deacon process (Prob. 17, Chap. Three) for making chlorine:

$$2HCl(g) + \tfrac{1}{2}O_2(g) \rightleftarrows H_2O(g) + Cl_2(g)$$

Using Gibbs energy data from Table 10-1 and the results of Problem 17, Chapter Three, derive an expression for lg K as a function of T for the Deacon reaction.

13. A cylinder provided with a piston contains the precise equivalent of $1N_2O_4(g)$, which is, to be sure, partially dissociated into $NO_2(g)$. Suppose that this gaseous mixture is expanded reversibly and isothermally from total pressure p_1 to total pressure p_2.

(A) Assuming equilibrium at all times, evaluate the Gibbs energy change by direct integration:

$$\Delta G = \int_{p_1}^{p_2} V dp$$

Suggested procedure: If a fraction, α, of N_2O_4 is dissociated, the equilibrium constant for the dissociation reaction

$$N_2O_4(g) \rightleftarrows 2NO_2(g)$$

will be

$$\frac{4\alpha^2 p}{1 - \alpha^2} = K$$

However, the total volume of the gas is given by the expression

$$V = \frac{(1 + \alpha)RT}{p}$$

By eliminating α, we express V as a function of p and T and then carry out the integration to obtain ΔG.

(B) Alternatively, we can evaluate ΔG by considering initial and final states only:

$$G' = n'_{NO_2}\mu'_{NO_2} + n'_{N_2O_4}\mu'_{N_2O_4}$$

$$G = n_{NO_2}\mu_{NO_2} + n_{N_2O_4}\mu_{N_2O_4}$$

Solving for the initial and final conditions, and rearranging, show that the answers to A and B are the same.

$$\text{Ans.} \quad \Delta G = 2RT \ln \left\{ \frac{\sqrt{K + 4p_2} - \sqrt{K}}{\sqrt{K + 4p_1} - \sqrt{K}} \right\}$$

11. Show that the over-all reaction

$$CO + H_2O + O_2 \longrightarrow CO_2 + H_2O_2$$

is thermodynamically favorable. Actually, there is little or no evidence of formation of peroxide when carbon monoxide is burned in the presence of water. Can the reader think of conditions under which this combination reaction would occur?

Statistical Mechanics; Maxwell–Boltzmann Statistics

Thermodynamic Probability

In Chapter Five we considered briefly the connection between entropy and probability. We shall now inquire into this relationship more systematically with the goal of establishing a basis for calculating thermodynamic quantities from molecular-structure data. The basic postulate that we shall use for this purpose is the relationship proposed by Planck—namely, that the entropy of a system is given by the equation

$$S = k \ln \Omega \qquad (11.1)$$

in which k is Boltzmann's constant and Ω is the thermodynamic probability. The plausibility of this equation was demonstrated in Chapter Five, but we hasten to add that it is still a postulate; it is, moreover, a most important one, for it constitutes the bridge that connects phenomenological thermodynamics with probability and molecular structure.

At this time we shall discuss further the nature of Ω, the thermodynamic probability, which was defined as the number of different ways in which a thermodynamic state can be realized. Consider a system of N objects or particles, which may be atoms, molecules, coins, dice, etc., each of which can exhibit any of a number of different aspects. A coin, for example, can turn up heads or tails; so we say it has two possible aspects. A die can turn up any of six faces; so it has six aspects. In the case of atoms or molecules, we shall be concerned with the various possible quantum states and energy levels; for some atomic or molecular systems the number of conceivable aspects or quantum states may be infinite.

Let us suppose that we have specified the number of objects or particles in our system that exhibit each conceivable kind of aspect. When this is done, we have also specified what is known as a *statistical state* of the system. If we have ten coins, for example, and say that six show heads and four show tails, we have specified a statistical state with respect to the coins. In defining the statistical state, we say nothing about which coins show heads and which ones show tails; we merely tell how many fall into each category. Similarly, if we have a number of molecules and assert that N_1 of them are in the first quantum level, N_2 in the second, and so on until we have accounted for all the molecules, we shall have specified a statistical state with respect to the system of molecules. Here again we say nothing about which molecules are in each quantum level; we say only how many are in each.

Having defined what we mean by "statistical state," we can now amplify our definition of the thermodynamic probability, Ω. The thermodynamic probability is defined as the number of different ways in which a statistical state can be attained. For illustrative purposes let us consider the tossing of distinguishable coins, say three in number. Let H_i equal the probability that the ith coin will turn up heads and T_i $(= 1 - H_i)$ the probability that it will turn up tails; if the coins are all balanced, these probabilities will equal $1/2$. (The following argument will be valid, however, even if the coins are loaded.) The various states that are possible as a result of tossing the three coins can be represented by the products appearing in the following shceme:

$$
\begin{aligned}
1 &= (H_1 + T_1)(H_2 + T_2)(H_3 + T_3) \\
&= H_1H_2H_3 + H_1H_2T_3 + H_1T_2H_3 + T_1H_2H_3 \\
&\quad + H_1T_2T_3 + T_1H_2T_3 + T_1T_2H_3 + T_1T_2T_3 \qquad (11.2)
\end{aligned}
$$

Now the probability that all the coins will turn up heads is $H_1H_2H_3$, the probability that the first two will be heads and the last tails is $H_1H_2T_3$, etc. Obviously, the sum of the probabilities of all combinations must be unity, as indicated by equation 11.2. Now let us stipulate that the coins are identical, so that $H_1 = H_2 = H_3$ and, of course, $T_1 = T_2 = T_3$. Let us further suppose that we do not care which particular coins turn up heads or tails but only how many turn up heads and how many turn up tails. Under these circumstances we can drop the subscripts that identify the coins and rewrite equation 11.2 as

$$1 = (H + T)^3 = H^3 + 3H^2T + 3HT^2 + T^3 \qquad (11.3)$$

Equation 11.3 tells us that the probability of getting three heads is H^3, the probability of two heads and one tail is $3H^2T$, etc. The factor 3 appears in front of H^2T (and HT^2) because there are three different ways of obtaining one tail and two heads; as indicated above, we do not care which of the three ways is actually realized. Here we have an illustration of what is meant by "thermodynamic probability"; the coefficients 1, 3, 3, and 1 appearing in equation 11.3 are precisely the thermodynamic probabilities of the states represented by all heads, two heads and a tail, two tails and a head, and all tails. If the coins are not loaded (that is, if $H = T = 1/2$), the absolute probabilities of the states will be 1/8, 3/8, 3/8, and 1/8 respectively.

The foregoing argument can now be generalized for a system of N equivalent coins in which H and T represent the probabilities that any particular coin will turn up heads or tails. The over-all probabilities for the various statistical states can easily be generated in the following way:

$$1 = (H + T)^N = \sum_{i=0}^{N} \frac{N!}{(N-i)!i!} H^{N-i}T^i \qquad (11.4)$$

A particular term in this summation represents the actual probability of realizing $N - i$ heads and i tails; the thermodynamic probability of such a state is $N!/(N-i)!i!$. Equation 11.4 can be rewritten in a more symmetrical form as follows:

$$1 = (H + T)^N = \sum \frac{N!}{N_1!N_2!} H^{N_1}T^{N_2} \qquad (11.5)$$

In this equation, N_1 and N_2 represent the numbers of heads and tails, with $N_1 + N_2 = N$.

Let us now apply the preceding methods to a system made of N objects each of which can exhibit six aspects (for example, dice). If $p_1, p_2, p_3, \ldots$ are the probabilities that the several aspects will show up, then

$$1 = (p_1 + p_2 + p_3 + p_4 + p_5 + p_6)^N$$

$$= \sum \sum \sum \sum \sum \frac{N!}{N_1!N_2!N_3!N_4!N_5!N_6!} p_1^{N_1}p_2^{N_2}p_3^{N_3}p_4^{N_4}p_5^{N_5}p_6^{N_6} \qquad (11.6)$$

in which expression N_i is the number of objects manifesting the ith aspect, subject, of course, to the constraint that $N_1 + N_2 + N_3 + N_4 + N_5 + N_6 = N$. The thermodynamic probability associated with a state characterized by $N_1, N_2, \ldots$ will be

$$\Omega = \frac{N!}{N_1! N_2! N_3! N_4! N_5! N_6!} \tag{11.7}$$

The extension of the foregoing to systems of N distinguishable objects distributed among s aspects is now obvious. If N_i denotes the number of objects in the ith category, the thermodynamic probability associated with a statistical state characterized by $N_1, N_2, N_3, \ldots N_s$ will be

$$\Omega = \frac{N!}{N_1! N_2! \ldots N_s!}$$

$$= \frac{N!}{\prod_i N_i!} \tag{11.8}$$

This equation is a very important one, and we shall use it frequently in our statistical mechanics. It should be borne in mind, however, that its validity rests upon the assumption that the objects (which might be atoms, molecules, electrons, etc.) are distinguishable from one another. For reasons that will be introduced later, this is not always so, and the use of equation 11.8 will therefore introduce errors in certain kinds of entropy calculations. The kind of statistics with which we shall be immediately concerned, and which rests upon the use of equation 11.8, is known as *Boltzmann statistics.*

Thermodynamic Probabilities of Systems in Equilibrium

Having established what we mean by "thermodynamic probability," we might think that substitution of equation 11.8 into equation 11.1 would give us the entropy of the system under consideration. Some ambiguity remains, however, for nothing has yet been said about what statistical state should be chosen to describe the actual system. We find the answer to this problem by recognizing that the entropy, and hence the thermodynamic probability, of an isolated system strives for a maximum; when it reaches that maximum, the system is said to be in a state of equilibrium. Thus, although equation 11.1 should give us the entropy corresponding to any conceivable state for which we know the thermodynamic probability, the entropy of an isolated system that has reached equilibrium will be that corresponding to the state of maximum thermodynamic probability. Our next task is therefore clearly defined:

we must turn to the problem of establishing the nature of the states
of maximum probability. When such a state has been determined, the
entropy of the system can be calculated by use of the equation

$$S = k \ln \; \Omega_{\text{max}} \tag{11.9}$$

in which Ω_{max} is the thermodynamic probability of that most probable
state.

For many statistical mechanical problems, it is possible to determine
the states of maximum probability by direct approaches requiring little
mathematics beyond elementary algebra. There are, however, some more
sophisticated methods, such as those involving the use of the calculus of
variations. Since variational techniques are generally used and since a
knowledge of such methods is desirable if one seeks to solve more complex problems, we shall initially carry out our derivations in that
manner. Subsequently (in Chap. Fifteen), we shall illustrate a different
and, in many respects, simpler method for deriving the important
statistical mechanical equations. Although mathematically simpler, the
methods to be employed in Chapter Fifteen require a greater recognition of the established principles of thermodynamics than do the abstract derivations that depend on the calculus of variations.

To determine when a function has a maximum (or a minimum)
value, we naturally think of the methods of the calculus, which are
readily applicable to continuous functions. In the case of thermodynamic probability, we observe from equation 11.8 that the function
is discontinuous because the factorial quantities are defined only for
whole numbers. When we deal with atomic and molecular systems, however, we invariably deal with such large samples that the numbers can
be considered continuous, from a practical point of view, since the relative increments represented by unit changes will be exceedingly small.
The problem of finding a maximum for the thermodynamic probability
would evidently be considerably simplified if only the factorial function
could be expressed as a continuous function that would be amenable
to differentiation. Fortunately, there is a simple, though approximate,
expression that will suffice for transforming the factorial into a continuous function.

Stirling's Approximation for *N*!

We shall now derive an approximate expression for $N!$ that can be
differentiated and manipulated like any ordinary function that might

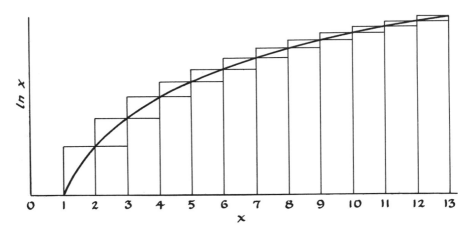

FIGURE 11-1   Plot of ln x against x with rectangular approximations to area.

be encountered. Consider Figure 11-1, in which ln x is plotted against x; in addition, steps are indicated, both above and below the smooth curve, with values equal to ln 2, ln 3, etc. The area under the smooth curve from $x = 1$ to $x = N$ will be given by the expression

$$\int_1^N \ln x \, dx = N \ln N - N + 1 \tag{11.10}$$

We can also approximate the same area by adding up the areas of the rectangles defined by the above-mentioned steps. If we choose the rectangles whose tops lie above the curve, we find for the area

$$\ln 2 + \ln 3 + \ldots \ln N = \ln N! \tag{11.11}$$

On the other hand, if we use the rectangles below the curve, we obtain

$$\ln 1 + \ln 2 + \ldots \ln (N - 1) = \ln (N - 1)!$$
$$= \ln N! - \ln N \tag{11.12}$$

The correct value of the area evidently lies between those given by equations 11.11 and 11.12; so we can assert, upon appropriate algebraic manipulation, that

$$N \ln N - N + 1 < \ln N! < (N + 1) \ln N - N + 1 \tag{11.13}$$

Now, if N is very large, we can neglect unity as compared with N without appreciably affecting the calculated values for the upper and lower limits of ln N!. With this approximation, the upper and lower limits become identical, from which we see that

$$\ln N! = N \ln N - N \tag{11.14}$$

This equation is the approximate form for $\ln N!$ that we shall use extensively when dealing with factorials. Because $\ln N$ changes very slowly with increasing N, equation 11.14 provides a good approximation for $\ln N!$ if N is large. However, a value of $N!$ calculated by using the exponential form of equation 11.14 will be in error by a considerable factor.

A better approximation for $\ln N!$, which we shall not derive here, is

$$\ln N! = \ln \sqrt{2\pi N} + N \ln N - N \qquad (11.15)$$

In nonlogarithmic form this becomes

$$N! = \sqrt{2\pi N} \left(\frac{N}{e}\right)^N \qquad (11.16)$$

which is fairly accurate for large N. It will be seen that the factor $\sqrt{2\pi N}$ can be considerable, although its logarithm will be nil compared with $N \ln N - N$. Equations 11.14 and 11.15 are known as Stirling's approximations to $\ln N!$. We shall use the simpler of these approximations, equation 11.14, to aid us in establishing states of maximum probability.

EXERCISE:

Assuming $\ln N!$ to be halfway between the upper and lower limits indicated by equation 11.13, show that

$$\ln N! = (N + 1/2) \ln N - N + 1$$

and hence that

$$N! = e\sqrt{N} \left(\frac{N}{e}\right)^N$$

Compare this result with equation 11.16.

EXERCISE:

Calculate values of 10!, 50!, and 100! and their logarithms to the base 10, using equations 11.14 and 11.15. Compare with the correct answers, which are 3.6288×10^6, 3.0414×10^{64}, and 9.3326×10^{157}, the common logarithms of which are 6.55976, 64.48307, and 157.97000.

States of Maximum Thermodynamic Probability

Consider N distinguishable objects each capable of exhibiting any of s different aspects. What is the state of maximum thermodynamic probability? We can obtain the answer by applying the calculus of variations to equation 11.8, which gives the thermodynamic probability of the system. Now a continuous function will exhibit a maximum under the same conditions as its logarithm; hence, if we maximize $\ln \Omega$, we simultaneously maximize Ω. Taking the logarithm of Ω, and making use of Stirling's approximation in the form of equation 11.14, we obtain the expression

$$\ln \Omega = N \ln N - N - \sum_{i=1}^{s} (N_i \ln N_i - N_i) \qquad (11.17)$$

Our variation problem now consists of adjusting the N_i's to make $\ln \Omega$ a maximum while keeping constant the total number of particles, N. In other words, we shall seek to find the functional dependencies of the dependent variables, $N_1, N_2, \ldots$, upon the independent variable, N, that will render $\ln \Omega$ a maximum. Since

$$N = N_1 + N_2 + \ldots + N_s$$

$$= \sum_{i=1}^{s} N_i \qquad (11.18)$$

and since the variation of an independent variable is equal to zero, it follows that

$$\delta N = \sum_{i=1}^{s} \delta N_i = 0 \qquad (11.19)$$

Equation 11.19 thus constitutes a constraint on the variations; in essence this means that there are only $s - 1$ functions to be found—let us say $N_1, N_2, \ldots N_{s-1}$—the last function, N_s, to be fixed by the constraint. If we now take the variation of $\ln \Omega$ from equation 11.17, bearing in mind that $\delta N = 0$, we find that

$$\delta \ln \Omega = - \sum_{i=1}^{s} \ln N_i \delta N_i$$

$$= - (\ln N_1 \delta N_1 + \ln N_2 \delta N_2 + \ldots \ln N_s \delta N_s) \qquad (11.20)$$

Making use of equation 11.19, we can eliminate one of the variations, say δN_s, which is equal to $-(\delta N_1 + \delta N_2 + \ldots \delta N_{s-1})$. Upon so substituting for δN_s, we find that

$$\delta \ln \Omega = \ln (N_s/N_1)\delta N_1 + \ln (N_s/N_2)\delta N_2$$

$$+ \ldots \ln (N_s/N_{s-1})\delta N_{s-1} = \sum_{i=1}^{s-1} \ln (N_s/N_i)\delta N_i \qquad (11.21)$$

Now, to make $\ln \Omega$ a maximum, we require that its variation be equal to zero. But we see from equation 11.21 that, if we are to guarantee that $\delta \ln \Omega = 0$, the coefficients of all of the unconstrained variations must be equal to zero. In other words,

$$\ln (N_s/N_i) = 0 \qquad (11.22)$$

for all values of i. This means that $N_i = N_s$ or, in other words, that

$$N_1 = N_2 \ldots = N_{s-1} = N_s \qquad (11.23)$$

Hence we see that the state of maximum thermodynamic probability is that in which an equal number of objects exhibits each of the different aspects. If we combine equation 11.23 with equation 11.18, it is obvious that

$$N_i = \frac{N}{s} \qquad (11.24)$$

This equation is the desired final result; it tells how each of the N_i's must depend on the independent variable N to make $\ln \Omega$ a maximum.

The simple result expressed by equation 11.24 can now be applied to a number of familiar examples. If a large number of coins (for which $s = 2$) are tossed, the state of maximum thermodynamic probability will be characterized by equal numbers of heads and tails. If the coins are not loaded, that same distribution will also define the state of maximum ordinary probability. Similarly, if a large number of dice are rolled, a maximum thermodynamic probability will be realized when equal numbers of dice present each of the six different aspects. Finally, if a large number of molecules are placed in a container that is subdivided by imaginary boundaries into a number of little regions of equal volume, the state of maximum thermodynamic probability will be that for which equal numbers of molecules are in each little volume—in other words, a state of uniform concentration.

Lagrangian Multipliers

We developed the preceding section in considerable detail to illustrate precisely what is involved when a variation problem is carried out. We shall now carry through the solution of the same problem by a different method, known as the method of Lagrangian multipliers. This method, the validity of which is not obvious, greatly simplifies problems involving more than one condition of constraint. We shall not attempt to justify this alternative procedure formally; rather, we trust that the reader will accept it through confidence gained by study of examples and exercises.

The procedure employing Lagrangian multipliers is carried out as follows. First, let us rewrite equation 11.20, at the same time setting $\delta \ln \Omega$ equal to zero in anticipation of maximizing the function:

$$\delta \ln \Omega = - \sum_{i=1}^{s} \ln N_i \delta N_i = 0 \qquad (11.25)$$

Next, let us multiply equation 11.19 by a parameter, α, which is known as a Lagrangian multiplier, to get the expression

$$\alpha \delta N = \sum_{i=1}^{s} \alpha \delta N_i = 0 \tag{11.26}$$

We shall now subtract equation 11.26 term by term from equation 11.25 to obtain

$$\delta \ln \Omega - \alpha \delta N = - \sum_{i=1}^{s} \{\ln N_i + \alpha\} \delta N_i = 0 \tag{11.27}$$

The essence of the method now is to regard the variations as mutually independent and not explicitly subject to the condition of constraint. The added degree of freedom thus assumed is compensated for by introduction of the multiplier, α, which is not yet determined, but which will subsequently be identified by use of the very constraint that we provisionally seek to avoid in explicit form. To maximize $\ln \Omega$, we set the coefficient of each of the variations in equation 11.27 equal to zero. Hence we write

$$\ln N_i + \alpha = 0 \tag{11.28}$$

or

$$N_i = e^{-\alpha} \text{ (a constant)} \tag{11.29}$$

But, since $\sum_{i=1}^{s} N_i = N$, it follows that

$$se^{-\alpha} = N \tag{11.30}$$

from which we conclude again, in agreement with equation 11.24, that $N_i = N/s$.

When there is only one condition of constraint, as in the example above, the use of multipliers is scarcely warranted. Indeed, if we were concerned only with the problem described above, we would not even bother to describe the method of Lagrangian multipliers. If there are two or more constraints, however, the multiplier method, as we shall see later, introduces simplification.

Thermodynamic Probabilities of Systems Involving Energy Levels

In the preceding sections it was shown that, if a large number of objects are distributed among a number of boxes, the state of maximum thermodynamic probability is attained when the same number of objects is placed in each of those boxes. This result comes about because each box is equally accessible, so far as thermodynamic probability is concerned, and hence a uniform distribution can be expected. Let us now imagine a somewhat different problem, related to the one above, but in addition involving energy considerations.

Suppose that we have a number of boxes placed at different levels in a gravitational field and that we now distribute among these boxes a number of marbles of uniform mass; the potential energy of any marble will obviously depend upon the height of the box in which it is placed. Suppose further that the total potential energy of the marbles is predetermined and that the distribution of these marbles among the various boxes is subject to such an over-all energy constraint. This mechanical analogue illustrates a very important statistical-mechanical problem—the distribution of atoms or molecules among various quantized energy levels subject to a fixed total energy.

Let us now think of a number of atoms or molecules distributed among energy levels with values denoted by E_1, E_2, E_3, etc. Sometimes it happens that certain of these levels are *degenerate*—that is to say, they can be realized in more than one way or through more than one combination of quantum numbers. Degeneracy of levels is equivalent to having several boxes of the mechanical analogue at the same height in the gravitational field. The number of such boxes at any one height, or the number of combinations of quantum numbers corresponding to any one level, is said to be the *degree of degeneracy* of the energy level. When a situation of this kind occurs, it is necessary that the energy be listed as many times as the level is degenerate if one uses equation 11.8 for the thermodynamic probability. If this is done, it is desirable, in the interest of clarity, to employ double subscripts for the N's. Thus, if i specifies the energy level of value E_i, then N_{ij} will denote the number of particles in the jth representation of that level if j takes on the values 1, 2, . . . g_i, with g_i equal to the degeneracy. With this understanding, equation 11.8 becomes

$$\Omega = \frac{N!}{\Pi\ \Pi\ N_{ij}!}_{i\ j} \tag{11.31}$$

and

$$\sum_{j=1}^{g_i} N_{ij} = N_i \tag{11.32}$$

where N_i is the total number of particles with energy E_i. At this point it should be clear that highly degenerate energy levels will tend to have higher total populations than levels of low degeneracy if we disregard the specific effect of the energy values. In other words, if we assume all other considerations to be equal, the particles will tend to go into those levels that have available the greatest number of "boxes." At the same time, however, the specific effect of the energy is exceedingly important, as will be disclosed by the result to be derived in the next section.

Maxwell-Boltzmann Distribution Law

Using equation 11.31 for the thermodynamic probability, we shall now derive the Maxwell-Boltzmann Distribution Law, which tells us how distinguishable atoms or molecules are distributed among their energy levels. Let us consider N particles with allowable energy levels E_1, E_2, E_3, etc. having degeneracies g_1, g_2, g_3, etc. Let us further suppose that the total energy of these particles is a fixed quantity, U. Since both the total number of molecules and the total energy are fixed, it is clear that our statistical problem involves two conditions of constraint. Bearing these in mind and using the method of Lagrangian multipliers, we shall now establish the most probable distribution of particles by maximizing the thermodynamic probability.

Taking the logarithm of the thermodynamic probability given by equation 11.31 and making use of Stirling's approximation, we obtain the expression

$$\ln \Omega = N \ln N - N - \sum_i \sum_j (N_{ij} \ln N_{ij} - N_{ij}) \qquad (11.33)$$

Moreover, we can assert that

$$N = \sum_i N_i = \sum_i \sum_j N_{ij} \qquad (11.34)$$

Finally, we impose the total energy constraint, which requires that

$$U = \sum_i N_i E_i = \sum_i \sum_j N_{ij} E_i \qquad (11.35)$$

Taking variations of equations 11.33, 11.34, and 11.35, we find that

$$\delta \ln \Omega = - \sum_i \sum_j \ln N_{ij} \delta N_{ij} \qquad (11.36)$$

$$\delta N = \sum_i \delta N_i = \sum_i \sum_j \delta N_{ij} = 0 \qquad (11.37)$$

$$\delta U = \sum_i E_i \delta N_i = \sum_i \sum_j E_i \delta N_{ij} = 0 \qquad (11.38)$$

Let us now multiply equation 11.37 by α and equation 11.38 by β; upon combining the results with equation 11.36, we obtain the following:

$$\delta \ln \Omega - \alpha \delta N - \beta \delta U = - \sum_i \sum_j (\ln N_{ij} + \alpha + \beta E_i) \delta N_{ij} \quad (11.39)$$

To render $\ln \Omega$ a maximum, we now set the coefficient of each of the variations of the right-hand side of equation 11.39 equal to zero. Hence we conclude that

$$\ln N_{ij} = -\alpha - \beta E_i \qquad (11.40)$$

or that

$$N_{ij} = e^{-\alpha - \beta E_i} \qquad (11.41)$$

Since N_{ij} depends only on i (through E_i) and not on j, we can further write for N_i, the total number of particles with energy E_i, the following:

$$N_i = \sum_{j=1}^{g_i} N_{ij} = g_i N_{ij} = g_i e^{-\alpha - \beta E_i} \qquad (11.42)$$

The two equations immediately above are statements of the *Maxwell-Boltzmann Distribution Law*. We can determine the two parameters, α and β, by making use of the original constraints, equations 11.34 and 11.35, in lieu of which the multipliers were introduced. Equation 11.41 tells how many molecules are likely to appear in a particular energy box, and equation 11.42 shows how many molecules will have a particular energy; as would be expected, the latter number is proportional to the degeneracy, g_i. Whenever one enumerates separately each of the states corresponding to a given energy, equation 11.41 should be used for describing the distribution of atoms or molecules among such states. It is often convenient, however, to lump together all particles with the same energy, in which case the factor g_i is introduced.

Alternative Treatment of Degenerate Systems

We carried out the derivation of the preceding section while focusing attention on the individual energy states and on the number of particles, N_{ij}, in each such state. Although the results so obtained are quite sufficient for the purpose we have in mind, we shall nevertheless find it instructive to consider an alternative approach, one that involves the thermodynamic probability expressed in the N_i's and g_i's instead of the N_{ij}'s.

In order to carry out the alternative derivation, we must first establish a new form for Ω. It is evident that N_i distinguishable particles can be placed in g_i energetically equivalent boxes in $g_i^{N_i}$ different ways. Therefore, if we start with the thermodynamic probability in the form of equation 11.8, which is valid for nondegenerate systems, we need only multiply that equation by $\prod_i g_i^{N_i}$ to convert it to the proper expression for degenerate systems. Hence we can write for the modified thermodynamic probability that

$$\Omega = N! \prod_i \frac{g_i^{N_i}}{N_i!} \tag{11.43}$$

We can also obtain equation 11.43 by an appropriate summation of thermodynamic probabilities, expressed by equation 11.31, over all of the states compatible with a specified set of N_i's. To do so, let us consider the quantity $g_i^{N_i}$, which can be written as $(1 + 1 + \ldots \text{ to } g_i \text{ terms})^{N_i}$. Upon expanding this sum of g_i unities by the multinomial theorem, we obtain the identity

$$g_i^{N_i} = \sum \sum \cdots \sum \frac{N_i!}{N_{i1}! \, N_{i2}! \ldots \ldots N_{ig_i}!} \tag{11.44}$$

in which the N_{ij}'s are subject to the constraint expressed by equation 11.32. Rearranging equation 11.44, we see that

$$\sum \sum \cdots \sum \frac{1}{\prod_j N_{ij}!} = \frac{g_i^{N_i}}{N_i!} \tag{11.45}$$

Let us now group together all of the Ω's as expressed by equation 11.31, that correspond to a particular value of i. Because of equation 11.45 such a grouping assumes precisely the form of equation 11.43.

The thermodynamic probabilities expressed by equations 11.31 and 11.43 are not generally equal to each other, and the two expressions are not precisely equivalent even though they are closely related. Equation 11.31 explicitly involves each quantum representation, whereas equation 11.43 deals with consolidated degenerate energy levels. For the *most probable* distribution, however, the values of Ω calculated from the two expressions are the same. (See the exercise below.)

EXERCISE:

Assuming the validity of Stirling's approximation in its simplest form, show that

$$\frac{g_i^{N_i}}{N_i!} = \frac{1}{\left[\left(\frac{N_i}{g_i}\right)!\right]^{g_i}}$$

Hence demonstrate the equivalence of equations 11.31 and 11.43 when applied to systems in which each subgroup of N_i particles is uniformly distributed among g_i energetically equivalent states. (Note that, in this exercise, it is not necessary to assume anything about how N_i depends on the magnitude of the energy; however, it is assumed that the most probable distribution is realized for the particles of a given energy.)

We can now derive the Maxwell-Boltzmann Distribution Law directly by maximizing Ω as expressed by equation 11.43 while subject to the two constraints involving the total number of particles and the total energy. Applying Stirling's approximation to the logarithm of Ω, we find that

$$\ln \Omega = N \ln N - N + \sum_i (N_i \ln g_i - N_i \ln N_i + N_i) \qquad (11.46)$$

Taking the variation of $\ln \Omega$ together with those derived from the aforementioned constraints, we see that

$$\delta \ln \Omega = \sum_i (\ln g_i - \ln N_i)\delta N_i \qquad (11.47)$$

$$\delta N = \sum_i \delta N_i = 0 \qquad (11.48)$$

$$\delta U = \sum_i E_i \delta N_i = 0 \qquad (11.49)$$

Multiplying equation 11.48 by α and equation 11.49 by β and then combining them with equation 11.47, we find that

$$\delta \ln \Omega - \alpha\delta N - \beta\delta U = \sum_i (\ln g_i - \ln N_i - \alpha - \beta E_i)\delta N_i \qquad (11.50)$$

To make $\ln \Omega$ a maximum, we shall now set the coefficients of δN_i equal to zero. Thus we obtain

$$\ln N_i = \ln g_i - \alpha - \beta E_i \qquad (11.51)$$

or

$$N_i = g_i e^{-\alpha - \beta E_i} \qquad (11.52)$$

Equation 11.52 is obviously equivalent to equation 11.42, derived earlier.

EXERCISE:

A large number of molecules are distributed among three different energy levels, which, with their degeneracies, are $E_1 = 0$, $g_1 = 1$; $E_2 = a$, $g_2 = 2$; $E_3 = 2a$, $g_3 = 3$. The average energy of the molecules is equal to $2a/3$. If the Boltzmann distribution holds, what fraction of the molecules has each of the different energies? If $a = 1.3$ kcal mol^{-1}, what are the values of β and $1/\beta$? (This problem illustrates how the average energy can be used to determine β. In practice, as we shall see in later sections, we generally know β, from which we calculate, among other things, the average energy.)

Identification of β

We can readily determine the significance of the multiplier β by considering the average energy, $\langle E \rangle$, of the particles in the system. This average, which is equal to U/N, will be

$$\langle E \rangle = \frac{U}{N} = \frac{\sum_i E_i N_i}{\sum_i N_i}$$

$$= \frac{\sum_i E_i g_i e^{-\beta E_i}}{\sum_i g_i e^{-\beta E_i}} \tag{11.53}$$

The factor $e^{-\alpha}$ obviously drops out when equation 11.52 is substituted into the expression for the quotient, U/N. It is evident, therefore, that β depends only on the average molecular energy and hence must be an intensive variable. We will now identify β more precisely by considering its relationship to the entropy.

As pointed out earlier, the entropy of a system should, according to equation 11.9, be given by $k \ln \Omega_{\max}$. Now we derived the Maxwell-Boltzmann Distribution Law by maximizing the thermodynamic probability. Hence, if we substitute the Maxwell-Boltzmann expression for N_i into the general equation for $k \ln \Omega$, we should be able to evaluate the entropy for a system in equilibrium. If we multiply equation 11.33 by k and replace $\ln N_{ij}$ by $-\alpha - \beta E_i$, we obtain the expression

$$S = Nk \ln N - Nk + k \sum_i \sum_j N_{ij}(\alpha + \beta E_i + 1) \tag{11.54}$$

Similarly, if we multiply equation 11.46 by k and replace $\ln N_i$ by $\ln g_i - \alpha - \beta E_i$, we find that

$$S = Nk \ln N - Nk + k \sum_i N_i(\alpha + \beta E_i + 1) \tag{11.55}$$

Upon carrying out the indicated summations, we obtain from each of these equations the following:

$$S = k \ln N! + Nk\alpha + k\beta U + Nk \qquad (11.56)$$

Let us now differentiate S while keeping N and all of the E_i's constant.* Upon doing so we see that

$$dS = Nkd\alpha + kUd\beta + k\beta dU \qquad (11.57)$$

But we also know that

$$N = \sum_i N_i = \sum_i g_i e^{-\alpha - \beta E_i} \qquad (11.58)$$

Differentiation of this equation under the specified conditions yields the result

$$dN = 0 = \sum_i (-d\alpha - E_i d\beta) g_i e^{-\alpha - \beta E_i}$$

$$= - \sum_i N_i (d\alpha + E_i d\beta)$$

$$= - Nd\alpha - Ud\beta \qquad (11.59)$$

Combining equation 11.59 with equation 11.57, we conclude that

$$dS = k\beta dU \qquad (11.60)$$

Since we obtained this differential of S by holding constant all independent variables except U, we conclude from equation 9.42 that

$$\left(\frac{\partial S}{\partial U}\right)_{V,N} = \frac{1}{T} = k\beta \qquad (11.61)$$

We have thus shown that β, one of the Lagrangian multipliers, is simply equal to $1/kT$. We can therefore rewrite the Maxwell-Boltzmann Distribution Law in the following commoner form:

$$N_i = C_i e^{-E_i/kT} \qquad (11.62)$$

In this equation the coefficient C_i is used to replace $e^{-\alpha} g_i$.

The Maxwell-Boltzmann Distribution Law is one of the most important relationships in theoretical physical chemistry. It is remarkably general in its applicability, and accordingly it frequently turns up in connection with special problems that do not explicitly involve statistical mechanics. For example, the pressure of an ideal gas in an isothermal

---

* If we are concerned with translational energy, the statement that E_i is constant implies, from quantum mechanics, that the dimensions of the container are fixed or that the volume is constant (see Chap. Twelve). Thus we are tacitly assuming that our independent variables are U, V, and N; for, although V is not explicitly involved, it is implicit in the E_i's. Our differentiation is, therefore, really a partial differentiation with respect to U.

It is also assumed, as a necessary condition for ideality, that the energy levels are unaffected by the concentration of particles.

atmosphere subject to the earth's gravitational field can be derived from ordinary thermodynamic considerations (see Chapter Seven). We readily obtain an equivalent result from equation 11.62 by replacing E_i by mgh, the potential energy of a molecule of mass m at height h above the earth's surface. Application of the Maxwell-Boltzmann Distribution Law to the translational energies of molecules leads to results fully compatible with the kinetic theory, as we shall see in the next chapter.

Problems

1. Two reporters, A and B, independently cover a certain event and report a particular detail in identical fashion. If A is known to be correct 4/5 of the time and B is known to be correct 3/4 of the time, what is the probability that their report is correct? If they disagreed with respect to the detail, what would be the probability that A's version was correct?

Ans.   12/13, 4/7

2. (A) Three cards, identical in size and shape, differ in the following details. One of the cards is white on both sides, one is black on both sides, and one is white on one side and black on the other. After the cards are mixed, one is drawn at random and placed on a table so that only one side is in view. If the visible side is white, what is the probability that the other side also is white? (There is a strong temptation to argue that the probability is 1/2, using the following faulty reasoning. Since the visible side is known to be white, it is obvious that one is not dealing with the black-black card. Since there was an equal chance of drawing the white-white card and the white-black card, there should be an even chance that the reverse side matches the visible white side. However, the correct answer to the question is 2/3. Explain why this is so.)

   (B) (This is a very simple problem, unless one has been confused by part A.) A family is known to have two children, and it is further established that at least one of them is a girl. What is the probability that the other is also a girl? [One might argue that the probability is 1/3 as follows. Let G and B denote girl and boy and let subscripts 1 and 2 denote the first and second born. Prior to knowing that one of the children is a girl, we would recognize four possible combinations, namely: G_1G_2, G_1B_2, B_1G_2, and B_1B_2. Upon ruling out B_1B_2, we have three possibilities remaining of which two represent brother-sister relationships and only one corresponds to two sisters. Hence we assert (erroneously) that the probability that the girl's sibling is a girl is 1/3. The correct answer is 1/2. Why?]

3. A weightless beam with three weightless pans is placed unsymmetrically on a fulcrum as shown in the accompanying figure. N uniform grains of

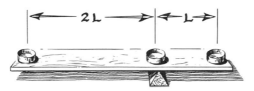

sand are distributed among the three pans in such a way that the beam remains in balance. What fraction of the sand should be placed in each of the three pans if the thermodynamic probability is to be a maximum?

Ans.   0.218, 0.346, 0.436

4. Consider a weightless bar of length $3L$ placed unsymmetrically over a ful-crum so that it is divided into portions of length $2L$ and L. N grains of sand are now distributed along the bar in a continuous manner subject to the constraint that the bar remain in balance. What distribution of sand along the bar has the highest thermodynamic probability?

5. Consider an isothermal atmosphere in a uniform gravitational field. Let the concentration of molecules (number per cm^3) at the earth's surface be c_0. Then the concentration at any height, h, above the earth's surface should be $c_0 \exp(-mgh/kT)$, according to the Boltzmann equation. Inte-gration of the concentration from $h = 0$ to $h = \infty$ then yields the total number of molecules above one square centimeter of earth's surface. This total number of molecules multiplied by the weight of each molecule, mg, equals the weight of air above one square centimeter of earth's surface, which should also equal the pressure, p_0, at the earth's surface. By carrying out the calculations indicated above, show that $p_0 = c_0 kT$ in agreement with the ideal gas-law equations. Was gaseous ideality introduced surrep-titiously, or does this problem constitute a derivation of the ideal-gas law?

6. (A) Consider a large number of diatomic molecules with electric dipole moment μ. When such a molecule is placed in an electric field of strength E, its potential energy will equal $-\mu E$ if the dipole is lined up with the field and $+\mu E$ if the dipole opposes the field. Suppose the molecules in the field are capable of exhibiting only two orientations, in the direction of the applied field and opposed thereto. At tempera-ture T, what fraction of the molecules will exhibit each kind of orien-tation, and what will be the average potential energy?

Ans. $\langle E \rangle = -\mu E \tanh(\mu E/kT)$

(B) The dipole moment of HCl is equal to 1.03×10^{-18} esu cm. If some hydrogen chloride at 300 K were placed in an electric field of 100,000 V cm^{-1}, what fraction of the molecules would be oriented in each direction? What would be the average energy?

(c) How strong a field would be necessary to cause 55 percent of the hydrogen chloride molecules to line up with the field at 300 K? What does this suggest as to the practical possibility of lining up a large fraction of molecular electric dipoles?

7. Consider a mixture of N molecules of one kind and N' molecules of another kind. Show that the thermodynamic probability of the mixture can be taken as the product of the separate thermodynamic probabilities—that is to say,

$$\Omega = \{N! \prod_i g_i^{N_i} / N_i!\} \{N'! \prod_i g_i'^{N_i'} / N_i'!\}$$

Using the method of Lagrangian multipliers, show that for the most probable distribution

$$N_i = g_i e^{-\alpha - \beta E_i}$$

and

$$N_i' = g_i' e^{-\alpha' - \beta E_i'}$$

where α and α' are different but β is the same for each expression. Why must β be the same for both kinds of molecules in a mixture?

8. Suppose that the energy of a particle can be represented by the expression $E = a\xi^n$ if ξ represents a coordinate or a momentum and n is a positive number. Let us further suppose that ξ can take on values from $-\infty$ to $+\infty$ if n is an even number or that ξ can assume only values from 0 to $+\infty$ if n is odd or nonintegral. Show that the average energy of a system of such particles subject to Boltzmann statistics will be

$$\langle E \rangle = \frac{kT}{n}$$

If n is equal to 2, as it would be for the kinetic energy of an object possessing one degree of freedom or for the potential energy of an oscillator subject to Hooke's Law, then $\langle E \rangle = kT/2$. Since this should apply for each quadratic term in the expression for the total energy of a particle possessing more than one degree of freedom, or possessing both kinetic and potential energies, we conclude that the average total energy of a particle will be

$$\langle E \rangle = \frac{skT}{2}$$

if s is the number of quadratic terms (coordinates or momenta) appearing in the energy expression. Thus the average energy of translation of a particle moving in three dimensions should be $3kT/2$. This proposition is known as the "principle of equipartition of energy." (See page 39.)

9. Derivation of Normal Error Function: Consider the random distribution of N unloaded coins. The probability of N_1 heads and N_2 tails will be

$$p(N_1, N_2) = \frac{N!}{2^N N_1! N_2!} \tag{A}$$

Let us define a new variable, x, such that $N_1 = (N + x)/2$ and $N_2 = (N - x)/2$; evidently x will be a measure of departure from the most probable distribution. (It should be observed that, if N is odd, x must be odd, and if N is even, x must be even. Moreover, for a given N, x can only take on values differing by increments of 2.) With this definition we can now write

$$p(N,x) = \frac{N!}{2^N \left(\dfrac{N + x}{2}\right)! \left(\dfrac{N - x}{2}\right)!} \tag{B}$$

Taking the logarithm of $p(N,x)$, assuming N to be a large number, and using the more accurate form of Stirling's approximation, equation 11.15, show, upon simplification, that

$$\ln p(N,x) = \ln \sqrt{\frac{2N}{\pi(N^2 - x^2)}} - \left(\frac{N + x}{2}\right) \ln \left(1 + \frac{x}{N}\right)$$
$$- \left(\frac{N - x}{2}\right) \ln \left(1 - \frac{x}{N}\right) \tag{C}$$

Assuming that x/N is always small compared with unity, expand the logarithms as power series in terms of x/N, rearrange, and neglect all terms of order of magnitude less than x^2/N. Then show that

$$\ln p(N,x) \approx \ln \sqrt{\frac{2}{\pi N}} - \frac{x^2}{2N} \tag{D}$$

or that

$$p(N,x) = \sqrt{\frac{2}{\pi N}} e^{-x^2/2N} \tag{E}$$

If this expression is correct, we can also assert that

$$\sum_x p(N,x) = 1 \tag{F}$$

If we represent the summation approximately by an integral, we can further write (assuming N to be large) that

$$\sum_x p(N,x) \approx \int_{-\infty}^{\infty} p(N,x) \frac{dx}{\Delta x} \tag{G}$$

in which expression $\Delta x = 2$, the interval between allowable values of x. (A graphical argument involving areas will show why the factor 2 must be introduced.)

Assuming now that x can take on continuous values, we observe that the probability of having a value of x in the range from x to $x + dx$ will be

$$P(N,x)dx = \frac{p(N,x)}{2}\,dx = \frac{1}{\sqrt{2\pi N}}\,e^{-x^2/2N}dx \qquad\qquad (\text{H})$$

The expression for $P(N,x)dx$ is known as the normal error function or Gaussian distribution. Although approximations were used in the derivation, the final result is correctly normalized to unity:

$$\int_{-\infty}^{\infty} P(N,x)dx = 1 \qquad\qquad (\text{I})$$

10. For $N = 10$, calculate exact values for $p(N,x)$, using equation B of Problem 9, and compare them with those calculated with equation E of the same problem. Make a graph with values of $P(N,x)$ erected as ordinates, and compare it with the smooth curve obtained from equation E.

12

Translational Properties
of Ideal Gases

Maxwell Distribution of Molecular Velocities

We shall now turn to the application of Maxwell-Boltzmann statistics to the translational properties of ideal gases. This kind of treatment will prove to describe quite completely the high-temperature behavior of monatomic gases: helium, neon, argon, etc. It will also be applicable to diatomic and polyatomic molecules, though it will provide only a partial description of them, since such molecules can undergo rotations and vibrations as well as translations. In this chapter, however, we shall regard our molecules as noninteracting point particles whose total energy is energy of translation.

Using the Maxwell-Boltzmann Distribution Law, we can readily derive the Maxwell distribution of molecular velocities, which was first obtained by kinetic-theory methods. Let us consider N molecules of an ideal gas, each of mass m. Because of the assumption concerning ideality, we can assert that the energy of any molecule will be independent of its spatial position relative to the other molecules. Moreover, as long as we are concerned only with translational effects, the energy of any molecule will be determined by its speed. If $\dot{x}$, $\dot{y}$, and $\dot{z}$ equal the components of velocity of a molecule, its energy will be

$$E = \tfrac{1}{2}m(\dot{x}^2 + \dot{y}^2 + \dot{z}^2) \tag{12.1}$$

Let us now characterize a particular way of realizing that energy by specifying that the velocity components be in the ranges from $\dot{x}$ to $\dot{x} + d\dot{x}$, from $\dot{y}$ to $\dot{y} + d\dot{y}$, and from $\dot{z}$ to $\dot{z} + d\dot{z}$. In accordance with the Maxwell-Boltzmann Distribution Law, the number of molecules, dN, in the specified class will be

$$dN = Ce^{\frac{-m(\dot{x}^2 + \dot{y}^2 + \dot{z}^2)}{2kT}} \, d\dot{x} \, d\dot{y} \, d\dot{z} \tag{12.2}$$

No degeneracy factor appears in this equation, for we are considering only one particular way in which a molecule can possess the specified energy and are not counting all the ways in which it can have the corresponding speed. We can evaluate the constant C by integrating the equation over all conceivable values of the velocity components—namely, from $-\infty$ to $+\infty$ for each. Hence

$$N = C \int_{-\infty}^{\infty} \int_{-\infty}^{\infty} \int_{-\infty}^{\infty} e^{\frac{-m(\dot{x}^2 + \dot{y}^2 + \dot{z}^2)}{2kT}} \, d\dot{x} \, d\dot{y} \, d\dot{z} \tag{12.3}$$

Using the results of Appendix B to evaluate the definite integrals, we find that

$$C = N \left(\frac{m}{2\pi kT}\right)^{3/2} \tag{12.4}$$

and therefore that

$$dN = N \left(\frac{m}{2\pi kT}\right)^{3/2} e^{\frac{-m(\dot{x}^2 + \dot{y}^2 + \dot{z}^2)}{2kT}} \, d\dot{x} \, d\dot{y} \, d\dot{z} \tag{12.5}$$

If we now replace $\dot{x}$, $\dot{y}$, and $\dot{z}$ by polar coordinates v, θ, and ϕ, so defined that $\dot{x} = v \sin\theta \cos\phi$, $\dot{y} = v \sin\theta \sin\phi$, and $\dot{z} = v \cos\theta$, and if we make the appropriate Jacobian transformation, equation 12.5 becomes

$$dN = N \left(\frac{m}{2\pi kT}\right)^{3/2} e^{-\frac{mv^2}{2kT}} v^2 \sin\theta \, dv \, d\theta \, d\phi \tag{12.6}$$

The quantity v is equal to the molecular speed, $\sqrt{\dot{x}^2 + \dot{y}^2 + \dot{z}^2}$, and the kinetic energy is equal to $\tfrac{1}{2}mv^2$. Since the energy does not depend upon θ or ϕ, we can readily integrate with respect to those variables over their ranges, which are $0 \leqslant \theta \leqslant \pi$ and $0 \leqslant \phi \leqslant 2\pi$. Upon so doing, we obtain a new distribution equation in the form

$$dN_v = N \left(\frac{m}{2\pi kT}\right)^{3/2} e^{-\frac{mv^2}{2kT}} 4\pi v^2 \, dv \tag{12.7}$$

This equation gives the number of molecules with speeds in the range from v to $v + dv$ regardless of the direction in which they may be moving. The factor $4\pi v^2$ is proportional to the degeneracy, which we should have been obliged to insert directly had we not chosen to go through the discussion involving velocity components.

The average value of a function of v, let us say $f(v)$, can now be computed by integration as follows:

$$\langle f(v) \rangle = \frac{1}{N} \int_{v=0}^{v=\infty} f(v)\, dN_v \qquad (12.8)$$

The average speed, $\langle v \rangle$, and the root-mean-square speed, $\sqrt{\langle v^2 \rangle}$, are thus shown to be given by the following formulas:

$$\langle v \rangle = \sqrt{\frac{8kT}{\pi m}} = 14{,}550 \sqrt{\frac{T}{M}} \text{ cm sec}^{-1} \qquad (12.9)$$

$$\sqrt{\langle v^2 \rangle} = \sqrt{\frac{3kT}{m}} = 15{,}790 \sqrt{\frac{T}{M}} \text{ cm sec}^{-1} \qquad (12.10)$$

In these equations, M is the molecular weight of the molecules and T the absolute temperature in degrees kelvin. To complete our picture, we can also evaluate the most probable speed, $\tilde{v}$, by finding when dN_v/dv is a maximum. The result is

$$\tilde{v} = \sqrt{\frac{2kT}{m}} = 12{,}900 \sqrt{\frac{T}{M}} \text{ cm sec}^{-1} \qquad (12.11)$$

In Figure 12-1 we have a plot of dN_v/dv against v in which $\tilde{v}$, $\langle v \rangle$, and $\sqrt{\langle v^2 \rangle}$ are indicated.

For a final variant of the Maxwell Distribution Law, we can specify how many molecules have energies in the range from E to $E + dE$. Upon making the appropriate change in variable, we obtain from equation 12.7 the result

$$dN_E = \frac{2N}{\sqrt{\pi}(kT)^{3/2}} e^{-\frac{E}{kT}} \sqrt{E}\, dE \qquad (12.12)$$

The distribution with respect to energies, you will observe, is independent of molecular weight.

FIGURE 12-1   Maxwell's distribution of molecular speeds. The number of molecules with speeds in the range v to $v + dv$ will be given by dN. The most probable speed, $\tilde{v}$, the average speed, $\langle v \rangle$, and the root-mean-square speed, $\sqrt{\langle v^2 \rangle}$, are all shown on the graph.

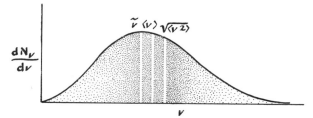

EXERCISE:

Equations 12.2–12.7 can equally well be written in terms of momenta instead of velocities. Rewrite the equations, using $p_x = m\dot{x}$, $p_y = m\dot{y}$, $p_z = m\dot{z}$, and $p = mv$ as variables in place of $\dot{x}$, $\dot{y}$, $\dot{z}$, and v.

EXERCISE:

(A) By carrying out the treatment of the preceding section, show for a two-dimensional gas that

$$dN_v = \frac{Nm}{kT} e^{-mv^2/2kT} v \, dv$$

(B) Derive expressions for $\tilde{v}$, $\langle v \rangle$, and $\sqrt{\langle v^2 \rangle}$ for a two-dimensional gas.

(C) Show for such a gas that

$$dN_E = \frac{N}{kT} e^{-E/kT} dE$$

EXERCISE:

(A) Derive the expressions given in equations 12.9, 12.10, and 12.11.

(B) Calculate the average speed, the root-mean-square speed, and the most probable speed in cm sec^{-1} for hydrogen and for oxygen at 0 and 1,000 °C.

EXERCISE:

Plot dN_E/dE against E and compare the result with Figure 12-1. Show that the most probable energy is $kT/2$. Why does this not correspond to the most probable speed?

Translational Energy of an Ideal Gas

The translational energy of an ideal gas can now be directly calculated from the results of the preceding section. For this purpose we might use any of the equations 12.5, 12.7, and 12.12. The procedure consists simply of multiplying dN_E by E, expressed in terms of the appropriate variables, and integrating the result over all conceivable values of those variables. If we choose equation 12.12 for this purpose, we see that

$$U = \int E \, dN_E$$

$$= \int_0^\infty \frac{2N}{\sqrt{\pi}(kT)^{3/2}} e^{-\frac{E}{kT}} E^{3/2} dE \tag{12.13}$$

Upon evaluating this integral, we find that

$$U = \tfrac{3}{2} NkT \tag{12.14}$$

If N equals Avogadro's number, then $Nk = R$; hence the translational energy of one mole of ideal gas is

$$U = \tfrac{3}{2}RT \tag{12.15}$$

a result compatible with the kinetic theory in combination with the ideal gas-law equation of state. It will be obvious from equation 12.15 that the heat capacity at constant volume attributable to translation effects will be simply $\tfrac{3}{2}R$ per mole of ideal gas.

In our calculations dealing with the translational motions of ideal gases, we have assumed that the possible energy values are practically continuous, thus justifying the processes of integration. Strictly speaking, however, translational energies, like other energies, are subject to quantization, and hence only discrete energy levels are allowed. This means that we should really carry out summations, such as those indicated in equation 11.53, in place of integrations like that of equation 12.13. It happens, however, that the separations between the translational energy levels in ordinary systems are so small (compared with kT) that for all practical purposes the energies can be considered continuous (except for temperatures near absolute zero). The result expressed by $U = \tfrac{3}{2}NkT$ is therefore quite exact for all temperatures at which known substances remain gaseous. In principle, however, that equation for translational energy would fail at exceedingly low temperatures if there were such a thing as a gas obeying Boltzmann statistics under those circumstances. Our reason for mentioning this is that the methods employed above will not work for rotations or vibrations of molecules except asymptotically for high temperatures. The correct procedure requires recognition of quantization, the precise effect of which will appear later.

Translational Quantum States

In deriving and discussing the Maxwell-Boltzmann Distribution Law, we made frequent reference to quantized energy levels and quantum states, but we did not give the details of any specific examples, nor did we show precisely how they might affect our thermodynamic functions. To be sure, we were able to calculate the translational energy of an ideal gas without knowledge of the quantized translational levels, but that calculation was successful only because of the exceedingly small separations between those levels. In any event, we now find ourselves in a position where little more can be developed, even about translational effects, without some knowledge of the quantum mechanics involved. Our purpose in postponing until now any specific reference to the results of quantum mechanics has been to enable the reader to appreciate just what can be done without such; from subsequent developments it should likewise become clear where quantum theory has its greatest impact on statistical mechanics.

It is not our intent to delve into the very important subject of quantum mechanics beyond presenting some of the conclusions that bear directly on the topics with which we are currently preoccupied. We shall only say that it is possible by various means—such as finding acceptable solutions to Schrödinger's wave equation—to establish the allowable energy levels for a number of atomic and molecular systems.

Let us consider, for example, a particle of mass m placed in a cubical box of edge length l. The quantum mechanics of this system is readily worked out, and the allowable energy levels are found to be expressed by the equation

$$E = \frac{h^2}{8ml^2} (n_x^2 + n_y^2 + n_z^2) \qquad (12.16)$$

in which h is Planck's constant and n_x, n_y, and n_z are integers different from zero. The lowest energy level will correspond to $n_x = n_y = n_z = 1$ and will have the value $3h^2/(8ml^2)$. The next-lowest energy level will have the value $6h^2/(8ml^2)$; this level is triply degenerate since it can be obtained by three combinations of the quantum numbers—namely, (2,1,1), (1,2,1), and (1,1,2)—for (n_x, n_y, n_z).

For convenience we shall now define a quantity n^2 by the equation

$$n^2 = n_x^2 + n_y^2 + n_z^2 \qquad (12.17)$$

Obviously, n^2 will not generally be a perfect square of an integer, but it is a whole number that determines the value of an energy level. (For convenience we may call it the square of the "total quantum number.") Recalling that we assumed a cubical shape for the box, we recognize that the volume, V, will equal l^3. Hence we can replace l^2 by $V^{2/3}$ and, in view of our definition of n^2, write that

$$E = \frac{n^2 h^2}{8m V^{2/3}} \qquad (12.18)$$

Degeneracy of Translational Energy Levels

It is not possible to write a simple formula for the degeneracy of translational energy levels, but we can derive an approximate formula, valid for large n, that will serve our purpose. Consider a simple cubic lattice made up of points whose coordinates are given by integers n_x, n_y, and n_z in the $(+++)$ octant of a Cartesian coordinate system (see Fig. 12-2). Each little cell in such a lattice will have unit volume since the edge lengths of the cubes will be unity. On the average, the number of points in any closed region of the lattice will be equal to the volume of the region.

If we now draw a vector from the origin to any point in this lattice, the square of the length of that vector will be equal to the sum of the squares of the components. If the components of the vectors are chosen as translational quantum numbers n_x, n_y, and n_z, the square of the length of the vector will equal n^2, which is the quantity to be placed in the expression for the energy of a particle in a box (equation 12.18). It should be clear, moreover, that different combinations of n_x, n_y, and n_z exhibiting the same sum of the squares will correspond to the same total energy, and the number of such combinations will equal the degeneracy of the level. To determine the degeneracy associated with a certain energy range, we need only ask how many points in the aforementioned lattice correspond to the specified energy range. Thus, if we think of n as being continuous, which we can do to a good approximation if n is large, the lattice volume corresponding to the range from n to $n + dn$ will equal one-eighth of the volume of a spherical shell of radius n and thickness dn. (The factor $1/8$ comes in because we consider only one octant in the coordinate system.) This volume, which equals $\frac{1}{2}\pi n^2 dn$, will, on the average, equal the number of lattice points contained in one-eighth of the

FIGURE 12-2 Part of a lattice of points (n_x, n_y, n_z) representing translational quantum numbers.

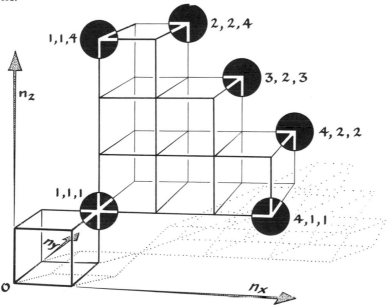

spherical shell. We can assert, therefore, that the degeneracy, dg, associated with total translational quantum numbers in the range from n to $n + dn$ will be

$$dg = \tfrac{1}{2}\pi n^2\, dn \tag{12.19}$$

If we now substitute for n its equivalent value in terms of E, using equation 12.18, we find that the degeneracy associated with the energy range from E to $E + dE$ will be

$$dg = \frac{4\sqrt{2}\pi m^{3/2} V E^{1/2}}{h^3}\, dE \tag{12.20}$$

The degeneracy of translational quantum states, you will note, is determined solely by the range of the net total quantum numbers and does not really depend upon the volume of the box. This statement may seem at variance with equation 12.20 since the volume does appear in that equation. The reason for this can be simply explained. Although the degeneracy is determined by the quantum numbers and their range of values, the energy depends upon both the quantum numbers and the volume. Hence, if the degeneracy is expressed in terms of the energy range, the volume must necessarily be introduced into the expression. This does not, however, alter the validity of the statement that the degeneracy of translational quantum states depends only upon the quantum numbers, as expressed by equation 12.19.* This fact will be utilized now as we turn to the derivation of the equation of state of an ideal gas.

Equation of State of an Ideal Gas

We are now in a position to establish the equation of state for an ideal gas by taking cognizance of the quantized energy levels and their dependence upon volume. To accomplish this end, let us return to equations 11.56, 11.57, and 11.58 and proceed as we did in Chapter Eleven except that we shall no longer require V to be constant. Specifically, we shall now take the total differential of S, keeping N constant as before but permitting the E_i's to vary in accordance with their dependence on V. With this understanding we obtain in place of equation 11.59 the following expression:

$$dN = 0 = \sum_i (-d\alpha - E_i d\beta - \beta dE_i) g_i e^{-\alpha - \beta E_i}$$

$$= \sum_i N_i \left(-d\alpha - E_i d\beta + \frac{2\beta E_i}{3V}\, dV \right)$$

---

* We could, of course, say that the degeneracy is a function of $EV^{2/3}$, which depends only on n.

$$= -N d\alpha - U d\beta + \frac{2\beta U}{3V} dV \qquad (12.21)$$

To obtain this result, we must remember that g_i depends only upon the quantum numbers and not upon the volume. E_i, on the other hand, does depend upon the volume, so its derivative with respect to that variable must be included in the calculation. Combining equation 12.21 with equation 11.57, we obtain the expression

$$dS = k\beta dU + \frac{2k\beta U}{3V} dV \qquad (12.22)$$

But we also know that

$$dS = \frac{1}{T} dU + \frac{p}{T} dV \qquad (12.23)$$

Hence we conclude, as before, that $k\beta = 1/T$ and that

$$\frac{p}{T} = \frac{2k\beta U}{3V} \qquad (12.24)$$

or, upon rearrangement and simplification, that

$$pV = \tfrac{2}{3}U \qquad (12.25)$$

Equation 12.25 describes the relation between the pressure-volume product of an ideal gas and the translational energy. In view of the established dependence of energy on temperature (equation 12.14), we further conclude that

$$pV = NkT \qquad (12.26)$$

which expression reduces, for one mole, to

$$pV = RT \qquad (12.27)$$

Our derivation of the equation of state for the ideal gas has made use of the dependence of the energy levels upon the volume of the system. This is reasonable, for we introduced the pressure by evaluating $(\partial S/\partial V)_U$, which necessarily called for volume dependence in some form.

Before leaving the equation of state for an ideal gas, we should reassure ourselves that quantization of translational energies produces no appreciable effect on the average translational energy expressed by equation 12.14, which was derived without recourse to quantum theory. Since our derivation involved replacement of summations by integrations, we should examine its validity by seeing whether or not the separations between energy levels are indeed small compared with kT. The spacings between translational energy levels are of the order of magnitude $h^2/(8mV^{2/3})$, so our test reduces to evaluating $h^2/(8mV^{2/3}kT)$. Using helium, for example, with a molecular weight of 4, and assuming that $V = 10^3$ cm^3 and $T = 300$ K, we find that the expression equals 2×10^{-19}, an exceedingly small number. The spacing of energy levels relative to kT is evidently so small, except at vanishingly low temperatures, that the exponentials of $-E_i/kT$ vary continuously, for all practical purposes, thus justifying the integrations that were used for calculating the translational energy.

Entropy of an Ideal Monatomic Gas

Our search for the entropy of an ideal gas is now almost complete. Examination of equation 11.56 discloses that the one quantity still required is α, one of the Lagrangian multipliers. To obtain α, we can rearrange equation 11.58 to get

$$e^\alpha = \frac{\sum_i g_i e^{-\beta E_i}}{N} \tag{12.28}$$

If we now replace the summation by an integration and substitute for g_i and E_i the values expressed by equations 12.19 and 12.18, we find that

$$e^\alpha = \frac{\int_0^\infty \tfrac{1}{2}\pi n^2 e^{-\frac{n^2 h^2 \beta}{8mV^{2/3}}} \, dn}{N} \tag{12.29}$$

Evaluation of the integral (see Appendix B) then yields

$$e^\alpha = \left(\frac{2\pi m}{\beta}\right)^{3/2} \cdot \frac{V}{Nh^3}$$

$$= \frac{(2\pi mkT)^{3/2} V}{Nh^3} \tag{12.30}$$

or

$$\alpha = \ln \frac{(2\pi mkT)^{3/2} V}{Nh^3} \tag{12.31}$$

Everything necessary for the evaluation of the translational entropy of an ideal gas is now available; substitution of the various quantities into equation 11.56 gives

$$S = k \ln N! + Nk \ln \frac{(2\pi mkT)^{3/2} V}{Nh^3} + \tfrac{5}{2}Nk$$

$$= k \ln N! + Nk \ln \frac{(2\pi m)^{3/2}(kT)^{5/2}}{ph^3} + \tfrac{5}{2}Nk \tag{12.32}$$

Having derived this equation for the entropy of an ideal monatomic gas, we now naturally ask if the result agrees with observation. We can test the equation by comparing the calculated entropy with that obtained experimentally, using the Third Law as a basis for obtaining an absolute value. For example, if we consider neon, a monatomic gas, at room temperatures, we can write the following for its absolute entropy:

$$S = \int_0^{T_{\text{fus}}} \frac{C(\text{c})}{T} dT + \frac{\Delta H_{\text{fus}}}{T_{\text{fus}}} + \int_{T_{\text{fus}}}^{T_{\text{vap}}} \frac{C(\text{liq})}{T} dT + \frac{\Delta H_{\text{vap}}}{T_{\text{vap}}} + \int_{T_{\text{vap}}}^{T} \frac{C(\text{g})}{T} dT \tag{12.33}$$

In this equation $C(\text{c})$, $C(\text{liq})$, and $C(\text{g})$ are the heat capacities of the solid, liquid, and vapor, and ΔH_{fus} and ΔH_{vap} are the heats of fusion and vaporization at temperatures T_{fus} and T_{vap} respectively.

If a comparison between theory and experiment is carried out, it is found that equation 12.32 is substantially correct except that the calculated entropy is too large by the amount $k \ln N!$. For one mole of gas the term $k \ln N!$ is very large indeed; yet one is reluctant to throw out the entire theory on that account, particularly since the theory appears to be satisfactory in all other respects. There is, however, a good reason why the term $k \ln N!$ should be dropped from the entropy. The reader will recall that, when we set up the expression for the thermo-dynamic probability in Boltzmann statistics, we assumed that the particles were all distinguishable. Now the number of different ways in which we can take N distinguishable particles without reference to statistical states is precisely $N!$, which appears as a factor in equations 11.31 and 11.43, the expressions for the thermodynamic probability, Ω. It is the assumed distinguishability that is responsible for the appearance of the term $k \ln N!$ in the entropy. Such distinguishability, if it existed, would imply, among other things, that the mixing of two identical samples of a particular gas would be accompanied by an entropy change. With this hypothesis we should be compelled to say, for example, that an entropy change would attend the mixing of one mole of helium at 0 °C and one atmosphere with another mole of helium at the same temperature and pressure to get two moles of helium at 0 °C and one atmosphere. But such a proposition cannot be accepted, for it would imply that the entropy of a gas is not an extensive property determined by the state of the system.*

In the light of these arguments, we shall suppress the term $k \ln N!$ from equation 12.32 and write that

$$S = Nk \ln \frac{(2\pi m)^{3/2}(ekT)^{5/2}}{ph^3} \tag{12.34}$$

e being the base of natural logarithms. This equation is a compact form of what is called the Sackur-Tetrode entropy of translation, the names being those of the men who contributed to the original derivation.

We can reduce equation 12.34 to more usable numerical form by putting in the numerical values for the constants and by replacing m

---

* If different isotopes were mixed, there would be an entropy change. This is perfectly reasonable, however, for different isotopes are certainly distinguishable from one another.

by the quotient of M, the molecular weight, and Avogadro's number. Upon changing the units of p to atmospheres, we obtain the expression

$$S = \tfrac{5}{2}R \ln T - R \ln p + \tfrac{3}{2}R \ln M - 1.1650\,R \qquad (12.35)$$

Equations 12.34 and 12.35, you will observe, are fully compatible with equation 6.23 in form; the coefficient of $\ln T$ (namely, $\tfrac{5}{2}R$) is just the heat capacity at constant pressure for a monatomic ideal gas. At 25 °C and one atmosphere, equation 12.35 further reduces to

$$S = 6.8635 \lg M + 25.991 \text{ cal deg}^{-1} \text{ mol}^{-1} \qquad (12.36)$$

EXERCISE:

By appropriate numerical calculations, reduce equation 12.34 to equation 12.35. By further substitution, obtain equation 12.36.

EXERCISE:

Calculate the entropies of neon, argon, and krypton at 25 °C and 1 atm, using equation 12.36. Compare your results with the following accepted values:

Substance	S cal deg^{-1} mol^{-1}
Ne	34.947
Ar	36.982
Kr	39.191

Identification of α

Just as we were able to establish that β is equal to $1/kT$, so we can now identify α, the other multiplier introduced in our variation treatment leading to the Boltzmann Distribution Law. To do so correctly, we must, for the reasons cited above, first suppress the term $k \ln N!$ from the entropy. With this understanding, we can rewrite equation 11.56 as

$$S = Nk\alpha + k\beta U + Nk \qquad (12.37)$$

We shall now deduce the total differential of the entropy, taking cognizance of all the independent variables, U, V, and N. Thus we obtain the expression

$$dS = Nk\,d\alpha + kU\,d\beta + k\beta\,dU + (k\alpha + k)\,dN \qquad (12.38)$$

We shall not now, as we did earlier, require that $dN = 0$, although we can still assert from equation 12.21 that

$$Nk\,d\alpha + kU\,d\beta = \frac{2k\beta U}{3V}\,dV - k\,dN \qquad (12.39)$$

By combining equations 12.38 and 12.39, we find that

$$dS = k\beta dU + \frac{2k\beta U}{3V} dV + k\alpha dN \qquad (12.40)$$

The first two terms are already familiar, but from the third we conclude that

$$k\alpha = \left(\frac{\partial S}{\partial N}\right)_{U,V} \qquad (12.41)$$

Multiplying both sides of equation 12.41 by Avogadro's number and thus replacing k by R and the number of molecules, N, by the number of moles, n, we conclude, in the light of equation 9.43, that

$$\alpha = \frac{1}{R}\left(\frac{\partial S}{\partial n}\right)_{U,V} = -\frac{\mu}{RT} \qquad (12.42)$$

μ being the chemical potential or, since we are dealing with a single substance, the molar Gibbs free energy. The molar Gibbs free energy of an ideal monatomic gas can now be written directly from equations 12.42 and 12.31:

$$G = -RT\alpha = -RT \ln \frac{(2\pi m k T)^{3/2} V}{N h^3} \qquad (12.43)$$

Introducing the ideal-gas equation of state, we find that

$$G = -RT \ln \frac{(2\pi m)^{3/2} (kT)^{5/2}}{p h^3} \qquad (12.44)$$

which is in complete agreement with the Sackur-Tetrode entropy given by equation 12.34.

Problems

1. Starting with the relationship $dN = e^{-\alpha - \beta E} dg$, with dg given by equation 12.20, derive equation 12.12. To eliminate $e^{-\alpha}$ observe that

$$\int_{E=0}^{E=\infty} dN = N$$

and hence that

$$dN = \frac{N e^{-\beta E} dg}{\int_0^\infty e^{-\beta E} dg}$$

from which expression the desired equation can be obtained.

2. Consider a two-dimensional ideal gas contained in a square box of area A. The allowable energy levels for particles in such a box will be given by the expression

$$E = \frac{n^2 h^2}{8mA}$$

in which

$$n^2 = n_x^2 + n_y^2$$

n_x and n_y being positive integers. Show that for large n the degeneracy will be $dg = \frac{1}{2}\pi n \, dn$ or

$$dg = \frac{2\pi m A}{h^2} \, dE$$

Hence show by the methods of the last problem that

$$dN = \frac{N}{kT} \, e^{-E/kT} dE$$

Show also that $U = NkT$ for a two-dimensional gas. (Compare this problem with the second exercise on p. 256.)

3. Derive an expression for $\langle v^n \rangle$ if n is a positive integer and v is the speed of an ideal gaseous molecule.

Ans. $\langle v^n \rangle = \left(\dfrac{kT}{m}\right)^{n/2} (n+1)(n-1) \ldots 5 \cdot 3 \cdot 1$ for n even

$\langle v^n \rangle = \sqrt{\dfrac{2}{\pi}} \left(\dfrac{kT}{m}\right)^{n/2} (n+1)(n-1)(n-3) \ldots 6 \cdot 4 \cdot 2$ for n odd

4. "Classical-quantum" Calculation of Translational Degeneracies: Equation 12.20 can be derived by classical-quantum considerations as follows. Consider the total volume in six-dimensional phase space (coordinates x, y, z and momenta p_x, p_y, p_z) associated with physical volume V and translational energies ranging from 0 to E. This generalized volume will be given by the expression

$$\phi = \int\!\!\int\!\!\int\!\!\int\!\!\int\!\!\int dx \, dy \, dz \, dp_x \, dp_y \, dp_z$$

One can carry out the triple integration over the coordinates directly to get

$$\phi = V \int\!\!\int\!\!\int dp_x \, dp_y \, dp_z$$

But, since the energy, E, is

$$E = \frac{1}{2m}(p_x^2 + p_y^2 + p_z^2) = \frac{1}{2m} p^2$$

the maximum value of the total momentum is $\sqrt{2mE}$. Hence the triple integral over the momenta will be precisely equal to the volume of a sphere of radius $\sqrt{2mE}$. Therefore

$$\phi = V\tfrac{4}{3}\pi(2mE)^{3/2}$$

Now the volume in phase space associated with the range of energies from E to $E + dE$ will be given if one differentiates ϕ as follows:

$$d\phi = 4\sqrt{2}\pi m^{3/2} V E^{1/2} dE$$

According to old quantum theory, the unit cell of phase space associated with a coordinate and its conjugate momentum is equal to h, Planck's constant. This is compatible with the uncertainty principle of Heisenberg, in which it is asserted that $\Delta x \Delta p_x \sim h$ if Δx and Δp_x are the uncertainties involved in the simultaneous measurement of x and p_x. For three coordinates and three momenta, the volume of a cell in phase space is h^3; hence the number of cells in phase space associated with the range of translational energies from E to $E + dE$ is equal to $d\phi/h^3$. This should also be the degeneracy:

$$dg = \frac{d\phi}{h^3} = \frac{4\sqrt{2}\pi m^{3/2} V E^{1/2} dE}{h^3} \qquad (12.20)$$

(The methods of this problem are of historical interest, for they were used prior to the more complete formulation of quantum mechanics.)

5. Calculate the fraction of ideal gaseous molecules with speeds in excess of ten times the most probable speed. For this purpose you may use the following approximate equation, valid asymptotically for large x:

$$\int_x^\infty e^{-x^2}\, dx = \frac{e^{-x^2}}{2x}\left(1 - \frac{1}{2x^2} + \cdots\right)$$

6. If a molecule of air near the earth's surface had a sufficiently high component of velocity upward, it could escape from the earth (provided it did not bump into other molecules on the way).

   (A) Derive an equation for the critical velocity component for escape from the earth. Let g_0 be the acceleration of gravity at the earth's surface and r_0 the radius of the earth.

   (B) What fraction of the molecules of air ($M \approx 29$) at 27 °C would have a vertical component of velocity equal to or greater than the escape velocity? (Here one must use the methods of Prob. 5.) Assume the radius of the earth to be 6,400 km. In making the calculation, bear in mind that only one component of velocity in a particular direction is involved.

   Ans.   (A) $v_{escape} = \sqrt{2g_0 r_0}$

7. The velocity of escape from the moon is about .21 of that for the earth. What fraction of air molecules at 27 °C would have a high enough velocity component to escape from the moon? In view of the smallness of this fraction, how might one explain the absence of an atmosphere on the moon?

8. (A) Assuming that 75 percent of the atoms of chlorine have mass number 35 and that 25 percent have mass number 37, what fraction of the molecules of chlorine will have masses 70, 72, and 74 if pairs of atoms are taken at random?

(B) For a total of 1 mole of molecular chlorine, what is the entropy of mixing of the three kinds of molecules?

(C) How does the answer to part B differ from the entropy of mixing of the same total number (2 moles) of atoms if we assume that no di-atomic molecules exist?

9. In calculating the Sackur-Tetrode entropy for a monatomic gas consist-ing of a mixture of isotopes, one can, to a good approximation, use the average atomic weight. (This does not, of course, take care of the entropy of mixing.)

(A) Derive an expression for the difference between the sum of the cal-culated entropies of two isotopes, of mole fractions x_1 and x_2, and the entropy calculated for the material assumed to have the average molecular weight.

(B) Assuming neon to consist of 90 percent isotope 20 and 10 percent isotope 22, calculate the above-defined difference in cal deg^{-1} mol^{-1}.

$$\text{Ans.} \quad (A) \quad \tfrac{3}{2}R \ln \left\{ \frac{M_1^{x_1} M_2^{x_2}}{x_1 M_1 + x_2 M_2} \right\}$$

10. Calculate the root-mean-square "total quantum number" of translation for argon at 27 °C contained in a cubical vessel of 1,000 cm^3. What is the degeneracy associated with the range from n to $n + \Delta n$ if $\Delta n = 1$ and n is the root-mean-square quantum number calculated above? Do these numbers impress you concerning the validity of treating the translational energy levels as being practically continuous?

11. Suppose one naively set the Sackur-Tetrode entropy, as expressed by equation 12.35, equal to zero and then solved for the pressure correspond-ing to a particular temperature. Would the answer be meaningful? Cal-culate the pressure in atmospheres for which the "Sackur-Tetrode entropy" of helium would equal zero at 300 K. What is the numerical significance of the result?

13

Partition Functions in Statistical Thermodynamics

Introduction

In the two preceding chapters we developed the subject of Boltzmann statistics and then applied it to the translational behavior of a monatomic ideal gas. The calculations that we made will actually provide a partial description of all ideal gases, whether they are monatomic or not. For a polyatomic molecule, however, the entropy, energy, and other related thermodynamic quantities will consist of several terms attributable not only to translation but also to rotation, vibration, and other kinds of possible molecular excitation. The Sackur-Tetrode entropy, for example, can be regarded as the translational contribution to the entropy of any ideal gas; to obtain the entire entropy, one must augment the translational portion by further contributions from modes of motion or kinds of excitation other than translation.

We have treated translational energy and entropy in some detail to enable the reader to gain an insight into the relationship between atomistic theory and statistical thermodynamics. Other manifestations of molecular energy can be handled in a similar manner, but it now becomes expedient to develop some general formulas that can be applied almost automatically to the calculation of thermodynamic quantities from molecular data. The procedures to be developed are reasonably foolproof, but there are pitfalls, and the reader is warned at the outset not to expect formulas into which substitutions can be made slavishly. The method involves the use of certain general functions known as partition functions, which we shall now define.

Consider a molecule or particle that can exist in any of a number of energy levels, E_i, with degeneracies g_i. For such a particle, the partition function, q, is defined by the equation

$$q = \sum_i g_i e^{-E_i/kT} \tag{13.1}$$

For a system of two or more noninteracting particles, the partition function, Q, is defined, quite naturally, as the product of the individual partition functions as follows:

$$Q = q_1 q_2 \ldots q_N \tag{13.2}$$

In particular, if the N noninteracting particles are identical, then the subscripts can be dropped and we see that

$$Q = q^N \tag{13.3}$$

For a mixture of N_A molecules of kind A, N_B molecules of kind B, etc., all free of intermolecular interactions, we can write

$$Q = q_A^{N_A} q_B^{N_B} \ldots \tag{13.4}$$

Equation 13.2 can also be used, at least approximately, for some systems of molecules subject to intermolecular interactions provided certain precautions are observed. This will be illustrated later in connection with our statistical mechanical treatment of nonideal gases.

EXERCISE:

Suppose a molecule, A, can exist in any of three energy levels of values a, $2a$, and $4a$ with degeneracies 1, 2, and 4, respectively. Another molecule, B, can exist in either of two levels, with energies $2a$ and $3a$ and corresponding degeneracies of 1 and 3.

(A) Set up partition functions, q_A and q_B, for the separate molecules.

(B) Regarding the two molecules as constituting a system, complete by direct count the table started below showing total energies and total degeneracies for the system.

E_A	E_B	$g_A g_B$	E_{total}	g_{total}
a	$2a$	1	$3a$	1
a	$3a$	3	$4a$	5
$2a$	$2a$	2		

After completing the table, set up the total partition function for the system and show that it is precisely equal to $q_A q_B$.

If we substitute translational energies for a molecule into equation 13.1, we obtain its translational partition function, denoted by q_{trans}. If we substitute rotational or vibrational energies, we get q_{rot} or q_{vib}. Strictly speaking, there is only one partition function for a molecule, namely, the one we obtain by putting into equation 13.1 all combinations of allowable energies with their degeneracies; however, if the different kinds of energy are noninteracting from an intramolecular point of view, the partition function for a molecule will be the product of the separate functions:

$$q = q_{trans} \cdot q_{rot} \cdot q_{vib} \cdots \qquad (13.5)$$

The rationale supporting the product form of equation 13.5 is similar to that in justification of equation 13.2 (see exercise on p. 270). Two different kinds of energy are said to be intramolecularly noninteracting when the excitation of one has no effect on the allowable levels of the other. Actually there often is some interaction, as between vibration and rotation, but the effect is usually small and is generally neglected.

If we compare equation 13.1, the definition of the partition function, with equation 11.58, we see that q is simply related to α, one of the Lagrangian multipliers, as follows:

$$q = Ne^\alpha \qquad (13.6)$$

In view of the established connection between α and the chemical potential, equation 12.42, it is evident that q is simply related to the Gibbs free energy; we defer specific discussion of this point. If we take the logarithm of the partition function and differentiate it partially with respect to T, we find that

$$\frac{\partial \ln q}{\partial T} = \frac{\sum_i g_i e^{-E_i/kT} E_i/kT^2}{\sum_i g_i e^{-E/kT}} \qquad (13.7)$$

In the light of equation 11.53 we conclude from the above that

$$U = NkT^2 \left(\frac{\partial \ln q}{\partial T}\right)_V \qquad (13.8)$$

This equation will evidently provide a convenient route for the direct calculation of the energy of a system. Obviously, we can obtain the heat capacity at constant volume, C_v, by evaluating $(\partial U/\partial T)_V$. The entropy of a system can likewise be easily computed. Substituting the energy given by equation 13.8 and α from equation 13.6 into equation 11.56, we find, upon simplification and use of Stirling's approximation, that the entropy can be expressed in the following simple closed form:

$$S = Nk \frac{\partial}{\partial T} (T \ln q) \qquad (13.9)$$

Moreover, the Helmholtz free energy is given by an even simpler equation:

$$A = -NkT \ln q \qquad (13.10)$$

Finally, the equation of state, which we can obtain by differentiating the Helmholtz partially with respect to V, can be expressed by the equation

$$p = -\left(\frac{\partial A}{\partial V}\right)_T$$

$$= NkT \left(\frac{\partial \ln q}{\partial V}\right)_T = kT \left(\frac{\partial \ln Q}{\partial V}\right)_T \qquad (13.11)$$

The form of dependence of U, S, and A on the logarithm of q means, in view of equation 13.5, that those functions are sums of terms attributable to translation, rotation, vibration, etc., so long as the assumption of no intramolecular energy interactions is valid. In other words

$$U = U_{\text{trans}} + U_{\text{rot}} + U_{\text{vib}} + \ldots \qquad (13.12)$$

with similar equations applicable to S, A, G, p, C_v, etc.

If we substitute the known translational energy levels and their degeneracies into the definition of q (equation 13.1), and if we evaluate the summation by taking recourse to integration, we obtain the following:

$$q_{\text{trans}} = \frac{(2\pi mkT)^{3/2} V}{h^3} \qquad (13.13)$$

Substitution of the translational partition function into equations 13.8 and 13.11 will give the correct translational energy and correct equation of state for an ideal gas. When the partition function is combined with equation 13.9, on the other hand, one obtains, for reasons explained on page 263, an entropy that is too large by $k \ln N!$; hence this amount of entropy must be subtracted from that given by the formula. (Similarly, the Helmholtz energy given by equation 13.10 will be too small by $kT \ln N!$ and must likewise be corrected.) However, since this particular correction applies only to translation and not to other modes of motion, we shall not explicitly indicate it in the general formulas.

To obtain the Gibbs free energy, one need only add the pressure-volume product to the Helmholtz function. Since $pV = NkT$ for an ideal gas,

$$G = -NkT \ln q + NkT \tag{13.14}$$

Substitution of q_{trans} into this equation will, of course, give a value of the Gibbs free energy that is too small by $kT \ln N!$. However, if we add this correction and rearrange in the light of Stirling's approximation, we find that

$$G = -NkT \ln (q/N) \tag{13.15}$$

This equation, when the translational partition function is substituted in it, gives the correct Gibbs energy of translation. Moreover, since the kind of indistinguishability correction described earlier is made only for translation, we can expect equation 13.15 to have general validity, provide q represents the total partition function. A combination of equations 13.15 and 13.6 will now give a result fully compatible with the identification of the Lagrangian multiplier, α, expressed by equation 12.42.

EXERCISE:

Derive equations 13.9 and 13.10 by the methods described in the text.

Thermodynamic Properties of Molecular Rotators

Consider a diatomic molecule such as H_2, Cl_2, or HCl. In addition to moving about with translational energy equal to $\frac{3}{2}RT$ per mole, these molecules can also rotate about their centers of gravity. Upon solving the quantum-mechanical problem of a rigid linear molecule rotating in space, we find that the allowable energy levels are given by expressions of the form

$$E_K = \frac{K(K + 1)h^2}{8\pi^2 I} \tag{13.16}$$

In this equation K is known as the rotational quantum number, a number that can assume the values 0, 1, 2, 3, etc. The other quantities in the expression include h (Planck's constant) and I, the moment of inertia of the molecule about its center of gravity. The rotational energy levels given by equation 13.16 are degenerate, for there is another quantum number, M, that can assume values 0, ± 1, ± 2, etc., up to and including $\pm K$. Although M does not appear explicitly in the expression for the energy, one can ascertain by direct count that there are $2K + 1$ possible values of M corresponding to a given energy level. The degeneracy g_K is therefore precisely equal to $2K + 1$. The rotational partition function for a molecule like HCl will then be

$$q_{\rm rot} = \sum_{K=0}^{\infty} (2K + 1)e^{-\frac{K(K+1)h^2}{8\pi^2 IkT}} \tag{13.17}$$

Now the spacings between rotational energy levels are not so small, compared with kT at room temperature, that we can always replace the summation by an integration. However, if we assume that this can be done, we can represent the degeneracy by the expression

$$dg = (2K + 1)dK = d(K^2 + K) \tag{13.18}$$

which then permits us to write that

$$q_{\rm rot} = \int_0^{\infty} e^{-\frac{(K^2+K)h^2}{8\pi^2 IkT}} \, d(K^2 + K) \tag{13.19}$$

Upon integration, we obtain for the rotational partition function the expression

$$q_{\rm rot} = \frac{8\pi^2 IkT}{h^2} \tag{13.20}$$

Substituting $q_{\rm rot}$ into equations 13.8, 13.9, and 13.10, we can calculate the rotational energy, the rotational entropy, and the rotational Helmholtz free energy of a rigid rotator in space. For one mole of material these turn out to be

$$U_{\rm rot} = RT \tag{13.21}$$

$$S_{\rm rot} = R \ln \frac{8\pi^2 IkT}{h^2} + R \tag{13.22}$$

$$A_{\rm rot} = -RT \ln \frac{8\pi^2 IkT}{h^2} \tag{13.23}$$

Evidently the heat capacity attributable to rotation is simply equal to R. This brings C_v for a diatomic molecule up to $\frac{5}{2}R$, which is roughly correct for all such molecules at room temperature. Moreover, since the rotational partition function does not depend upon the volume of the container, its partial derivative with respect to V is equal to zero. Hence the rotational contribution to the pressure of a system must be zero, and the Gibbs energy of rotation is equal to the Helmholtz rotational energy.

As pointed out above, the results obtained for linear rotators in space, assuming that the partition-function summations can be replaced by integrals, are actually valid only at high temperatures—that is, temperatures for which kT is large compared with the spacings between rotational levels. From the expression for the energy it is clear that molecules with high moments of inertia will have smaller spacings between their rotational energy levels than will molecules with low moments of inertia. Actually it turns out that, for molecules like NO, CO_2, and Cl_2, a fairly complete excitation of molecular rotation will be realized even below room temperatures. On the other hand, the moments of inertia of molecules like hydrogen or the hydrogen halides are sufficiently small so that ordinary room temperatures are just barely high enough to justify replacing the summation by integration. For this reason a more exact evaluation of the partition function must be carried out by direct summation if the results are to be used over any appreciable range of reduced temperatures. In the case of hydrogen a further complication arises because of the existence of two kinds of hydrogen, ortho and para, which we shall discuss later.

EXERCISE:

Show that the number of molecules having rotational energies in the range from E to $E + dE$ is given by the expression

$$dN - \frac{N}{kT} e^{-E/kT} dE$$

in which N is the total number of molecules. (Compare with the second exercise on p. 256.)

EXERCISE:

The total angular momentum of a rotator without fixed axis is

$$p = \sqrt{K(K+1)}\, h/2\pi$$

Show that the number of molecules with angular momenta in the range from p to $p + dp$ will equal

$$dN = \frac{N}{IkT} e^{-p^2/2IkT} p\, dp$$

Effect of Molecular Symmetry on Rotational Partition Functions

You will recall that the assumption of distinguishability of molecules led to an entropy that was too great by $k \ln N!$. As pointed out earlier, this assumed distinguishability gives rise to a hypothetical entropy change when a gas is mixed with itself. Since this is an untenable point of view, we subtracted the term $k \ln N!$ from the translational entropy— a procedure equivalent to dividing the thermodynamic probability by $N!$, which is precisely the number of different ways in which N distinguishable objects can be taken.

A somewhat similar situation exists in connection with rotation. In this case, however, we are concerned with the indistinguishability that like atoms within a molecule have when they occupy equivalent positions with respect to rotational symmetry. To illustrate, let us consider a simple diatomic molecule, X_2. If this molecule is rotated about a twofold symmetry axis through $180°$, the resulting configuration is identical to that initially exhibited. If the atoms are truly indistinguishable, the new configuration will not be different from the old. Hence the thermodynamic probability, which is a count of the number of distinguishable configurations exhibited by a molecule, will be only half as great for a molecule of type X_2 as it would be for the diatomic molecule XY, in which the atoms are unlike. (See Fig. 13-1.) In other words, the process of rotating molecules of type X_2 about their twofold axes will result in no change in rotational entropy for the same reason that the process of interchanging locations of pairs of identical molecules results in no change of translational entropy. The rotational partition function ex-

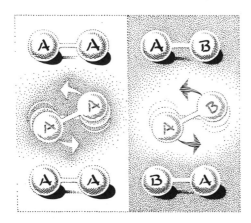

FIGURE 13-1 When a symmetrical molecule, A-A, is rotated through $180°$, the new configuration is indistinguishable from the old; on the other hand, when an unsymmetrical molecule, A-B, is similarly rotated, a distinguishable new configuration is obtained. Thus the total number of distinguishable configurations for A-A is only half as great as for A-B.

pressed by equations 13.17 and 13.20 must therefore be divided by 2 if the molecule is a symmetrical one of the type X_2. This same factor would occur for molecules like CO_2 and C_2H_2, which have two equivalent positions related to each other by simple rotation. The same correction would not apply to nitrous oxide, however, for it does not have a twofold symmetry axis perpendicular to the line of the molecule.

Other molecules, such as methane, have even greater symmetry numbers. Although we shall not here discuss in detail the rotational energy levels of nonlinear molecules, we shall point out that the methane molecule, which has the shape of a regular tetrahedron, has the symmetry number 12 because it can exhibit twelve rotationally related but essentially indistinguishable aspects.

The extra rotational entropy that symmetrical diatomic molecules would apparently have if we failed to take atomic indistinguishability into account would be of little consequence to us except in connection with the Third Law. Consider molecules of X_2 in a perfect crystal at absolute zero. If the two ends of the molecule were distinguishable from each other, but equivalent so far as interactions with their neighbors were concerned, the solid crystal would have, at absolute zero, an entropy equal to $Nk \ln 2$, for there are precisely 2^N ways of orienting N double-ended molecules in a crystalline lattice. (See Prob. 1, Chap. Five.) However, if we assign an entropy of zero to the crystal at absolute zero, we are compelled, in order to reduce the gaseous entropy by the necessary amount, to take into account the symmetry number in the gaseous state.

Taking cognizance of indistinguishability is a convenient although oversimplified method of accounting for an anomalous entropy. From the point of view of quantum mechanics, indistinguishability is really an outgrowth of certain symmetry requirements for the wave functions used to describe a molecule. (Similar arguments can also be advanced for translational phenomena, but the net result is the same as that obtained by stipulating indistinguishability of particles.) As a result of such considerations, the rotational partition function at high temperatures should be divided by a symmetry number, σ, which may equal 1 or 2 for linear molecules. This will, of course, diminish the calculated entropy by $R \ln \sigma$ and will correspondingly change the Helmholtz and Gibbs free energies. The energy, heat capacity, and equation of state, on the other hand, will be unaffected by the symmetry factor. Putting the symmetry factor into equation 13.22 and substituting a numerical value for $8\pi^2 k/h^2$, we obtain for the molar rotational entropy of a linear rotator the expression

$$S_{rot} = R \ln (IT/\sigma) + 89.4076\ R \qquad (13.24)$$

in which I is expressed in g cm^2. This can be reduced further, at 25 °C to

$$S_{rot} = 4.5757\ \lg (I/\sigma) + 188.993\ \text{cal deg}^{-1}\ \text{mol}^{-1} \qquad (13.25)$$

Although we shall not treat in detail the behavior of nonlinear rotators, we shall mention at this point some of the conclusions pertaining to them. If the three principal moments of inertia of a rigid nonlinear molecule are I_1, I_2, and I_3, the high-temperature partition function will be

$$q_{rot} = \frac{\sqrt{\pi}}{\sigma} \left(\frac{8\pi^2 I_1 kT}{h^2}\right)^{1/2} \left(\frac{8\pi^2 I_2 kT}{h^2}\right)^{1/2} \left(\frac{8\pi^2 I_3 kT}{h^2}\right)^{1/2} \qquad (13.26)$$

From this equation it can be readily established that the rotational energy is $\frac{3}{2}RT$ and the rotational heat capacity $\frac{3}{2}R$. If the values of the moments of inertia are expressed in c.g.s. units, the expression for the molar entropy becomes

$$S_{rot} = \frac{3}{2}R \ln T + R \ln \frac{(I_1 I_2 I_3)^{1/2}}{\sigma} + 134.684\ R \qquad (13.27)$$

At 25 °C this equation further reduces to

$$S_{rot} = 4.5757\ \lg \frac{(I_1 I_2 I_3)^{1/2}}{\sigma} + 284.627\ \text{cal deg}^{-1}\ \text{mol}^{-1} \qquad (13.28)$$

Ortho and Para Hydrogen

We mentioned above that hydrogen can exist in two different forms known as the ortho and para varieties. This difference arises because the nuclear spins of the two protons can be paired to give either a symmetric combination (spins parallel) or an antisymmetric combination (spins opposed). The total Schrödinger wave function for the description of protons must, however, assume an antisymmetric form. (This means that, if the coordinates of two protons are interchanged, the total wave function associated with them must change sign.) Now all the rotational levels that have even quantum numbers—$K = 0$, 2, 4, etc.—are symmetric, whereas those with odd quantum numbers—$K = 1$, 3, 5, etc.—are antisymmetric. The symmetric rotational functions must accordingly be combined with antisymmetric nuclear-spin functions to provide overall antisymmetry; on the other hand, the odd rotational levels must be combined with symmetric nuclear-spin functions. It turns out that there are three symmetric spin functions for the hydrogen molecule and only one antisymmetric combination; hence we see that the odd levels will have a statistical weight of three to one against the even levels.

The effect of nuclear spins on the magnitudes of rotational energy levels is entirely negligible, but the effect on quantum weights is profound. Hydrogen molecules in even rotational states are said to be para hydrogen, and those in odd states are called the ortho variety. In the absence of a catalyst such as charcoal, hydrogen molecules in odd rotational levels (ortho) can undergo transitions only to other odd levels. Similarly, molecules in even states (para) can experience transitions only to other even states. Taking into account the statistical weights, we can accordingly write the rotational partition functions for the two kinds of hydrogen as follows:

$$q_{para} = \sum_{K \; even} (2K + 1)e^{-\frac{K(K+1)h^2}{8\pi^2 IkT}} \qquad (13.29)$$

$$q_{ortho} = 3 \sum_{K \; odd} (2K + 1)e^{-\frac{K(K+1)h^2}{8\pi^2 IkT}} \qquad (13.30)$$

If the two kinds of hydrogen are in equilibrium with each other, a state that one can realize by keeping the material in the presence of a catalyst, the rotational partition function for all hydrogen will be the sum of the two functions above. At high temperatures, because of the weight factor of three in favor of the ortho variety, the equilibrium will result in three parts of ortho to one part of para. At low temperatures, the para variety will be favored, for all the molecules, upon sufficient cooling in the presence of a catalyst, will go to the zeroth rotational level, which is a particular state of para hydrogen.

From the foregoing discussion we see that we can speak about four kinds of hydrogen, pure para, pure ortho, the equilibrium mixture, and the three-to-one mixture. The three-to-one mixture is what we normally have; if it is cooled in the absence of a catalyst, it will remain a three-to-one mixture. The energy and heat capacity of that mixture will be the weighted averages of the separate quantities for the two kinds of hydrogen. To obtain the total entropy, on the other hand, one must augment the weighted average of the separate entropies by the entropy of mixing (of the ortho and para forms). The equilibrium mixture is, of course, what we should have if we kept hydrogen in contact with a catalyst that promoted the transition from one variety to the other. We can make pure para hydrogen by cooling hydrogen to low temperatures in the presence of such a catalyst. We might then, after removing the catalyst, heat the pure para hydrogen to room temperature. In Figure 13-2 we see graphs of the rotational heat capacities of the different kinds of hydrogen.

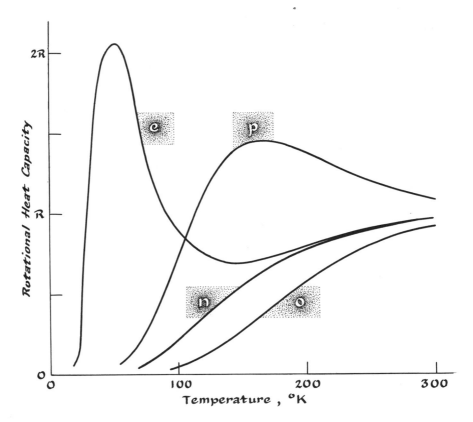

FIGURE 13-2   Rotational heat capacities for different kinds of hydrogen. The letter *o* denotes ortho hydrogen, *p* the para variety, *e* the equilibrium mixture, and *n* the normal 3-to-1 ortho-para mixture.

Deuterium also can exist in ortho and para forms; but, because the spin of a deuteron is different from that of a proton and because the total wave function for deuterons must be symmetric instead of anti-symmetric, the high-temperature ratio of ortho to para deuterium is not three to one. Computations verified by experiment show that the ortho-to-para ratio of deuterium at elevated temperatures is one to two.

At moderately high temperatures the effect of nuclear spins can be dealt with for most gases by introduction of appropriate symmetry factors. The only practical effect of nuclear spins on thermodynamic functions is on the quantum weights to be attached to the different levels, and this can generally be compensated for by geometrical symmetry considerations.

Thermodynamic Properties of Molecular Vibrators

Another type of energy possessed by diatomic and polyatomic molecules is energy of vibration. The vibration of a simple diatomic molecule is easily pictured; complex polyatomic molecules, on the other hand, can undergo numerous modes of vibration, each of which can be represented by a different pattern for the relative displacements of the atoms from their equilibrium positions. All such vibrations, however, can be reduced approximately to the equivalent of the vibration of a simple harmonic oscillator, which is the model that we shall discuss. A simple harmonic oscillator is most conveniently represented by a mass attached to a large immovable object by a weightless spring that obeys Hooke's Law. Diatomic molecules whose atoms can vibrate in relation to each other can, to a certain degree of approximation, be regarded as simple harmonic oscillators, provided one uses for the mass of the oscillator the so-called "reduced mass" of the pair of atoms.

The simple harmonic representation of the vibration of a diatomic molecule can be justified as follows. Let us consider the potential energy (Fig. 13-3) of a molecule AB in a particular electronic state as a function of the distance, r, between the atoms. It is assumed in the figure that

FIGURE 13-3   Potential energy of a diatomic molecule as a function of internuclear separation, r. D is the dissociation energy and D_0 the energy required for dissociation from the lowest vibrational level.

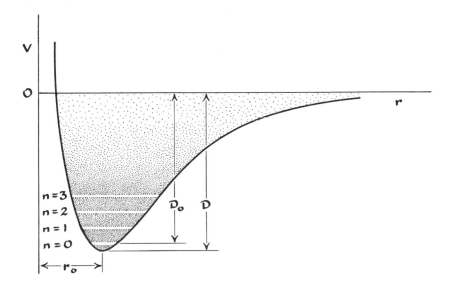

the potential energy is zero when the atoms are infinitely distant from each other. The stable configuration of the molecule is represented by the point corresponding to the minimum in the potential energy, which occurs at separation r_0, and the value of that minimum potential energy is $-D$ if D is the dissociation energy of the molecule. The vibration of a molecule can now be pictured as the rocking motion of a particle rolling up and down each side of the potential curve in the neighborhood of r_0, the equilibrium position. The higher the vibrational energy, the greater will be the amplitude of the vibration, or, in other words, the greater will be the periodic displacements from the equilibrium position.

If the energy of vibration is not too great (that is, if the molecule is not displaced a great deal from its equilibrium configuration), the potential energy curve will be approximately a parabola whose equation is

$$V = -D + \tfrac{1}{2}k(r - r_0)^2 \tag{13.31}$$

In this expression k is the Hooke's Law constant for the hypothetical spring holding the two particles together. From classical mechanics it can be shown that the natural vibrational frequency, ν_0, of a mass, m, subjected to Hooke's Law forces by a spring of constant k will be

$$\nu_0 = \frac{1}{2\pi}\sqrt{\frac{k}{m}} \tag{13.32}$$

In molecular-structure work it is usual to speak of characteristic vibrational frequencies rather than of Hooke's Law constants, but the two are obviously related.

When we deal with vibrational energy levels, it is customary to adopt a new reference point by setting the energy equal to zero for the atoms in their equilibrium positions; this is equivalent to dropping the dissociation energy, D, from equation 13.31. For simplicity's sake the quantity $r - r_0$, which measures displacement from equilibrium, is usually replaced by a new symbol that can be regarded as the vibrational coordinate. For polyatomic molecules there may be several vibrational frequencies, in which case the relative displacements of the various atoms are related to certain "normal coordinates," of which there will be one for each mode of vibration.

As long as $r - r_0$ is not too large, the parabolic representation of the potential energy can be used without serious error. If we assume the validity of such a potential function, it is possible to determine by quantum mechanics the allowable energy levels for a simple harmonic oscillator. These energy levels, measured from the bottom of the potential energy curve, are given by the expression

$$E_n = (n + 1/2)h\nu_0 \tag{13.33}$$

In this equation n must take on values 0, 1, 2, 3, etc.; h is Planck's constant; and ν_0 is the classical frequency of vibration. The term $\frac{1}{2}h\nu_0$ is called the "zero-point energy," for it equals the energy of the oscillator in its lowest vibrational state. The zero-point vibrational energy is a remarkable manifestation of quantum mechanics, for it has no basis for existence in the classical view. Because of the zero-point energy, the effective dissociation energy for diatomic molecules, usually denoted by D_0, is equal to $D - \frac{1}{2}h\nu_0$, this quantity representing the work that must be done in completely dissociating a diatomic molecule from its lowest vibrational state. Several vibrational levels for a typical diatomic molecule are represented by the horizontal lines in Figure 13-3. For a one-dimensional oscillator the vibrational levels are all nondegenerate; that is, $g_n = 1$.

With a knowledge of the vibrational energy levels of a simple harmonic oscillator it is easy to evaluate the partition function for a vibrating molecule. Making use of equation 13.33, we obtain the expression

$$q_{\text{vib}} = \sum_{n=0}^{\infty} e^{-(n+\frac{1}{2})h\nu_0/kT} \tag{13.34}$$

If we now represent the quantity $h\nu_0/kT$ by the letter x, the partition function can be rewritten as follows:

$$q_{\text{vib}} = e^{-x/2}\{1 + e^{-x} + e^{-2x} + e^{-3x} + \ldots\} \tag{13.35}$$

Since e^{-x} must always be less than unity, the infinite series appearing above can be represented exactly in the following closed form:

$$q_{\text{vib}} = \frac{e^{-x/2}}{1 - e^{-x}} = \frac{e^{-h\nu_0/2kT}}{1 - e^{-h\nu_0/kT}} \tag{13.36}$$

Introducing the vibrational partition function into equations 13.8, 13.9, and 13.10, we find for the energy, entropy, and Helmholtz energy of vibration the following expressions, valid for one mole of material:

$$U_{\text{vib}} = RT\left(\frac{x}{2} + \frac{x}{e^x - 1}\right)$$

$$= \frac{N_A h\nu_0}{2} + \frac{RTx}{e^x - 1} \tag{13.37}$$

$$S_{\text{vib}} = -R\ln(1 - e^{-x}) + \frac{Rx}{e^x - 1} \tag{13.38}$$

$$A_{\text{vib}} = RT\left[\frac{x}{2} + \ln(1 - e^{-x})\right]$$

$$= \frac{N_A h\nu_0}{2} + RT\ln(1 - e^{-x}) \tag{13.39}$$

For the vibrational contribution to the heat capacity we need only differentiate the energy with respect to temperature to obtain the expression

$$C_{vib} = \frac{Rx^2 e^x}{(e^x - 1)^2} \tag{13.40}$$

At room temperatures the entropy and heat capacity of vibration will be negligible for all but very low frequencies; however, the total zero-point energy, $N_A h\nu_0/2$, will be appreciable. Although the zero-point energy appears in the expressions for the energy and the Helmholtz function, it does not affect the entropy. At absolute zero, for example, the total energy equals the zero-point energy, but the entropy is zero because the thermodynamic probability is equal to unity. Finally, since the vibrational partition function does not depend upon the volume, we see that vibrations will make no contribution to the equation of state; the vibrational Gibbs energy therefore equals the vibrational Helmholtz energy.

We were able to evaluate the vibrational partition function exactly by assuming the correctness of the energy levels expressed by equation 13.33, and we did not need to replace the summation by an integral. If we approximated to q_{vib} by integration, however, as we did for translation and vibration, we should obtain the following:

$$q_{vib} \approx \frac{kT}{h\nu_0} e^{-h\nu_0/2kT} = \frac{1}{x} e^{-x/2} \tag{13.41}$$

We can also obtain this result from equation 13.36 by taking the limiting form of the expression for large values of $kT/h\nu_0$. From equation 13.41 it is possible to establish high-temperature values for the energy, entropy, Helmholtz energy, and heat capacity; in this way it can be shown that the molar heat capacity of vibration approaches R at high temperatures. The approximation represented by equation 13.41 is very poor for most molecular vibrations at room temperatures, even though its equivalent was all right for rotations and translations.

In the preceding treatment of a simple harmonic oscillator we found it convenient to measure the energy from the minimum of the potential energy curve. When we make thermodynamic calculations, however, especially of equilibrium constants, it is important that we adopt a consistent reference level for the energies of all the molecules involved. We can accomplish this most easily by assigning zero energy to the stationary atoms that are obtained by complete dissociation of the molecules. For a diatomic molecule this calls for restoration of the dissociation energy, D, into our expressions; we accomplish this readily by replacing the zero-point energy by $-D + \frac{1}{2}h\nu_0$. If the effective dissociation energy, $D - \frac{1}{2}h\nu_0$, is represented by D_0, the partition function and the energy then become

$$q_{\text{vib}} = \frac{e^{D_0/kT}}{1 - e^{-x}} \tag{13.42}$$

$$U_{\text{vib}} = U_0 + \frac{RTx}{e^x - 1} \tag{13.43}$$

where U_0 ($= -N_A D_0$) is the negative of the molar bond energy or dissociation energy of the molecule at 0 K. The Helmholtz and Gibbs free energies will be affected in a similar manner, but the entropy and heat capacity will, of course, be unchanged by this shift in reference level.

For a polyatomic molecule, the total bond energy must be introduced and a partition function set up for each of the characteristic vibrations. The total vibrational partition function will then be a product of the separate functions, and the total energy, entropy, etc. will be the sum of the separate contributions.

Electronic Partition Functions

Our applications of statistical mechanics so far have all involved energies attributable to atomic motions or, more precisely, to motions characterized by the velocities of atomic nuclei. We shall now turn to the question of energies within atoms, specifically to electronic energies, which to a certain extent can be treated without reference to any nuclear motions. You will recall that the possible energy levels of translation are so close together that translational excitation is realized even at very low temperatures. For rotation, too, the excitation proved to be complete for most substances at room temperature, although molecules like hydrogen show some dependence of heat capacity on temperature at moderately low temperatures. Most vibrational energy levels, on the other hand, are sufficiently spread out so that there is little vibrational excitation at room temperatures; as a result, the vibrational contribution to heat capacity is almost negligible except for very low frequencies.

We now come to the problem of electronic excitation, and here we find that the energy spacings are of the order of an electron volt, which is equivalent to 23 kilocalories per mole. Since this is very much larger than RT at ordinary temperatures, we can, when calculating the electronic partition function, usually neglect atomic populations in all except the ground state. Of course, if one had a gas at a temperature of several thousand degrees Kelvin, a significant fraction of the atoms might be found in excited states; at such temperatures any ordinary molecules would dissociate.

If E_0 is the energy and g_0 the corresponding degeneracy of the electronic ground state of an atom, and if the next higher level differs from that of E_0 by an amount considerably greater than kT, we can approximately express the electronic partition function by the equation

$$q_{el} = g_0 e^{-E_0/kT} \qquad (13.44)$$

Substitution of this partition function into the appropriate equations for the thermodynamic functions of interest will confirm the obvious conclusion that the energy of N such atoms will be precisely NE_0 and that the entropy will be $Nk \ln g_0$. The heat capacity of electronic excitation will, of course, be zero under the assumptions made. If we choose to measure electronic levels from the ground state (in other words, if we set E_0 equal to zero), the total electronic energy will be zero, but the molar entropy will remain equal to $R \ln g_0$. From a practical point of view, so far as thermodynamics is concerned, all we need to know about electronic levels is g_0, the degeneracy of the ground state. This quantity is determined by the total angular momentum of the electrons, which is represented by a vector whose magnitude is proportional to a number, j, and which can assume $2j + 1$ different orientations in space. This number of possible orientations, $2j + 1$, is precisely the degeneracy of the level. The value of j is specified as a subscript in the Russell-Saunders term representation employed by spectroscopists to characterize atoms in their various states. The ground state of atomic hydrogen, for example, is represented by the term $^2S_{1/2}$, from which it will be seen that j equals $1/2$. Hence the degeneracy of a hydrogen atom in its ground state will be equal to 2. The ground state of chlorine, on the other hand, is represented by the term $^2P_{3/2}$; since j is equal to $3/2$, the degeneracy equals 4.

The preceding discussion provides a sufficient basis for carrying out most of the calculations that are ordinarily required. The investigator should assure himself on every occasion, however, that all the assumed conditions are fulfilled. If, for example, close-lying electronic energy levels are accidentally encountered, or if the calculations are made for exceedingly high temperatures, a more complete partition function should be set up along the general lines established earlier. More than two terms would rarely be required; and, of course, one could never hope in practice to approximate to the electronic partition function by integration.

Calculation of Equilibrium Constants

Having established methods of calculating thermodynamic quantities from molecular-structure data, we are now in a position to extend the methods toward the evaluation of equilibrium constants for chemical reactions involving ideal gases. All that is required, in essence, is the calculation of standard Gibbs free energies or chemical potentials, appropriate combinations of which will yield the desired constants. To illustrate, let us consider a simple dissociation reaction such as

$$A_2(g) \rightleftharpoons 2A(g)$$

The reaction potential for this dissociation will be

$$\Delta\tilde{\mu} = 2\mu_A - \mu_{A_2} \tag{13.45}$$

in which μ_A and μ_{A_2} are chemical potentials. Under the assumed conditions of gaseous ideality, these partial molar quantities will be equal to the molar Gibbs energies of the pure gases at the partial pressures of interest. In view of equation 13.15, this reaction potential can be rewritten as

$$\Delta\tilde{\mu} = -RT \ln \frac{(q_A/N)^2}{(q_{A_2}/N)} \tag{13.46}$$

provided the q's are the partition functions and N is Avogadro's number. Now q_A will simply equal the translational partition function of atom A multiplied by g_0, the degeneracy of its electronic ground state. In the light of equation 13.13 and the equation of state for an ideal gas, therefore, we see that

$$\frac{q_A}{N} = \frac{(2\pi m_A)^{3/2}(kT)^{5/2}g_0}{p_A h^3} \tag{13.47}$$

if m_A is the mass of one atom of A. Q_{A_2}, on the other hand, will involve contributions from rotation and vibration as well as translation. (We shall assume that the degeneracy of the electronic ground state for the molecule is unity, as it usually is.) It follows that

$$\frac{q_{A_2}}{N} = \frac{(4\pi m_A)^{3/2}(kT)^{5/2}}{p_{A_2} h^3} \cdot \frac{8\pi^2 I k T}{\sigma h^2} \cdot \frac{e^{D_0/kT}}{1 - e^{-h\nu_0/kT}} \tag{13.48}$$

in which expression the various symbols represent quantities defined earlier and σ, the symmetry number, equals 2. A combination of equations 13.47 and 13.48 with 13.46 will now yield

$$\Delta\tilde{\mu} = -RT \ln \frac{g_0^2(m_A kT)^{3/2}(1 - e^{-h\nu_0/kT})e^{-D_0/kT}}{4\sqrt{\pi}Ih}$$

$$+ RT \ln \frac{p_A^2}{p_{A_2}} \tag{13.49}$$

If the partial pressures are set equal to unity, the reaction potential will assume its standard value, which is equal to $-RT \ln K_{eq}$. Therefore

$$K_{eq} = \frac{g_0^2 (m_A k T)^{3/2} (1 - e^{-h\nu_0/kT}) e^{-D_0/kT}}{4\sqrt{\pi} Ih} \qquad (13.50)$$

Since K_{eq}, as written, has the dimensions of pressure, it will be expressed, in the c.g.s. system of units, in dynes per square centimeter; one must introduce an appropriate factor to convert it to atmospheres.

Unless the vibrational frequency, ν_0, of the diatomic molecule is quite low, the term $e^{-h\nu_0/kT}$, compared with unity, can be neglected at moderate temperatures. If we neglect vibrational excitation, we can write for $\ln K_{eq}$ the following:

$$\ln K_{eq} = \ln \frac{g_0^2 (m_A k)^{3/2}}{4\sqrt{\pi} Ih} + \tfrac{3}{2} \ln T - \frac{\Delta U_0}{RT} \qquad (13.51)$$

In this expression we have replaced D_0/kT by $\Delta U_0/RT$, thus converting to molar units. (In this instance ΔU_0 is simply equal to the molar bond energy of the diatomic molecule.) Upon differentiating $\ln K_{eq}$ with respect to T, we obtain the expression

$$\frac{d \ln K_{eq}}{dT} = \frac{\Delta \tilde{H}}{RT^2} = \frac{3}{2T} + \frac{\Delta U_0}{RT^2} \qquad (13.52)$$

The enthalpy of the dissociation reaction will therefore be

$$\Delta \tilde{H} = \tfrac{3}{2} RT + \Delta U_0 \qquad (13.53)$$

This enthalpy change is really made up of a combination of the following terms: (a) ΔU_0, the energy of dissociation at 0 K; (b) $2 \cdot \tfrac{3}{2} RT$, the energy of translation of $2A(g)$; (c) $-(\tfrac{3}{2}RT + RT)$, the energies of translation and rotation of $A_2(g)$; and (d) $\Delta(pV) = RT$, which is necessary for the conversion of ΔU to ΔH.

As another example for the calculation of an equilibrium constant, consider the reaction

$$A_2(g) + B_2(g) \rightleftharpoons 2AB(g)$$

For this reaction we can evidently write

$$\Delta \tilde{\mu} = -RT \ln \frac{q_{AB}^2}{q_{A_2} q_{B_2}} \qquad (13.54)$$

Proceeding as before, we find that

$$K_{eq} = \frac{M_{AB}^3 I_{AB}^2 \sigma_{A_2} \sigma_{B_2}}{M_{A_2}^{3/2} M_{B_2}^{3/2} I_{A_2} I_{B_2} \sigma_{AB}^2} e^{-\Delta U_0/RT} \qquad (13.55)$$

In writing this equation, we have again neglected any vibrational excitation; $\sigma_{AB} = 1$, $\sigma_{A_2} = \sigma_{B_2} = 2$, and the M's are molecular weights.

Since $\Delta(pV) = 0$ for the reaction and K_{eq} is therefore nondimensional, only ratios of masses and ratios of moments of inertia are required in the expression. ΔU_0 is the net change in total bond energy for the reaction, and ΔH obviously equals ΔU_0 under the conditions assumed. For a more precise calculation, of course, one should include the effect of temperature on vibration.

For a general reaction of the type

$$aA + bB + \ldots \rightleftharpoons eE + fF + \ldots$$

in which A, B, $\ldots$ and E, F, $\ldots$ represent any gaseous molecules, monatomic or polyatomic, the equilibrium constant can be written as

$$K_{eq} = \frac{q_E^e q_F^f \ldots N^{-\Delta\nu} e^{-\Delta U_0/RT}}{q_A^a q_B^b \ldots} \qquad (13.56)$$

in which expression N is Avogadro's number, $\Delta\nu = (e + f + \ldots) - (a + b + \ldots)$, ΔU_0 is the change in total bond energy, and the q's are the molecular partition functions evaluated at unit pressures but excluding those portions of the vibrational contributions that are attributable to the U_0's, which are taken care of by the term $e^{-\Delta U_0/RT}$.

Statistical Mechanics of Nonideal Gases; Excluded Volume

The preceding statistical-mechanical treatment of gases is based on the assumption that the molecules are not subject to any intermolecular interactions; in other words, the molecules act like point particles that exert no forces on their neighbors. To be sure, we took into account such matters as rotation and vibration, but these were assumed to be strictly intramolecular in nature and to have no effect on the translational properties. It should be recognized, however, that actual molecules do exert forces on other molecules, such forces affecting, more or less, all modes of motion. In an effort to describe partially the behavior of actual, nonideal gases, we shall now consider the effect of intermolecular forces on the translational partition function of a gas, thus establishing a basis for calculating a modified equation of state.

To illustrate our method, we shall first consider only the effect of molecular size on the equation of state; this is approximately equivalent to finding van der Waals' b. Our molecular model will consist of N hard spheres of radius r_0, which interact with each other only when they touch. It is also assumed that the molecules are strictly nondeformable and that collisions between them are perfectly elastic.

In arriving at the equation of state of an ideal gas, using equation 13.11, we saw that the only part of the partition function that contributed to the final answer was the volume. As far as the spatial coordinates are concerned, the partition function for a single molecule can simply be taken to be the volume, V, which represents that portion of the partition function attributable to integration over the coordinates, but not including that arising from the momenta. Moreover, if the coordinate part of the partition function for a single molecule is V, then the partition function for N molecules will be V^V since, assuming that we deal with point particles, each molecule will have the full volume available.

What will be the effect of finite molecular size? To answer this question, let us imagine placing molecules one by one into a container of volume V and then determining the effect of each addition on the partition function. We may assume that the first molecule put into the container has available to it the entire volume if we overlook the constant surface effect that makes it impossible for the center of a molecule to get closer to the surface than the distance r_0. For this reason we can say that the partition function, q_1, for the first molecule will simply equal V, the ideal value. The second molecule, on the other hand, will not have the full volume available, since the presence of the first molecule excludes some spatial possibilities. If the spheres are perfectly hard, the center of the second molecule can never get closer than $2r_0$ to the center of the first. Therefore, insofar as the second molecule is concerned, the first molecule appears to exclude a volume $8v$, provided $v = (\frac{4}{3})\pi r_0^3$. Hence q_2 will equal $V - 8v$ and the partition function for two molecules will be $V(V - 8v)$. Regardless of the order in which the two molecules are placed in the container, the answer will be the same. It is important to recognize, however, that the partition function for two molecules is not the square of any simple function for one molecule.

When we get to the third molecule, we find that it will be excluded from $2 \times 8v$, or $16v$. (This neglects a small correction for possible overlap of the spheres of influence of the two first molecules. If the first two molecules have their centers separated by a distance in the range $2r_0$ to $4r_0$, then $16v$ is too large, but this error would only involve a higher order correction that we shall neglect.) Hence q_3 equals $V - 16v$ and the partition function for three molecules becomes $V(V - 8v)(V - 16v)$. Continuing this process indefinitely, we see that the partition function for N molecules will be

$$Q = V(V - 8v)(V - 16v) \ldots \ldots (V - [N - 1]8v)$$

$$= \prod_{i=0}^{N-1} (V - 8iv) \tag{13.57}$$

For a gaseous system, the term $8iv$ will be small compared to V even for the largest value of i, which is $N - 1$. At most, the correction term will be $8[N - 1]v$, or roughly eight times the volume of the condensed material, which should be small compared to the volume of the gas.

To obtain the equation of state for a gaseous system of hard spheres, we shall employ equation 13.11 in the form involving the total partition function Q. (See also equation 13.2.) Evidently

$$\ln Q = \sum_{i=0}^{N-1} \ln (V - 8iv) \tag{13.58}$$

Therefore

$$p = kT \left(\frac{\partial \ln Q}{\partial V}\right)_T$$

$$= kT \sum_{i=0}^{N-1} \frac{1}{V - 8iv} \tag{13.59}$$

But since $8iv$ is small compared to V, we can write that, approximately

$$\frac{1}{V - 8iv} = \frac{1}{V}\left(1 + \frac{8iv}{V}\right) \tag{13.60}$$

Substituting equation 13.60 into equation 13.59, we obtain

$$p = kT \sum_{i=0}^{N-1} \left(\frac{1}{V} + \frac{8iv}{V^2}\right) \tag{13.61}$$

But we know that $\sum_{i=1}^{N-1} i = N(N - 1)/2$. Hence

$$p = \frac{NkT}{V}\left(1 + \frac{4[N - 1]v}{V}\right) \tag{13.62}$$

Neglecting unity compared to N, we see that the relative correction term for nonideality is simply four times the ratio of the total volume of the hard spheres to the total volume of the gas. Now, if we omit the constant a from van der Waals' equation, we see that for one mole of a van der Waals gas,

$$p = \frac{RT}{V - b} \approx \frac{RT}{V}\left(1 + \frac{b}{V}\right) \tag{13.63}$$

Comparing equations 13.62 and 13.63, we conclude that van der Waals' b is just four times the volume of the molecules.

The foregoing treatment, although limited to the effect of volume only, was carried out to illustrate the method. In particular, the reader should observe that the volume part of the partition function, equation 13.57, is not simply derived by raising to the power N a function that is assignable to a single molecule. Having demonstrated this method for building up the partition function, we shall now extend it to the more general case in which intermolecular forces occur at distances greater than that of nearest approach; in this way we hope to learn something about van der Waals' a.

Statistical Mechanics of Nonideal Gases; Effect of Intermolecular Forces in Addition to Excluded Volume

In dealing with nonideal gases subject to intermolecular forces operating at distances greater than that of nearest approach, we shall assume once again that the molecules are spherically symmetric. We shall further assume that a pair of molecules, whose centers are separated by a distance r, are subject to an intermolecular potential energy as follows:

$$E = c/r^m \text{ for } r \geqslant 2r_0$$
$$E = \infty \text{ for } 0 < r < 2r_0 \tag{13.64}$$

This potential energy implies that the molecules behave like hard spheres when $r = 2r_0$, the distance of nearest approach. For larger distances, however, the potential energy decreases as an inverse power of r, the exponent m being so large that the energy becomes negligible when r equals several times the distance of nearest approach (see Fig. 13-4). Actually, for reasons that will appear later, m must be greater than 3; probably it is close to 6, which is the value for so-called London forces. If the value of the potential energy at the distance of nearest approach is $-E_0$, then evidently $-E_0 = c/(2r_0)^m$, so that for $r \geqslant 2r_0$

$$E = -E_0(2r_0/r)^m \tag{13.65}$$

The partition function for N molecules in a volume V subject to the forces described above can now be developed by envisioning the successive addition of molecules to the container. Evidently the first molecule that is introduced will contribute the factor $q_1 = V$ to the partition function, since it will encounter no interactions. Upon adding the second molecule, however, the partition function will be multiplied by

$$q_2 = \int e^{-E/kT} dV \tag{13.66}$$

in which E is the intermolecular potential energy and the integration is carried out over the total volume V. Since E is infinite for $0 < r < 2r_0$, it follows that a volume $8v$ will be automatically excluded as demon-

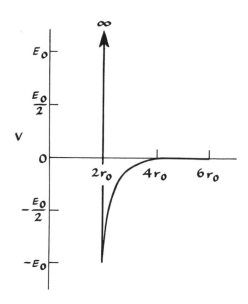

FIGURE 13-4   Potential energy of interaction as function of distance between two
molecules according to equations 13.64 and
13.65 with $m = 6$.

strated in the preceding section; the variability of E with distance for
$r > 2r_0$ will, however, alter the nature of the integral from that for
simple hard spheres.

To provide for the limits of integration in equation 13.66, let us
rewrite the equation as follows:

$$q_2 = \int dV + \int (e^{-E/kT} - 1)\, dV \qquad (13.67)$$

The first integral simply yields the total volume V, which would, of
course, be the entire contribution if E were zero everywhere. To evaluate
the second integral, let us place the first molecule at the origin of the
coordinate system and replace dV by $4\pi r^2 dr$. Then, since E is assumed to
be infinite for $r < 2r_0$, the second integral will yield $-32\pi r_0^3/3$ or $-8v$
for the range $0 < r < 2r_0$. Thus we obtain

$$q_2 = V - 8v + \int_{2r_0}^{\infty} (e^{-E/kT} - 1)4\pi r^2 dr \qquad (13.68)$$

The lower limit of the remaining integral is $2r_0$, since we have already
taken into account the excluded volume effect, and the upper limit can

be considered infinite since the integrand vanishes long before r is large enough to approach the limits of the container. Let us now assume that E_0, which represents the largest absolute value for E, is small compared with kT. Then, if we retain only the first two terms in the expansion of the exponential function, we can write

$$q_2 = V - 8v - \int_{2r_0}^{\infty} \frac{E}{kT} 4\pi r^2 dr \qquad (13.69)$$

or, upon substituting for E from equation 13.65,

$$q_2 = V - 8v + \frac{4\pi E_0 (2r_0)^m}{kT} \int_{2r_0}^{\infty} r^{-m+2} dr \qquad (13.70)$$

$$= V - 8v + \frac{24vE_0}{(m-3)kT}$$

It should now be clear why m must be greater than 3, for if it were not so, the integral would diverge.

When the third molecule is placed in the container, it will be subject to the excluded volume of two molecules, or $16v$, and outside the excluded volume it will interact with the first two molecules to make the temperature-dependent part of the volume correction equal to $48vE_0/(m-3)kT$. Generalizing this argument we see that the ith molecule will contribute to the partition function a factor, q_i, given by

$$q_i = V - 8iv + \frac{24ivE_0}{(m-3)kT} \qquad (13.71)$$

The total partition function will be

$$q = \prod_{i=0}^{N-1} q_i = \prod_{i=0}^{N-1} \{V - 8iv[1 - 3E_0/(m-3)kT]\} \qquad (13.72)$$

The equation of state can now be directly determined. Proceeding as we did in deriving equation 13.62, it will be seen that

$$p = kT \sum_{i=0}^{N-1} \frac{1}{V - 8iv[1 - 3E_0/(m-3)kT]}$$

$$\approx kT \sum_{i=0}^{N-1} \frac{1}{V} \left\{ 1 + \frac{8iv[1 - 3E_0/(m-3)kT]}{V} \right\}$$

$$\approx \frac{NkT}{V} \left\{ 1 + \frac{4(N-1)v[1 - 3E_0/(m-3)kT]}{V} \right\} \qquad (13.73)$$

Let us now compare equation 13.73 with van der Waals' equation expanded through its second virial term, namely

$$p = \frac{RT}{V} \left[1 + \frac{(b - a/RT)}{V} \right] \qquad (13.74)$$

Evidently $4Nv$ can be identified once again as b, neglecting unity compared to N; moreover we also recognize that a will be given by

$$a = \frac{12N^2vE_0}{m-3} = \frac{3NbE_0}{m-3} \qquad (13.75)$$

(N must equal Avogadro's number if a and b are applicable to one mole of gas.) From these results, we conclude that a/b, which should equal the heat of vaporization of a van der Waals liquid (see eq. 8.44), will be given by

$$\Delta H_{\text{vap}} = \frac{a}{b} = \frac{3NE_0}{m-3} \qquad (13.76)$$

In particular, if $m = 6$ (London forces), then the heat of vaporization is simply equal to NE_0, which is certainly of reasonable magnitude. We also see, from equations 8.33 in conjunction with the conclusion of this section that

$$p_c = \frac{E_0}{36(m-3)v}$$

$$V_c = 12\,Nv$$

$$kT_c = \frac{8E_0}{9(m-3)} \qquad (13.77)$$

The preceding treatment, although only approximate, does give an insight into the nature of gaseous nonideality. That it gives rise to a familiar form for the second virial coefficient is gratifying, but it must be remembered that the derivation involved a great many approximations. Nevertheless, application of the theory to actual systems should give some understanding of the order of magnitude of the potential energies of molecular interactions in relation to molecular dimensions. At best, however, it represents only a starting point for a most complicated kind of calculation.

Einstein's Theory of Heat Capacities of Solids

The methods of statistical mechanics that we have found useful in describing the thermodynamic behavior of ideal gases can also be employed with considerable success in connection with solid crystalline substances. Let us consider, for example, a crystal containing N atoms of an element arranged in a definite lattice structure. A simple dynamic model for describing that system was proposed by Einstein, who suggested that such a crystal be regarded as consisting of N three-dimensional isotropic harmonic oscillators, each with the same characteristic vibrational frequency, ν. As long as the vibration of an atom in a direction parallel to a particular axis is unperturbed by vibrations in directions perpendicular to the axis, the partition function for an isotropic three-dimensional oscillator will simply equal the cube of the partition function for a one-dimensional oscillator. The partition function for a one-dimensional oscillator is given by equation 13.36, so we can write for one atom (or ion) undergoing three-dimensional oscillations that

$$q = \frac{e^{-3x/2}}{(1 - e^{-x})^3} = \frac{e^{-3h\nu/2kT}}{(1 - e^{-h\nu/kT})^3} \qquad (13.78)$$

provided $x = h\nu/kT$. From equation 13.78 we deduce that the several thermodynamic functions of interest (energy, entropy, heat capacity, etc.) will be exactly $3N$ times what we should expect for a one-dimensional oscillator of frequency ν. Specifically, we can write for one mole that

$$U = 3RT \left(\frac{x}{2} + \frac{x}{e^x - 1} \right)$$
$$= \frac{3N_A h\nu}{2} + \frac{3RTx}{e^x - 1} \qquad (13.79)$$

The term $3N_A h\nu/2$ occurs once again as the zero-point energy, a quantity that will have no effect on the heat capacity or entropy. The heat capacity, C_v, will be

$$C_v = \left(\frac{\partial U}{\partial T} \right)_v = \frac{3Rx^2 e^x}{(e^x - 1)^2} \qquad (13.80)$$

As the absolute temperature approaches zero, C_v likewise approaches zero, as it must if the entropy is to remain finite at absolute zero. For high temperatures, on the other hand, $x \longrightarrow 0$, $U \longrightarrow 3RT$, and $C_v \longrightarrow 3R$. The limiting value of $3R$ for the molar heat capacity is in essential agreement with the observations of Dulong and Petit, who found atomic heat capacities (at constant pressure) to be about 6.2 cal deg^{-1}.

EXERCISE:

Show that, as $x \longrightarrow 0$, the expression of equation 13.79 becomes equal to $3RT$ plus zero point energy.

Debye's Theory of Heat Capacities of Solids

Although the Einstein formula for the heat capacity of a crystal possesses most of the desired qualitative features, it fails to describe adequately the behavior of crystals near absolute zero. To improve upon Einstein's model, Debye proposed that a crystal be treated like a giant molecule possessing many different vibrational frequencies ranging almost continuously from zero up to a maximum value, ν_{max}. The degeneracy associated with a frequency will equal the number of different ways in which standing waves of the corresponding wavelength can be realized within the crystal of interest. It follows that low frequencies, which are associated with long wavelengths, will have low degeneracies; the longer the waves, the more difficult it is to place a half-integral number of them in a crystal. High frequencies, on the other hand, will correspond to low standing wavelengths, and such waves can be arranged in a crystal in numerous ways. Making use of classical vibration theory, Debye recognized that the degeneracies associated with a range of frequencies from ν to $\nu + d\nu$ will be proportional to $\nu^2 d\nu$.* The total number of vibrational degrees of freedom in a crystal consisting of N atoms will be $3N$, and this must be precisely the same as the sum total of all of the frequencies, weighted by their degeneracies. Hence, if we let dg equal $C\nu^2 d\nu$, we can assert that

$$3N = \int_0^{\nu_{max}} C\nu^2 d\nu = \frac{C\nu^3_{max}}{3} \qquad (13.81)$$

From this we see that the coefficient C must equal $9N/\nu^3_{max}$, from which we conclude that

$$dg = \frac{9N\nu^2 d\nu}{\nu^3_{max}} \qquad (13.82)$$

Unlike Einstein, who assumed a single vibrational frequency, Debye recognized that a crystal can exhibit numerous modes of vibration with the degeneracies specified above. Hence the total energy of a crystal, according to Debye, will be obtained by adding together the energies associated with all of the various vibrational frequencies, properly weighted by their degeneracies. From equation 13.37 we see that the average energy of an oscillator of a single degree of freedom and a frequency ν is

$$\langle E_\nu \rangle = \frac{h\nu}{2} + \frac{h\nu}{e^{h\nu/kT} - 1} \qquad (13.83)$$

Multiplying $\langle E_\nu \rangle$ by dg and integrating, we obtain the total vibrational energy of a crystal as follows:

---

* You will recall that the degeneracy associated with translational energy levels for a particle in a box is equal to $\frac{1}{2}\pi n^2\, dn$ if n represents the total quantum number. To find an acceptable solution of Schrödinger's wave equation for a particle in a box, one had to fit the equivalent of standing waves into a box; the degeneracy for a particle in a box turns out to have the same kind of dependence on the wavelength of the Schrödinger wave function as does the degeneracy of vibrational waves in a crystal. The quantum number n is proportional to the frequency of a de Broglie (Schrödinger) wave.

$$U = \int_0^{\nu_{max}} \left(\frac{h\nu}{2} + \frac{h\nu}{e^{h\nu/kT} - 1} \right) \frac{9N\nu^2}{\nu_{max}^3} \, d\nu \qquad (13.84)$$

The first term of the integrand can be readily integrated to give the total zero-point energy of the crystal, $9Nh\nu_{max}/8$. We can simplify the second term, as we did earlier, by letting x equal $h\nu/kT$. Thus we see that

$$U = \frac{9Nh\nu_{max}}{8} + \frac{9N(kT)^4}{(h\nu_{max})^3} \int_0^{\frac{h\nu_{max}}{kT}} \frac{x^3 \, dx}{e^x - 1} \qquad (13.85)$$

We shall now define what is known as a "characteristic temperature," θ, by the simple equation

$$\theta = \frac{h\nu_{max}}{k} \qquad (13.86)$$

In view of this definition, we see, with N equal to Avogadro's number, that

$$U = \frac{9R\theta}{8} + \frac{9RT^4}{\theta^3} \int_0^{\theta/T} \frac{x^3 \, dx}{e^x - 1} \qquad (13.87)$$

The integral appearing in this equation cannot be expressed in simple closed form, except approximately for small values of T/θ; for large values of T/θ it can be evaluated by recourse to a series expansion. If T/θ is sufficiently small, the upper limit of the integral will be so large that the contribution of the integrand for values of x in excess of θ/T will be negligible. Hence for small values of T/θ we can approximate to the integral by letting the upper limit become infinite. If we denote that integral by the letter I, we see that

$$I = \int_0^\infty \frac{x^3 \, dx}{e^x - 1} = \int_0^\infty \frac{x^3 e^{-x} \, dx}{1 - e^{-x}}$$

$$= \int_0^\infty x^3 (e^{-x} + e^{-2x} + e^{-3x} + \ldots) dx$$

$$= 6 \left(\frac{1}{1^4} + \frac{1}{2^4} + \frac{1}{3^4} + \ldots \right) \qquad (13.88)$$

It can be shown, moreover, by consideration of certain Fourier series, that the infinite series, $\sum_{i=1}^{\infty} 1/i^4$, converges to $\pi^4/90$. Hence

$$I = \frac{6\pi^4}{90} = \frac{\pi^4}{15} \qquad (13.89)$$

Substituting this value of I for the integral in equation 13.87, we find for small values of T/θ that

$$U = \frac{9R\theta}{8} + \frac{3\pi^4 R T^4}{5\theta^3} \tag{13.90}$$

Differentiating the energy with respect to T, we see that

$$C_v = \frac{12\pi^4 R}{5}\left(\frac{T}{\theta}\right)^3 = 464.6 \left(\frac{T}{\theta}\right)^3 \text{ cal deg}^{-1}\text{ mol}^{-1} \tag{13.91}$$

This equation was given earlier (as equation 5.13) for use in connection with Third Law calculations, and a brief discussion of its applicability appears in Chapter Five.

In letting the upper limit of the integral of equation 13.87 become infinite, we introduced an approximation that was convenient but really not necessary. Without much difficulty we can use the finite upper limit in evaluating the integral indicated in equation 13.88. In practice, however, because of the computational problems involved, the exact expression so obtained is of little use for high-temperature calculations; for low temperatures, of course, the approximation is a good one.

To obtain a useful expression for the calculation of heat capacities of solids at high temperatures, we shall evaluate the integral of equation 13.87 by another means. First, we observe that for large values of T/θ the upper limit of integration becomes small; hence, if we expand the integrand as a power series in x, we can expect to get a series that will converge rapidly under the assumed conditions. The easiest way to proceed, for our purpose, is to expand $e^x - 1$ in a power series and to divide it into x^3 by long division. Thus we obtain

$$\frac{x^3}{e^x - 1} = x^2 - \frac{x^3}{2} + \frac{x^4}{12} - \frac{x^6}{720} + \cdots \tag{13.92}$$

Inserting this result into equation 13.87, we find, upon integration, that

$$U = 3RT\left[1 + \frac{1}{20}\left(\frac{\theta}{T}\right)^2 - \frac{1}{1,680}\left(\frac{\theta}{T}\right)^4 + \cdots\right] \tag{13.93}$$

Upon differentiating partially with respect to T, we see that

$$C_v = 3R\left[1 - \frac{1}{20}\left(\frac{\theta}{T}\right)^2 + \frac{1}{560}\left(\frac{\theta}{T}\right)^4 - \cdots\right] \tag{13.94}$$

An extended calculation will disclose that the next term in the series is equal to $-3R(\theta/T)^6/18,144$. For $T > \theta/2$, the abbreviated expression given above is adequate for most purposes. Observe that, as $T \longrightarrow \infty$, the heat capacity approaches $3R$, as predicted from the Einstein theory.

It is evident, from examination of the above developments, that the heat capacity of a solid, according to either the Debye or the Einstein theory, is a function of T/θ only while θ, the characteristic temperature,

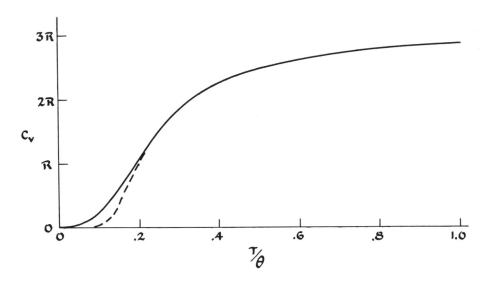

FIGURE 13-5   Heat capacity of a solid, according to the Debye theory (solid line) and
the Einstein theory (broken line), as a function of T/θ if θ is the Debye characteristic
temperature. The Einstein curve was adjusted to coincide with the Debye curve at
high temperatures.

is a parameter whose value depends upon the substance. (The characteristic temperature for the Einstein theory is simply $h\nu/k$, in which
expression ν is the single characteristic frequency. It will not have the
same value as the Debye characteristic temperature, which is related
to the maximum frequency.) If one plots the heat capacity of a solid
against ln T, one should therefore get, for different materials, curves
that are displaced from one another along the abscissa but are otherwise
superimposable. This is indeed found to be so. In Figure 13-5 is a plot
of C_v against T/θ for the Debye and Einstein theories. The most significant difference between the two is at low temperatures; the Debye
expression for C_v approaches zero much less rapidly as $T \longrightarrow 0$ than
does the Einstein prediction. It is in this respect that Debye's theory is
superior for describing actual crystals.

Problems

1. A rigid rotator with fixed axis has allowable energy levels expressed by the equation

$$E_M = \frac{M^2 h^2}{8\pi^2 I}$$

in which $M = 0, \pm 1, \pm 2, \ldots$. Replacing the summation by an integration, show that the partition function for a rigid rotator with fixed axis is

$$q_{rot} = \left(\frac{8\pi^3 I k T}{h^2}\right)^{1/2}$$

Hence show that the energy of a mole of such rotators will be equal to $\frac{1}{2}RT$.

2. Consider a particle in a one-dimensional box. The allowable energy levels will be given by the expression

$$E_x = \frac{n_x^2 h^2}{8m l_x^2}$$

in which l_x is the length of the box and n_x is a positive integer. Show that the partition function for such a particle will be

$$q_x = \frac{(2\pi m k T)^{1/2} l_x}{h}$$

From this result show that the partition function for a particle in a rectangular parallelepiped box will be a product of the partition functions for the several dimensions:

$$q_{trans} = q_x \cdot q_y \cdot q_z = \frac{(2\pi m k T)^{3/2} V}{h^3}$$

In this expression $V = l_x l_y l_z$. (This method of describing a three-dimensional system would not necessarily be valid if the energy associated with motion in one direction depended upon the energies attributable to perpendicular motions. As long as the three components are mutually independent, this technique is valid. See the discussion on p. 270.)

3. Consider N molecules with allowable energy levels E_i of degeneracy g_i. Suppose we arbitrarily divide the energy levels into two groups, designated A and B, for which we set up separate partition functions, q_A and q_B.
   (A) Show that the total partition function q_{AB} is equal to $q_A + q_B$.
   (B) Show that the equilibrium constant, K, for the equilibrium $A \rightleftharpoons B$ will be

$$K = \frac{q_B}{q_A}$$

   (c) Show also that the numbers of molecules, N_A and N_B, in the two classes will be

$$N_A = N q_A / q_{AB}$$
$$N_B = N q_B / q_{AB}$$

(D) Show further that the total energy of the system will be

$$U = N_A k T^2 \frac{\partial \ln q_A}{\partial T} + N_B k T^2 \frac{\partial \ln q_B}{\partial T}$$

$$= n_A U_A + n_B U_B$$

if n_A and n_B denote numbers of moles.

(E) Finally, by use of partition functions, show that the total heat capacity of the equilibrium system at constant volume will be

$$C_v = n_A C_{vA} + n_B C_{vB} + \frac{n_A n_B (\Delta \tilde{U})^2}{(n_A + n_B) R T^2}$$

if $\Delta \tilde{U}$ is the "energy of reaction" in the passage from A to B. (Compare with Prob. 4, Chap. Ten, and Prob. 17, Chap. Eight.)

4. Suppose that a molecule could exist in either of two energy levels, E_1 and E_2, with degeneracies g_1 and g_2. Derive expressions for U, S, A, and C_v for N molecules of the substance. Taking cognizance of the number of molecules, N_1 and N_2, in each of the two states, show that the heat capacity will be

$$C_v = \frac{N_1 N_2}{N k T^2} (E_2 - E_1)^2$$

Compare this result with the third term in the answer to Problem 3, part E, and explain.

5. Prove, for a mole of substance obeying Boltzmann statistics, that the heat capacity at equilibrium will be

$$C_v/R = \sum_i \sum_{j > i} f_i f_j (E_j - E_i)^2 / (kT)^2$$

if f_i and f_j are the fractions of the molecules with energies of E_i and E_j. In the light of this result, show further that

$$C_v/R = (\langle E^2 \rangle - \langle E \rangle^2)/(kT)^2$$

if $\langle E^2 \rangle$ is the average squared energy and $\langle E \rangle^2$ the square of the average energy. Illustrate the correctness of this formula by evaluating $\langle E^2 \rangle$ and $\langle E \rangle^2$ for the translational motion of molecules, using the Maxwell-Boltzmann distribution of velocities. (This problem shows that, if the molecular energies are narrowly distributed, the heat capacity will be small. If all the molecules, at equilibrium, were in one energy level, the heat capacity would vanish. According to Boltzmann statistics, this could be realized at absolute zero.)

6. (A) A two-dimensional isotropic harmonic oscillator has energy levels given by the expression

$$E_n = (n + 1) h \nu_0$$

in which $n = n_x + n_y$ and n_x and n_y take on the values 0, 1, 2, ...

Show that the degeneracy, g_n, is equal to $n + 1$. Hence show directly that the partition function will be

$$q = \frac{e^{-h\nu_0/kT}}{(1 - e^{-h\nu_0/kT})^2}$$

(B) For a three-dimensional isotropic harmonic oscillator the energy levels are given by the expression

$$E_n = (n + 3/2)h\nu_0$$

in which $n = n_x + n_y + n_z$. Show that the degeneracy is $(n + 1)(n + 2)/2$ and that

$$q = \frac{e^{-3h\nu_0/kT}}{(1 - e^{-h\nu_0/kT})^3}$$

(C) Compare these partition functions with that of the one-dimensional harmonic oscillator, and explain.

(D) How do the thermodynamic functions for multidimensional oscillators compare with those for the one-dimensional oscillator?

7. The moment of inertia of hydrogen iodide is 4.272×10^{-40} g cm^2. Calculate the rotational energy, entropy, and Helmholtz free energy for 1 mole of hydrogen iodide at 0 °C and at 100 °C.

8. The vibrational frequency of hydrogen iodide is equal to 2309.5 cm^{-1}. Calculate the vibrational energy, entropy, and Helmholtz free energy for 1 mole of hydrogen iodide at 0 °C and at 100 °C.

9. In its ground state the chlorine atom has the term symbol $^2P_{3/2}$, in its next higher level the designation $^2P_{1/2}$. If the two levels are 881 cm^{-1} apart, what fraction of chlorine atoms will be in each state at 0, 500, and 1,000 °C? What are the values of U, S, A, and C_v attributable to electronic excitation for a mole of chlorine atoms?

10. Consider the equilibrium

$$I_2(g) \rightleftharpoons 2I(g)$$

(A) Set up an expression for the equilibrium constant for this reaction, using the following data: the bond energy, ΔU_0, is equal to 35,603 cal; the moment of inertia of the I_2 molecule is equal to 750.2×10^{-40} g cm^2; the vibrational frequency of the diatomic molecule is 213.67 cm^{-1}; the molecule has an electronic degeneracy equal to unity, but the atom, in a $^2P_{3/2}$ state, has a degeneracy equal to 4.

(B) Calculate numerical values for K_p, and compare them with the following experimental results:

t	K_p (observed)
800 °C	1.14×10^{-2} atm
1,000	1.65×10^{-1}
1,200	1.23

11. In Problem 8 of Chapter Twelve it was shown that the entropy of mixing of isotopic chlorine atoms is not the same as the entropy of mixing of the equivalent number of atoms in their molecular forms. Show that the difference, $2x_1x_2R \ln 2$ for 2 moles of atoms, arises from the difference between the symmetry numbers of $Cl^{35}Cl^{37}$ and Cl_2^{35} or Cl_2^{37}. Hence observe how the indistinguishability of atoms must be taken into account in connection with rotation.

12. Estimate the standard Gibbs energy change in calories for the following reaction at 25 °C:

$$Cl_2(g) + Br_2(g) \rightleftarrows 2BrCl(g)$$

The value of ΔU_0 for this reaction can be taken as the heat of reaction, -0.387 kcal. As for the effect of rotation, use for the covalent radii of Br and Cl the values 1.14 and 0.99 A, from which the moments of inertia can be calculated. [Be sure to use the appropriate reduced masses, such as $m_{Br}m_{Cl}/(m_{Br} + m_{Cl})$ for the molecule BrCl.] Finally, you may neglect any vibrational excitation (or assume that such effects cancel). Compare your answer with the accepted value of -1.211 kcal.

13. Using the Debye equations for the heat capacity of a solid, calculate and plot C_v against T/θ. Use both the low- and the high-temperature formulas, with a moderate range for overlap, and observe the extent to which the two functions meet.

14. (A) Using the method of Debye, derive an expression for the heat capacity of a two-dimensional solid. For this purpose, the degeneracy associated with frequencies in the range ν to $\nu + d\nu$ will be proportional to $\nu d\nu$.

    (B) Repeat part A for a one-dimensional system. Here the degeneracy will be proportional to $d\nu$.

Ans.   (A) $C_v = \text{const} \cdot T^2$

       (B) $C_v = \text{const} \cdot T$

14

Bose-Einstein and
Fermi-Dirac Statistics

Distinguishability of Particles

The preceding three chapters have been concerned with Boltzmann statistics and with its applications to chemical systems. For most chemical purposes, Boltzmann statistics is quite adequate, even though it is necessary to correct rather artificially for the implied distinguishability of like atoms to make the calculated entropies compatible with the Third Law. But although patched-up Boltzmann statistics is satisfactory for many chemical purposes, there remain a number of problems for which it is unsatisfactory. Moreover, there is always uncertainty attending the introduction of artificial corrections to accommodate such matters as indistinguishability; hence, it is desirable to look for more appropriate kinds of statistics.

When discussing indistinguishability of like atoms with reference to rotational partition functions (see p. 277), it was pointed out that a more fundamental way of taking account of indistinguishability involves a quantum-mechanical approach. This approach deals with the symmetry of wave functions and rests on a basic underlying concept which requires that wave functions for different particles be either completely symmetric or completely antisymmetric with respect to interchange of coordinates. Specifically, the statistics applicable to particles whose total wave functions must be completely symmetric is known as Bose-Einstein statistics. In the calculation of thermodynamic probabilities, this kind of statistics can be considered equivalent to counting the number of ways in which indistinguishable particles can be placed into distinguishable boxes (or energy levels) without limiting the number of particles placed in any one box. A second kind of statistics, known as Fermi-Dirac statistics, is applicable to systems of particles whose total wave functions must be antisymmetric with respect to interchange of coordinates. Applied to thermodynamic probabilities, Fermi-Dirac statistics counts the number of ways in which indistinguishable particles can be placed in a number of distinguishable boxes, subject to the condition that not more than one particle be placed in any one box. This stipulation, when applied to electrons, becomes the equivalent of the Pauli principle, which, when applied to the old Bohr theory of atomic structure, amounts to saying that no two electrons can have the same complete set of quantum numbers (including electron spin as one of the numbers).

Fermi-Dirac statistics is applicable to so-called *fermions*, which have spins of $\frac{1}{2}$, $\frac{3}{2}$, $\frac{5}{2}$, ... The most important application of Fermi-Dirac statistics is to electrons, including the electronic theory of metals. Bose-Einstein statistics is applicable to *bosons*, which are particles with integral spins, 0, 1, 2, ... For example, Bose-Einstein statistics can be applied to helium, with particular reference to its low-temperature behavior. It also has an important special application to radiation, which can be treated as a gas of photons.

Bose-Einstein Statistics

To establish an expression for the thermodynamic probability, Ω, of a Bose-Einstein system, we shall modify the method used to arrive at Ω for Boltzmann statistics, as expressed by equation 11.43. In particular, we must first ask ourselves how many ways N_i indistinguishable particles, each of energy E_i, can be placed in g_i distinguishable boxes. If N_i particles were lined up in a row, and if $g_i - 1$ partitions were inserted in the line of particles, the N_i particles would be divided into g_i separate groups. If both the particles and the partitions were distinguishable, there would be $(g_i - 1 + N_i)!$ different ways of composing such a lineup. But if we are to recognize true indistinguishability, this quantity must be divided by $N_i!$, the number of ways N_i distinguishable particles could be arranged. Moreover, we must also divide by $(g_i - 1)!$, since that represents the number of different orders in which numbered partitions could be inserted, for the order in which they are considered does not matter so long as each is considered separately. In other words, dividing by $(g_i - 1)!$ is equivalent to arranging the boxes in some stipulated order, let us say 1, 2, . . . , up to g_i. The first box must be to the left of the first partition, the second box immediately to the left of the second partition, and so on, until we reach the last box, which must be to the right of the $(g_i - 1)$th partition. Hence, Ω_i, the contribution of N_i particles of energy E_i to the total thermodynamic probability will be given by

$$\Omega_i = \frac{(g_i - 1 + N_i)!}{(g_i - 1)!N_i!} \tag{14.1}$$

The total thermodynamic probability will be given by an appropriate continued product of the Ω_i's namely:

$$\Omega = \prod_i \frac{(g_i - 1 + N_i)!}{(g_i - 1)!N_i!} \tag{14.2}$$

in which i takes on values that serve to denote all of the energy levels.

EXERCISE:

Consider five particles to be placed in three boxes. Show by direct count how many different ways the particles can be distributed among the boxes assuming that the boxes are distinguishable but that the particles are not. Proceed by establishing arrays like 5, 0, 0; 0, 5, 0; 0, 0, 5; 4, 1, 0; 4, 0, 1; and so on, down to 2, 2, 1; 2, 1, 2; and 1, 2, 2. Compare your count with that calculated by use of equation 14.1.

Having established the necessary expression for the thermodynamic probability, we can now derive the Bose-Einstein distribution law, using a method almost identical to that employed in our second derivation of the Maxwell-Boltzmann law (p. 245).

If we neglect unity compared to g_i and make use of Stirling's approximation, we find from equation 14.2 that

$$\ln \Omega = \sum_i \{(g_i + N_i) \ln (g_i + N_i) - g_i \ln g_i - N_i \ln N_i\} \tag{14.3}$$

The variation of $\ln \Omega$ can also be obtained to give

$$\delta \ln \Omega = \sum_i \{\ln (g_i + N_i) - \ln N_i\} \delta N_i \tag{14.4}$$

Moreover, by making use of the familiar constraints on the total number of particles, N, and the total energy, U, we assert that

$$\delta N = \sum_i \delta N_i = 0 \tag{14.5}$$

and that

$$\delta U = \sum_i E_i \delta N_i = 0 \tag{14.6}$$

Multiplying equation 14.5 by α and equation 14.6 by β and then combining them with equation 14.4, we obtain

$$\delta \ln \Omega - \alpha \delta N - \beta \delta U = \sum_i \{\ln (1 + g_i/N_i) - \alpha - \beta E_i\} \delta N_i \tag{14.7}$$

To make $\ln \Omega$ a maximum, the coefficients of δN_i are set equal to zero, thereby obtaining

$$\ln (1 + g_i/N_i) = \alpha + \beta E_i \tag{14.8}$$

Rearrangement of equation 14.8 by solving for N_i now yields the Bose-Einstein distribution law:

$$N_i = \frac{g_i}{e^{\alpha + \beta E_i} - 1} \tag{14.9}$$

Notice that equation 14.9 differs from the Maxwell-Boltzmann expression only by the presence of the term -1 in the denominator. For gases at high temperatures this term is of little consequence, but for gases at low temperatures and for radiation it is important.

Entropy of a Bose-Einstein Gas

We can now calculate the entropy of a Bose-Einstein gas by inserting the values of N_i given by equation 14.9 into equation 14.3, which, when multiplied by k, gives us $k \ln \Omega_{max}$, or the entropy, S. Before making an explicit substitution, we can rearrange $k \ln \Omega$ to obtain

$$S = k \sum_i \{N_i \ln (1 + g_i/N_i) + g_i \ln (1 + N_i/g_i)\} \tag{14.10}$$

Substituting $\ln (1 + g_i/N_i)$ from equation 14.8 into equation 14.10, we obtain upon further rearrangement

$$S = k \sum_i N_i \{\alpha + \beta E_i + \ln (1 + N_i/g_i)^{g_i/N_i}\} \tag{14.11}$$

Let us now denote N_i/g_i by z; thus the argument of the logarithm in equation 14.11 becomes $(1 + z)^{1/z}$, a function that approaches e, the base of natural logarithms, as its limit when z approaches zero. If g_i is much larger than N_i, it follows that z will be small. Accordingly, we can say that $(1 + N_i/g_i)^{g_i/N_i}$ is approximately equal to e, since for the higher translational energy levels z will be small. Hence, upon evaluating the indicated summations in equation 14.11, we can write that

$$S = Nk\alpha + k\beta U + Nk \tag{14.12}$$

This expression for the entropy is identical with that derived from Boltzmann statistics, equation 11.56, except for the term $k \ln N!$. Since the Boltzmann entropy is in fact too large by the amount $k \ln N!$, we see that the Bose-Einstein calculation renders unnecessary any arbitrary correction. It should be emphasized, however, that the absence of $k \ln N!$ in the Bose-Einstein entropy is a result of prior elimination of the distinguishability of molecules when formulating the thermodynamic probability.

It can readily be shown that α and β, the Lagrangian multipliers used in deriving the Bose-Einstein distribution law are identical in significance to the same parameters in the Maxwell-Boltzmann expression. That is, $\beta = 1/kT$ and $\alpha = -\mu/RT$, in which μ, the chemical potential, equals the molar Gibbs energy (see p. 265). It can also be shown that $k \sum_i g_i \ln$ $(1 + N_i/g_i)$, which appears in equation 14.10, is precisely equal to pV/T and, of course, to $2U/(3T)$. (See Prob. 7 of this chapter.) This conclusion is valid even if N_i is not small compared with g_i.

At elevated temperatures, a Bose-Einstein gas can be treated as a Boltzmann gas, and most of the applications of statistical mechanics already discussed can be carried through unchanged. At very low temperatures, a Bose-Einstein gas, such as helium, departs markedly from the behavior of a Boltzmann gas, and exhibits a special kind of condensation. The treatment of radiation, however, provides a most striking example of the applicability of Bose-Einstein statistics.

Radiation

Before using Bose-Einstein statistics to describe radiation, we shall see what might be deduced by more ordinary thermodynamic methods employing only the elementary concepts of kinetic theory. Although an elaborate theory of electromagnetic waves has been worked out to describe radiation, it is also possible to consider radiation as a photon gas—a gas consisting of particles each of rest mass zero moving about with the velocity of light. According to the kinetic theory of gases, the pressure-volume product of a gas is directly related to the number of particles, N, the particle mass, m, and their average square velocity, $\langle v^2 \rangle$, by the equation

$$pV = \tfrac{1}{3}Nm\langle v^2 \rangle \tag{14.13}$$

In a photon gas, all particles are moving at the velocity of light, c, at which speed each will possess a finite mass, m, different from zero, such that the total energy of a photon, mc^2, will equal $h\nu$, in which ν is the frequency of the radiation. (In accordance with the special theory of relativity, the rest mass must be presumed to be zero if the mass at velocity c is to be finite.) It follows, therefore, that the pressure-volume product of radiation will be given by

$$pV = \tfrac{1}{3}U \tag{14.14}$$

in which U is the total energy. Alternatively,

$$p = \tfrac{1}{3}u \tag{14.15}$$

in which u is the energy density, a function of temperature only. (That pV for radiation equals one-third instead of two-thirds of the kinetic energy, as is true for gases, is simply a consequence of the fact that the energy of a photon is mc^2 instead of the familiar $\tfrac{1}{2}mv^2$.)

We are now in a position to make some useful deductions about radiation in equilibrium with its surroundings (i.e., black-body radiation at some temperature T). Let us consider the general expression for dS, in which S is considered to be a function of U and V:

$$dS = \frac{dU}{T} + \frac{pdV}{T} \tag{14.16}$$

Recognizing that

$$U = uV \tag{14.17}$$

we see that

$$dS = \frac{V}{T}\frac{du}{dT}\,dT + \frac{4u}{3T}\,dV \tag{14.18}$$

But since dS is an exact differential, it follows that

$$\frac{\partial^2 S}{\partial V \partial T} = \frac{1}{T}\frac{du}{dT} = \frac{4}{3T}\frac{du}{dT} - \frac{4u}{3T^2} \tag{14.19}$$

Upon rearranging equation 14.19 it will be seen that

$$\frac{du}{dT} = \frac{4u}{T} \tag{14.20}$$

which upon integration yields

$$u = aT^4 \tag{14.21}$$

in which a is a constant. Equation 14.21 is closely related to the Stefan-Boltzmann law, a physical law that states that the energy flux of radiation emitted by a black body is proportional to the fourth power of the absolute temperature.

From equation 14.21, we find, upon substituting into equation 14.18, that

$$dS = 4aVT^2\,dT + \frac{4aT^3}{3}\,dV \tag{14.22}$$

Direct integration of equation 14.22 yields

$$S = \tfrac{4}{3}aVT^3 \tag{14.23}$$

The constant of integration was set equal to zero in equation 14.23, since zero volume of radiation must have zero entropy. Evidently the total energy of a volume V of black-body radiation will be given by

$$U = aVT^4 \tag{14.24}$$

and the pressure by

$$p = \tfrac{1}{3}aT^4 \tag{14.25}$$

Finally, it is interesting to note that the Gibbs free energy, G, of black-body radiation will be zero, since $G = U + pV - TS$.

Planck Distribution Law

In the preceding section it was shown that certain over-all thermodynamic properties of radiation could be deduced by some rather general considerations that did not require a knowledge of the distribution of radiation energy. We shall now turn to the question of that distribution, making use of Bose-Einstein statistical mechanics.

For the thermodynamic probability we shall again use equation 14.2, and we shall immediately suppress the term -1 compared to g_i. In maximizing the thermodynamic probability, however, it is necessary to introduce only one constraint instead of two; thus only one Lagrangian multiplier is required. Since the total energy, U, must be conserved, equation 14.6 and the multiplier β will be employed as before. But the number of particles (or photons), N, need not be constant, since photons can be readily created and destroyed. (From a relativistic point of view, conservation of energy and mass are simultaneously realized, so that one multiplier suffices for photons.) Apropos of not requiring conservation of the number of photons, we recall that α, the Lagrangian multiplier corresponding to the number of particles, is proportional to the Gibbs free energy of the system (eq. 12.42); and since the Gibbs energy of radiation is zero at equilibrium, we might have anticipated no need for α. To obtain the radiation distribution law, we need only rewrite equation 14.9, omitting α, as follows:

$$N_i = \frac{g_i}{e^{\beta E_i} - 1} \tag{14.26}$$

It can be demonstrated once again that $\beta = 1/kT$; however, we cannot use equation 14.12 for the entropy of radiation, even with $\alpha = 0$, since it is not valid to assume that N_i/g_i is small. Accordingly, we must consider more precisely the nature of g_i.

For g_i, we shall use an expression determined by how many waves of a given wavelength can fit in a cubical box of volume V. In essence, this calculation was done when we considered the degeneracy of translational levels for a particle in a box. In the quantum mechanical problem, the waves are the so-called de Broglie waves, and the total quantum number, n (of eq. 12.19), is directly proportional to the frequency of the wave. To fit an integral number of half waves into a box of edge length l (or $V^{1/3}$), we require that $n\lambda/2 = l$, and since $\lambda\nu = c$ for radiation, it follows that

$$n = \frac{2l\nu}{c} = \frac{2V^{1/3}\nu}{c} \tag{14.27}$$

Substituting n from equation 14.27 into equation 12.19, we obtain $dg = 4\pi V\nu^2 d\nu/c^3$. Radiation can, however, be polarized; hence this expression for dg must be multiplied by 2 to take into account the fact that each of two kinds of polarized waves can be fit in a given place. Therefore we write

$$dg = \frac{8\pi V\nu^2}{c^3} d\nu \tag{14.28}$$

We are now in a position to write the Planck distribution law for radiation. Replacing g_i in equation 14.26 by dg of equation 14.28 and recognizing that the energy, E, of a photon equals $h\nu$, we obtain for dN, the number of photons with frequencies in the range ν to $\nu + d\nu$, the following expression:

$$dN = \frac{(8\pi V\nu^2/c^3)d\nu}{e^{h\nu/kT} - 1} \tag{14.29}$$

To obtain the total energy in the range ν to $\nu + d\nu$, it is necessary to multiply dN by $h\nu$; to convert this result to energy densities, we must further divide by V. In this way we obtain

$$du = \frac{(8\pi h\nu^3/c^3)d\nu}{e^{h\nu/kT} - 1}$$ (14.30)

which is the Planck distribution law expressed in terms of frequencies. Replacing ν by c/λ, the result can be rewritten for energy densities of radiation in the range λ to $\lambda + d\lambda$ as follows:

$$du = \frac{8\pi chd\lambda}{\lambda^5(e^{hc/\lambda kt} - 1)}$$ (14.31)

This is the usual form in which the Planck distribution law is expressed. Values of $du/d\lambda$ are plotted vs. λ for various temperatures in figure 14-1. It will be observed that the energy density of radiation passes through a maximum at each temperature and that it goes to zero both as $\lambda \longrightarrow 0$ and as $\lambda \longrightarrow \infty$. The wavelength corresponding to the maximum can be obtained by differentiating the function with respect to λ and setting the derivative equal to zero. Upon appropriate numerical analysis, it is found that the corresponding wavelength, λ_{max}, is related to the temperature as follows:

$$\lambda_{max} T = \text{const}$$ (14.32)

This conclusion is known as Wien's displacement law.

The total energy density of radiation can be obtained by integrating du, using equation 14.30, from $\nu = 0$ to $\nu \longrightarrow \infty$. When we define a new variable, x, equal to $h\nu/kT$, the integral becomes

$$u = \frac{8\pi k^4 T^4}{c^3 h^3} \int_0^\infty \frac{x^3 \, dx}{e^x - 1}$$ (14.33)

The definite integral appearing in equation 14.33 is precisely the same as that appearing in equation 13.88, which was evaluated as follows:

$$\int_0^\infty \frac{x^3 \, dx}{e^x - 1} = \int_0^\infty \frac{x^3 e^{-x}}{1 - e^{-x}} dx = \int_0^\infty x^3(e^{-x} + e^{-2x} + e^{-3x} + \ldots) \, dx$$

$$= 6\left(\frac{1}{1^4} + \frac{1}{2^4} + \frac{1}{3^4} + \ldots\right)$$ (14.34)

It was also pointed out that the infinite series $\sum_{i=1}^{\infty} 1/i^4$ converges to $\pi^4/90$, so that the integral equals $\pi^4/15$ (eq. 13.89). Substituting back into equation 14.33 we see that

$$u = \frac{8\pi^5 k^4 T^4}{15c^3 h^3}$$ (14.35)

This is, of course, the same as equation 14.21, except that the constant a is now identified as $8\pi^5 k^4/15c^3 h^3$. The flux of radiation, s, coming from a black body will be given by

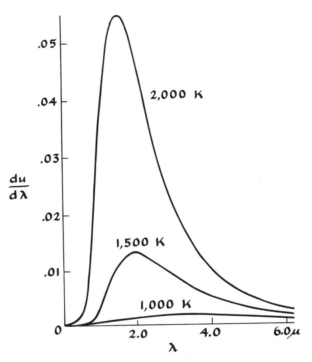

FIGURE 14-1 Graphs of energy density, $du/d\lambda$, in ergs cm^{-3} micron^{-1} versus λ for blackbody radiation according to the Planck distribution law.

$$s = \sigma T^4 = (ca/4) T^4 \qquad (14.36)$$

in which σ, Stefan's constant, is equal to $2\pi^5 k^4/15h^3c^2$, or 5.6698×10^{-5} ergs cm^{-2} sec^{-1} deg^{-4}.

Through use of kinetic-theory arguments in conjunction with classical thermodynamics, we already deduced relationships between the entropy, pressure, and energy of radiation. The statistical-mechanical approach, however, should be self-sufficient, so we shall now complete our treatment by calculating the entropy and pressure. If we substitute the Planck distribution formula, equation 14.26, into the expression for entropy, equation 14.10, we obtain upon rearrangement

$$S = k\beta U - k \sum_i g_i \ln (1 - e^{-\beta E_i}) \qquad (14.37)$$

Replacing the summation by an integration, taking cognizance of equation 14.28, letting $E = h\nu$ and $\beta = 1/kT$, we see that

$$S = \frac{U}{T} - \frac{8\pi Vk}{c^3} \int_0^\infty \nu^2 \ln (1 - e^{-h\nu/kT}) \, d\nu \qquad (14.38)$$

Once again, let $x = h\nu/kT$, to give

$$
\begin{aligned}
S &= \frac{U}{T} - \frac{8\pi Vk^4 T^3}{h^3 c^3} \int_0^\infty x^2 \ln\left(1 - e^{-x}\right) dx \\
&= \frac{U}{T} + \frac{8\pi Vk^4 T^3}{h^3 c^3} \int_0^\infty x^2 \left[e^{-x} + \frac{e^{-2x}}{2} + \frac{e^{-3x}}{3} + \dots\right] dx \\
&= \frac{U}{T} + \frac{8\pi Vk^4 T^3}{h^3 c^3} \cdot 2 \left[\frac{1}{1^4} + \frac{1}{2^4} + \frac{1}{3^4} + \dots\right] \\
&= \frac{U}{T} + \frac{U}{3T} \\
&= \frac{4U}{3T}
\end{aligned}
\tag{14.39}
$$

The Helmholtz free energy, $A = U - TS$, will be given by

$$
A = -\frac{U}{3} = -\frac{uV}{3} \tag{14.40}
$$

so that the pressure can be calculated to be

$$
p = -\left(\frac{\partial A}{\partial V}\right)_T = \frac{u}{3} \tag{14.41}
$$

which is the same as the earlier result. Naturally, the Gibbs free energy will turn out to be zero.

Fermi-Dirac Statistics

As mentioned near the beginning of this chapter, Fermi-Dirac statistics can be regarded as statistics dealing with indistinguishable particles placed in distinguishable boxes subject to the constraint that not more than one particle can be in any one box. To obtain the thermodynamic probability for this kind of statistics, consider g_i boxes, all corresponding to an energy E_i. If N_i particles are placed in these boxes, there must remain $g_i - N_i$ unfilled boxes. Now the number of ways g_i entities can be divided into two categories of sizes $g_i - N_i$ and N_i is precisely

$$
\Omega_i = \frac{g_i!}{(g_i - N_i)!N_i!} \tag{14.42}
$$

This expression for Ω_i can be taken as a factor in the thermodynamic probability. Evidently the total thermodynamic probability for a Fermi-Dirac system will be

$$
\Omega = \prod_i \frac{g_i!}{(g_i - N_i)!N_i!} \tag{14.43}
$$

EXERCISE:

Consider three particles to be placed in five boxes. Show by direct count how many different ways the particles can be distributed among the boxes assuming that the boxes are distinguishable but that the particles are not and that not more than one particle can be placed in any one box. Proceed by establishing arrays like 11100, 11010, 11001, etc. Compare your count with that calculated by use of equation 14.42.

To establish the Fermi-Dirac distribution law, we now proceed along familiar lines by first taking the logarithm of Ω, and then making use of Stirling's approximation to obtain

$$\ln \Omega = \sum_i g_i \ln g_i - (g_i - N_i) \ln (g_i - N_i) - N_i \ln N_i \quad (14.44)$$

Evidently the variation of $\ln \Omega$ will be given by

$$\delta \ln \Omega = \sum \{\ln (g_i - N_i) - \ln N_i\} \delta N_i \quad (14.45)$$

Once again the derivation requires imposing the two constraints on δN and δU, equations 14.5 and 14.6. Multiplying δN by α and δU by β and combining with equation 14.45, we see that

$$\delta \ln \Omega - \alpha \delta N - \beta \delta U = \sum_i \{\ln (g_i/N_i -- 1) - \alpha - \beta E_i\} \delta N_i \quad (14.46)$$

Setting each of the coefficients of δN_i equal to zero, we find that

$$\ln (g_i/N_i - 1) = \alpha + \beta E_i \quad (14.47)$$

or, upon solving for N_i,

$$N_i = \frac{g_i}{e^{\alpha+\beta E_i} + 1} \quad (14.48)$$

This is the Fermi-Dirac distribution law; like that of Bose-Einstein, it differs in appearance only slightly from the Maxwell-Boltzmann equation. Except at very high temperatures, however, the effect of the term $+1$ in the denominator is very important.

The entropy of a Fermi-Dirac gas can now be directly calculated. Upon multiplying $\ln \Omega$ (eq. 14.44) by k and rearranging, we find that

$$S = k \sum_i N_i \ln (g_i/N_i - 1) - g_i \ln (1 - N_i/g_i) \quad (14.49)$$

provided N_i is given by the distribution law. Using that law and substituting equation 14.47 into equation 14.49, we obtain

$$S = k \sum_i N_i\{(\alpha + \beta E_i) - \ln (1 - N_i/g_i)^{g_i/N_i}\} \quad (14.50)$$

Upon letting $z = N_i/g_i$, the argument of the logarithm becomes $(1 - z)^{1/z}$. Assuming that (at high temperatures) z is exceedingly small, the quantity $(1 - z)^{1/z}$ is substantially equal to $1/e$. Hence

$$S = Nk\alpha + k\beta U + Nk \qquad (14.51)$$

an expression identical to that of equation 14.12. Once again we recognize this as the correct expression for translational entropy at elevated temperatures. The indistinguishability of particles is, of course, implicit in the treatment and no further correction for a term like $k \ln N!$ is necessary. Similar to the situation for Bose-Einstein statistics, the latter term in equation 14.49, $-k \sum_i g_i \ln (1 - N_i/g_i)$, can be shown to equal pV/T or $2U/(3T)$ even if $N_i \longrightarrow g_i$. (See Prob. 7.)

Fermi-Dirac Gas at Absolute Zero

A most important application of Fermi-Dirac statistics is in the treatment of electrons in metals. A metal, each of whose atoms contributes one or more free conducting electrons, can be regarded more or less as an electron gas. The total energy of such electrons, even at absolute zero, can be considerable because of the limitation on cell population characteristic of Fermi-Dirac statistics. This is not unlike the situation encountered in the approximate description of the normal state of an atom, in which one can think of the electrons successively filling up the lowest energy levels available until the total number of electrons are assigned to their orbitals. Similarly, in the case of a metal, the free electrons first fill the lowest translational levels, then the next lowest, and so on until all the electrons are in place. The number of energy levels in the range E to $E + dE$ for particles in a box of volume V was given by equation 12.20 when we considered translational properties of gases. For electrons, however, that number should be multiplied by 2 since each electron* can have a spin of either plus or minus $\frac{1}{2}$. Hence we can write for electrons that

$$dg = \frac{8\sqrt{2}\pi m^{3/2} V E^{1/2}}{h^3} dE \qquad (14.52)$$

---

* Such doubling of the number of boxes is equivalent to saying that two electrons, but no more than two, can occupy each of the original boxes. This produces no substantive effect on the statistics involved nor on the nature of the distribution equation.

If there are N electrons in the metal and if the highest energy of an electron is E_0, with full occupancy implied for levels in the range $0 < E < E_0$, then

$$N = \int_0^{E_0} \frac{8\sqrt{2}\pi m^{3/2} V E^{1/2}}{h^3}\, dE = \frac{16\sqrt{2}\pi m^{3/2} V E_0^{3/2}}{3h^3} \qquad (14.53)$$

The total energy of these electrons will be given by

$$U = \int_0^{E_0} \frac{8\sqrt{2}\pi m^{3/2} V E^{3/2}}{h^3}\, dE = \frac{16\sqrt{2}\pi m^{3/2} V E_0^{5/2}}{5h^3} \qquad (14.54)$$

From the ratio, U/N, we see immediately that the average energy is precisely equal to $\frac{3}{5}E_0$. The quantity E_0 can, of course, be eliminated between equations 14.53 and 14.54; if this is done, we obtain the following expression for the energy:

$$U = \frac{3Nh^2}{40m}\left(\frac{3N}{\pi V}\right)^{2/3} \qquad (14.55)$$

The foregoing calculation of the energy of an electron gas at absolute zero (in this connection a metal at room temperature is practically at absolute zero) is quite compatible with the Fermi-Dirac distribution law, which can be written as

$$dN_E = \frac{dg_E}{e^{\alpha + E/kT} + 1} \qquad (14.56)$$

As $T \longrightarrow 0$, it is evident that $dN_E \longrightarrow dg_E$ provided $E < -\alpha k T$; on the other hand, if $E > -\alpha k T$, then $dN_E \longrightarrow 0$. Evidently $-\alpha k T$ is precisely the same as E_0 and, since $G = -\alpha N k T$ for one mole, it follows that

$$G = NE_0 = \tfrac{5}{3}U \qquad (14.57)$$

We would, of course, expect the entropy to be equal to zero at absolute zero. This is easily verified from the statistical point of view through examination of the thermodynamic probability factor, Ω_i, of equation 14.42. Evidently if $E < E_0$, then $N_i = g_i$ making $\Omega_i = 1$; similarly if $E > E_0$, then $N_i = 0$ again making $\Omega_i = 1$. Therefore $\Omega = 1$ and $S = 0$.

Finally, since $pV = G - U + TS$, we conclude that

$$pV = \tfrac{2}{3}U \qquad (14.58)$$

a result expected from kinetic theory.

Fermi-Dirac Gas at Temperatures Near Absolute Zero

The treatment of an electron gas for temperatures greater than zero is rather complicated. However, the following approximate equations, each representing the beginning of an infinite series in T, can be derived:

$$U = \frac{3Nh^2}{40m}\left(\frac{3N}{\pi V}\right)^{2/3} + \frac{2N\pi^2 mk^2}{h^2}\left(\frac{\pi V}{3N}\right)^{2/3}T^2 \qquad (14.59)$$

$$S = C_v = \frac{4N\pi^2 mk^2}{h^2}\left(\frac{\pi V}{3N}\right)^{2/3}T \qquad (14.60)*$$

$$A = \frac{3Nh^2}{40m}\left(\frac{3N}{\pi V}\right)^{2/3} - \frac{2N\pi^2 mk^2}{h^2}\left(\frac{\pi V}{3N}\right)^{2/3}T^2 \qquad (14.61)$$

and

$$G = \frac{Nh^2}{8m}\left(\frac{3N}{\pi V}\right)^{2/3} - \frac{2N\pi^2 mk^2}{3h^2}\left(\frac{\pi V}{3N}\right)^{2/3}T^2 \qquad (14.62)$$

Bose-Einstein and Fermi-Dirac Gases at High Temperatures

It has already been mentioned that at high temperatures both the Bose-Einstein and Fermi-Dirac statistics approach that of Maxwell-Boltzmann. In this connection, a "high temperature" is not the same for each of the two special statistics; room temperature is high for Bose-Einstein systems but low for those subject to Fermi-Dirac statistics. Boltzmann statistics, of course, gives rise to the ideal gas law equation, $pV = NkT$, for point particles free of intermolecular forces. Under standard conditions of pressure and temperature, a gas of similar particles subject to Bose-Einstein statistics would exhibit negligible deviations from ideality, so far as non-Boltzmann behavior is concerned. Under the conditions specified, actual gases will deviate somewhat from ideal behavior because of intermolecular forces that are very much more important than the effect of Bose-Einstein statistics.

In Fermi-Dirac systems, ordinary room temperature is still a low temperature, at least for electrons in metals. To be sure, as the temperature becomes infinite, the equation of state of a Fermi-Dirac gas will approach the familiar ideal gas law equation, but from a practical point of view, Fermi-Dirac statistics is generally applied to systems presumed to be at low temperatures. In any event, we shall not deal with the problem of gaseous degeneracy attributable to departures from Boltzmann behavior at high temperatures—whether those temperatures are easily realizable or whether they are envisioned only in an asymptotic way.

---

* The equality of S and C_v is correct, of course, only for the linear term in T and not for higher order terms that are not shown.

Problems

1. If $h\nu/kT$ is small compared to unity, show from equation 14.30 that

$$du = \frac{8\pi\nu^2 kT}{c^3}\, d\nu$$

This says that the energy density is just kT times the number of ways that waves with frequencies in the range ν to $\nu + d\nu$ can be placed in a unit volume. This classical result is fully in accord with the principle of equipartition of energy (see p. 39), and holds for the long wavelength end of the Planck distribution formula.

2. If $h\nu/kT$ is large compared to unity, show from equation 14.30 that

$$du = \frac{8\pi h\nu^3}{c^3}\, e^{-h\nu/kT} d\nu$$

This result would describe "Boltzmann photons" and is valid for the short wavelength end of the Planck formula.

3. Calculate the pressure in dyn cm^{-2} and in Torr of black-body radiation at 2000 K.

4. (A) Derive an expression for the equilibrium number of photons, of all frequencies, per unit volume of space at a temperature T using the Planck formula.

   (B) Show that the average energy of the photons is given by

$$\langle E \rangle = h\langle \nu \rangle = 3kT \frac{\displaystyle\sum_{i=1}^{\infty} \frac{1}{i^4}}{\displaystyle\sum_{i=1}^{\infty} \frac{1}{i^3}} = 2.7012kT$$

   (We have seen that $\displaystyle\sum_{i=1}^{\infty} 1/i^4 = \pi^4/90$, which is numerically equal to 1.08232. No simple closed form for $\displaystyle\sum_{i=1}^{\infty} 1/i^3$ has yet been reported but the numerical value of the sum has been determined to be 1.20206.)

5. Letting $x = hc/\lambda kT$, show that $du/d\lambda$ (from equation 14.31) is a maximum when

$$x = 5(1 - e^{-x})$$

Evidently x is approximately equal to 5. A better approximation is obtained by setting $x = 5(1 - e^{-5})$, and so on. Calculate the value of Wien's displacement constant appearing in equation 14.32.

6. Starting with the equation for the energy of a Fermi-Dirac gas at low temperatures (equation 14.59), derive equations 14.60, 14.61, and 14.62 by thermodynamic manipulations without recourse to statistical theory.

7. Equations 14.10 and 14.49 can be written in the following form:

$$S = k \sum_i N_i \ln (g_i/N_i \pm 1) \pm g_i \ln (1 \pm N_i/g_i)$$

provided the upper sign (wherever $\pm$ occurs) holds for Bose-Einstein statistics and the lower for that of Fermi-Dirac. Moreover, since

$$N_i = \frac{g_i}{e^{\alpha+\beta E_i} \mp 1}$$

it can be simply shown that

$$S = Nk\alpha + k\beta U \pm k \sum_i g_i \ln (1 \pm N_i/g_i)$$

Using this equation, and assuming only translational energy, form the total differential of S, recognizing that N, α, β, U, and N_i are all variable. Hence show that

$$dS = Nk d\alpha + kU d\beta + k\alpha dN + k\beta dU + k \sum_i \frac{g_i dN_i}{g_i \pm N_i}$$

Now form dN_i by differentiating the second equation given in this problem. (Remember that E_i depends on V. Substituting dN_i so obtained into the equation immediately above, show upon rearrangement that

$$dS = k\beta dU + \frac{2k\beta U}{3V} dV + k\alpha dN$$

Using the basic equations of thermodynamics, demonstrate that $\beta = 1/kT$, $p = 2U/3V$, and $\alpha = -\mu/kT$, if μ is given on a molecular basis. Hence prove that

$$U = 3pV/2 = \pm (3kT/2) \sum_i g_i \ln (1 \pm N_i/g_i)$$

But it is also true that

$$U = \sum_i \frac{g_i E_i}{e^{\alpha+\beta E_i} \mp 1}$$

so the two expressions for U can be equated.

8. Show by replacing summations by integrations, that

$$\sum_i \frac{g_i E_i}{e^{\alpha+\beta E_i} \mp 1} = \pm (3kT/2) \sum_i g_i \ln (1 \pm N_i/g_i)$$

provided E_i represents translational energy and Bose-Einstein or Fermi-Dirac statistics apply. Proceed by expanding both sides as power series in $e^{-\alpha-\beta E_i}$ and then integrate, assuming the validity of equations 12.18 and 12.19 where n is the total quantum number.

15

Alternative Derivations of Distribution Laws

Introduction

In Chapter Eleven we delved into the matter of statistical mechanics by considering probabilities and their connection with the entropy function. We started out with the basic postulate that

$$S = k \ln \Omega \qquad (15.1)$$

where Ω is the thermodynamic probability. Remembering that the entropy strives for a maximum in an isolated system of constant energy and volume, we then proceeded to determine the most probable distributions, since maximum probability would give rise to maximum entropy.

Now the maximization of entropy for an adiabatically enclosed system is not the only way of establishing equilibrium. We have seen, for example, that the Gibbs free energy strives for a minimum when temperature and pressure are constant and that the Helmholtz free energy likewise strives for a minimum under constant volume and temperature constraints. (See expressions 7.9 and 7.4.) Indeed, if we could keep the entropy and volume constant, the energy would seek a minimum; so would the enthalpy, provided the entropy and pressure were fixed.

That the maximization of entropy is by no means the only way of establishing equilibrium was also pointed out in connection with reaction potentials, summarized by equations 10.19. All this suggests that there may be other ways of deriving the statistical mechanical distribution laws—ways that would not require focusing primary attention on the entropy. In this chapter we shall do so by minimizing the Helmholtz free energy subject to constant temperature and volume constraints.

While we are deriving the distribution laws by looking at the Helmholtz free energy, we shall also avoid the use of the calculus of variations, Lagrangian multipliers, and Stirling's approximation. In short, we shall achieve the desired results by methods quite different from those used in Chapters Eleven and Fourteen. In my opinion, these methods are actually easier, but in any event they should prove instructive by providing another insight into the general subject.

States of Maximum Thermodynamic Probability

First of all, let us consider the simple problem, first treated on page 238, of finding the maximum thermodynamic probability of N particles distributed among s equally accessible boxes. The only constraint is that the number of particles, N, is fixed. Now let us suppose that we have found the solution to the problem, and that a set of values N_1, N_2, N_3, . . . , N_s renders Ω, as expressed by equation 11.8, a maximum. In other words,

$$\Omega_{\max} = \frac{N!}{N_1!N_2!N_3! \ldots N_s!} \tag{15.2}$$

Now if the set of values N_1, N_2, N_3, . . . , N_s does indeed make Ω a maximum, then any redistribution of the N_i's cannot possibly increase Ω. In particular, let us take one particle from box 2 and add it to box 1. We then obtain a new Ω, to be denoted by Ω', as follows:

$$\Omega' = \frac{N!}{(N_1 + 1)!(N_2 - 1)!N_3! \ldots N_s!} \tag{15.3}$$

Dividing equation 15.3 by 15.2, we obtain

$$\frac{\Omega'}{\Omega_{\max}} = \frac{N_2}{N_1 + 1} \leqslant 1 \tag{15.4}$$

We assert that the ratio, Ω'/Ω_{max}, must be equal to or less than unity because, by hypothesis, we started with a maximum Ω. Hence

$$N_2 \leqslant N_1 + 1 \tag{15.5}$$

Now let us perturb the system represented by equation 15.2 by removing one particle from box 1 and adding it to box 2, thus obtaining Ω'':

$$\Omega'' = \frac{N!}{(N_1 - 1)!(N_2 + 1)!N_3! \ldots N_s!} \tag{15.6}$$

In a manner similar to that employed above, we can assert that

$$\frac{\Omega''}{\Omega_{max}} = \frac{N_1}{N_2 + 1} \leqslant 1 \tag{15.7}$$

or

$$N_1 \leqslant N_2 + 1 \tag{15.8}$$

Now the only ways in which expressions 15.5 and 15.8 can be simultaneously satisfied is for

$$N_1 = N_2$$

or

$$N_1 = N_2 \pm 1 \tag{15.9}$$

Since the same argument can be advanced for any pair of boxes, we conclude in general either that the populations of all boxes are equal or that no two boxes can differ in population by more than one.

If only the inequality portions of expressions 15.5 and 15.8 were imposed, then the only solution would be for all N_i's to be equal. This situation could be realized if N were an integral multiple of s. On the other hand, if N is not equal to s times an integer, then the boxes will have populations equal to the integers immediately above and below the average population, subject to the total number being N. Thus, if one distributed 93 particles in 10 boxes, the most probable distribution would have 9 particles in each of 7 boxes and 10 particles in each of 3. Similarly, if one tossed $2n + 1$ coins (n being an integer), then the most probable distribution would have n or $n + 1$ heads.

The problem just completed is so simple that it scarcely warrants a detailed discussion. However, the method described will now be used for a more important problem, namely, that of deriving the statistical mechanical distribution laws.

Boltzmann Distribution Law

The methods of the last section will first be employed to derive an already familiar equation, the Boltzmann distribution law. This will be done by minimizing the Helmholtz free energy under constant volume and temperature constraints. It will be presumed, at the outset, that we know what we mean by temperature and that we can make use of it directly. The Helmholtz free energy, A, is, of course, defined by the equation

$$A = U - TS \qquad (15.10)$$

For a statistical system consisting of N_1 particles of energy E_1, N_2 particles of energy E_2, etc., we can write

$$U = \sum_i N_i E_i \qquad (15.11)$$

bearing in mind that

$$N = \sum_i N_i \qquad (15.12)$$

The entropy, S, is still assumed to be connected with the thermodynamic probability, Ω, through equation 15.1. For a Boltzmann system, that is to say, one consisting of distinguishable particles distributed among distinguishable energy levels, the thermodynamic probability given earlier by equation 11.43 is

$$\Omega = N! \prod_i \frac{g_i^{N_i}}{N_i!} \qquad (15.13)$$

where g_i is the degeneracy of the ith level, corresponding to energy E_i. It follows, therefore, that

$$A = \sum_i N_i E_i - kT \ln N! \prod_i \frac{g_i^{N_i}}{N_i!} \qquad (15.14)$$

Let us now make a simple algebraic transformation that will render the expression for A easier to handle. Specifically, let us write

$$\sum_i N_i E_i = -kT \ln \prod_i e^{-N_i E_i / kT} \qquad (15.15)$$

Equation 15.15 is easily verified as an identity; when it is introduced into equation 15.14 one obtains

$$A = -kT \ln N! \prod_i \frac{\left(g_i e^{-E_i / kT}\right)^{N_i}}{N_i!} \qquad (15.16)$$

Let us now ask what the N_i values should be to render A a minimum, which is quite equivalent to finding the maximum argument of the logarithm in equation 15.16. To be meaningful for our purpose, A must be minimized as temperature, T, and volume, V, are held constant. The temperature appears explicitly in equation 15.16, but the volume is only implicit; as we saw in Chapter Twelve, the energy levels for translational motion depend upon the dimensions of the container and thus upon the volume. Obviously, the only things in A that need be varied are the N_i's keeping N constant. This is a simple task to which we shall now turn.

For the sake of convenience, let us define a quantity f_i as follows:

$$f_i = g_i e^{-E_i/kT} \tag{15.17}$$

Our task now reduces to finding the distribution of N_i's that will maximize a function χ defined by

$$\chi = \frac{f_1^{N_1} f_2^{N_2} f_3^{N_3} \cdots}{N_1! N_2! N_3! \cdots}$$

$$= \prod_i \frac{f_i^{N_i}}{N_i!} \tag{15.18}$$

Let us suppose that we have found the set of N_i's, namely, N_1, N_2, N_3, . . . , that constitutes the solution. If expression 15.18 as written is already a maximum, then transferring one particle from level 2, let us say, to level 1 cannot increase the value of χ but must lead to some value, χ', as follows:

$$\chi' = \frac{f_1^{N_1+1} f_2^{N_2-1} f_3^{N_3} \cdots}{(N_1 + 1)!(N_2 - 1)!N_3! \cdots} \ll \chi \tag{15.19}$$

Similarly, if we perturb the most probable distribution by moving a particle from level 1 to level 2, we shall again alter the expression for χ to another, denoted by χ'', such that

$$\chi'' = \frac{f_1^{N_1-1} f_2^{N_2+1} f_3^{N_3} \cdots}{(N_1 - 1)!(N_2 + 1)!N_3! \cdots} \ll \chi \tag{15.20}$$

If we divide χ' and χ'' each by χ, we obtain two simple inequalities involving N_1 and N_2, namely

$$\frac{\chi'}{\chi_{max}} = \frac{f_1}{N_1 + 1} \cdot \frac{N_2}{f_2} \ll 1 \tag{15.21}$$

and

$$\frac{\chi''}{\chi_{max}} = \frac{N_1}{f_1} \cdot \frac{f_2}{N_2 + 1} \ll 1 \tag{15.22}$$

Upon combining and rearranging the two inequalities we find that

$$\frac{1}{1 + 1/N_2} \ll \frac{f_1}{N_1} \cdot \frac{N_2}{f_2} \ll 1 + 1/N_1 \qquad (15.23)$$

Now if both N_1 and N_2 are large—and this assumption will generally be valid in our statistical mechanical applications—then f_1/N_1 must be substantially equal to f_2/N_2, since the expression $(f_1 N_2)/(f_2 N_1)$ must approach unity as N_1 and N_2 become infinite. Moreover, since it does not matter between what pair of levels we transfer particles, it follows in general that

$$\frac{N_1}{f_1} = \frac{N_2}{f_2} = \frac{N_3}{f_3} \ldots = \lambda \qquad (15.24)$$

where λ is a parameter independent of the energy levels. Therefore,

$$N_i = \lambda f_i$$
$$= \lambda g_i e^{-E_i/kT} \qquad (15.25)$$

Equation 15.25 expresses the well-known Boltzmann distribution law.

It will be noted that the foregoing derivation avoided, as promised, any use of the formal methods of the calculus of variations, Lagrangian multipliers, or Stirling's approximation. It was assumed, to be sure, that we were dealing with large numbers of particles, but this assumption is inherent in other derivations as well. The underlying reason why the present derivation proved to be so simple is the fact that we focused on the Helmholtz free energy instead of on the entropy. The temperature appears naturally, and no need arises to identify an initially unidentified Lagrangian multiplier with the temperature.

The assumption that the N_i's are large is really not important or necessary. If equations 15.24 are even approximately correct, then in view of the inequalities 15.21 and 15.22 we can assert that

$$-\frac{1}{N_i} \approx -\frac{1}{\lambda f_i} \ll \frac{N_i}{\lambda f_i} - \frac{N_j}{\lambda f_j} \ll \frac{1}{\lambda f_j} \approx \frac{1}{N_j} \qquad (15.26)$$

If anywhere in the system there exists at least one ratio $N_j/(\lambda f_j)$ equal to unity, then $N_i/(\lambda f_i)$ must be in the range $1 - 1/N_i$ to $1 + 1/N_j$. If N_i and N_j are of the same order of magnitude, then the range of allowable values for the most probable N_i must be confined to $\lambda f_i \pm 1$. If the magnitude of N_j is very large, then $\lambda f_i - 1 \ll N_i \ll \lambda f_i$ for the most probable value of N_i. Actually, there can be fluctuations outside the range, for it is by no means certain that a system will exhibit its most probable distribution.

Chemical Potential and Activity

Let us now turn to establishing the significance of the parameter λ. Before proceeding, however, we should take cognizance of the error in equations 15.13, 15.14 and 15.16 introduced by reason of the assumed distinguishability of Boltzmann particles. Because the particles are really indistinguishable, the term $N!$, which equals the number of distinguishable ways in which N particles can be arranged in order, should be eliminated; accordingly we shall simply omit that factor. This act on our part should not be deemed a weakness of the present derivation, since the difficulty is inherent in Boltzmann statistics by whatever means it is developed (see p. 263). Later in this chapter we shall deal with Fermi-Dirac and Bose-Einstein statistics, which give rise to correct results directly.

If we omit the factor $N!$ from equation 15.16 and substitute from equation 15.25 into 15.16, we should arrive at a proper expression for the Helmholtz free energy, A. From such it is then easy, by ordinary thermodynamic manipulations, to derive expressions for a host of other thermodynamically important functions, including the equation of state, heat capacity, etc. The evaluation of A can, of course, be readily accomplished by using Stirling's approximation for the factorial; however, since we have already made a point about how we can avoid that approximation, we will continue to do so in this chapter.

In the light of the preceding discussion, we can express A_{N-1}, the Helmholtz free energy for $N - 1$ particles, as follows:

$$A_{N-1} = - kT \ln \prod_i \frac{f_i^{N_i}}{N_i!} \tag{15.27}$$

provided the N_i's all add up to $N - 1$. Let us now ask what A_N would be. If one particle is added at constant volume and temperature to a large system of particles in statistical equilibrium, it must, of course, end up in some particular level. Since by hypothesis the system is in equilibrium, any perturbation produced by adding one particle to the system should do no more than increase the population of some level, say the jth level, by one. This is true because any additional net changes would imply nonequilibrium before the perturbation.* Therefore, assuming that the population of the jth level increases by one, the change in A

---

* There can, of course, be fluctuations, but they do not change the substance of the argument. For our purpose, we can assume that the added particle goes into the level most fractionally deficient in population but yet in the range stipulated by inequalities 15.23.

attending the increase in number of particles from $N - 1$ to N will be given by

$$A_N - A_{N-1} = -kT \ln f_j/N_j \tag{15.28}$$

But we know from equation 15.24 that $N_j = \lambda f_j$ for all j. Therefore

$$A_N - A_{N-1} = kT \ln \lambda \tag{15.29}$$

It is significant that j does not appear in equation 15.29; this means that A increases by an amount independent of the level into which the added particles goes. Moreover, if N is large, it is clear that $A_N - A_{N-1}$ must be essentially equal to $(\partial A/\partial N)_{T,V}$, since constancy of T and V have been stipulated throughout. But that partial derivative is also the chemical potential, μ, per molecule. In other words,

$$\mu = kT \ln \lambda \tag{15.30}$$

By definition, λ is known as the absolute activity of the material.* It is important to note that adding (or removing) one particle changes the Helmholtz free energy by an amount independent of the energy level into which the particle is placed (or from which it is removed). In other words, the chemical potential of the substance must be the same for each energy level and each level can, in a sense, be regarded as representing a different form of the same substance. Accordingly, a high energy level at equilibrium must correspond to a high entropy increment accompanying the addition or removal of a particle from such a level; similarly low energy levels at equilibrium are so populated that the entropy will change by only a small amount when a particle is added or removed. This concept is summarized by equation 15.25 rewritten as follows:

$$\lambda = \frac{N_i e^{E_i/kT}}{g_i} \tag{15.31}$$

The absolute activity, which must be the same for each level, is proportional to the population of the level and to the exponential function of the energy divided by kT; it is also inversely proportional to the number of ways in which a particle (distinguishable from others in Boltzmann statistics) can be added to the level. These factors are all so balanced that the chemical potentials associated with all levels at equilibrium must be equal.

---

* The general concept of activity will be dealt with in more detail in Chapter Nineteen. If μ in equation 15.30 is expressed on a molar basis, then k should be replaced by R, leaving λ unchanged.

Finally, it should be observed from equation 15.25 and from the definition of the partition function (equation 13.1) that

$$\lambda = \frac{N}{q} \qquad (15.32)$$

But from equation 15.30 we also see that the chemical potential (per molecule) will be given by

$$\mu = -kT \ln (q/N) \qquad (15.33)$$

Obviously for N molecules we should write

$$G = -NkT \ln (q/N) \qquad (15.34)$$

which is identical to equation 13.15. This result requires no further corrections since the $N!$ term has already been taken care of. Further elaboration is unnecessary, because the connection with earlier treatments is now complete.

The methods employed heretofore in this chapter for deriving the Boltzmann distribution law can also be used, with a little awkwardness, to derive the Bose-Einstein and Fermi-Dirac distribution laws. (Such derivations will be left for problems at the end of the chapter.) There is, however, an even simpler, yet highly general, method for handling all statistical thermodynamic distribution laws, a subject to which we shall now turn.

Distribution Laws and Microscopic Equilibrium

If we reflect on the discussion of chemical potential in the last section, we can recognize still another approach to statistical mechanics. Instead of maximizing or minimizing certain thermodynamic functions, subject to appropriate constraints, it is quite equivalent to stipulate that at equilibrium the chemical potential of the particles making up the system must be the same for all energy levels. This follows because a state of equilibrium must be identical, for example, with a state of minimum Helmholtz free energy for specified temperature and volume. If equilibrium and a minimum Helmholtz free energy were not simultaneously realized, the very foundations of classical thermodynamics would be untenable. Accordingly, we can treat general statistical thermodynamics from the point of view of microscopic equilibrium between energy levels.

Let us first inquire into the magnitude of the chemical potential by looking at the change in Helmholtz free energy attending the addition of one particle to a system in equilibrium. Suppose initially that there are $N - 1$ particles, with N_j particles in the jth energy level for all j except for the ith level, which we shall assume has a population of $N_i - 1$. We shall now add one particle to the ith level, thereby increasing the total energy by E_i and the population of that level up to N_i. If $\Omega(N - 1, N_i - 1)$ and $\Omega(N, N_i)$ are the thermodynamic probabilities before and after addition of the particle, then the change in entropy associated with adding the particle will be $k \ln [\Omega(N, N_i)/\Omega(N - 1, N_i - 1)]$. Therefore, the chemical potential (per molecule) will be given by

$$\mu = E_i - kT \ln [\Omega(N,N_i)/\Omega(N - 1, N_i - 1)] \qquad (15.35)$$

If we now define an absolute activity, λ, in accordance with equation 15.30, we assert that

$$\frac{\Omega(N - 1, N_i - 1)}{\Omega(N, N_i)} = \lambda e^{-E_i/kT} \qquad (15.36)$$

Equation 15.36 is a general statistical mechanical distribution law, valid for any system of counting, whether Boltzmann, Bose-Einstein, or Fermi-Dirac. Upon putting in the appropriate factor for Ω for Boltzmann systems, one obtains the Boltzmann law directly, provided the term $N!$ is suppressed.

For Bose-Einstein statistics, the expression for Ω is given by equation 14.2, which after neglecting unity compared to g_i becomes

$$\Omega = \prod_i \frac{(g_i + N_i)!}{g_i! N_i!} \qquad (15.37)$$

Upon substituting equation 15.37 into 15.36 we obtain

$$\frac{N_i}{g_i + N_i} = \lambda e^{-E_i/kT} \qquad (15.38)$$

which, upon rearrangement, yields the familiar form

$$N_i = \frac{g_i}{(1/\lambda)e^{E_i/kT} - 1} \qquad (15.39)$$

This equation is equivalent to equation 14.9 provided

$$\mu = -\alpha kT = kT \ln \lambda \qquad (15.40)$$

The expression for the thermodynamic probability of a Fermi-Dirac system is given by equation 14.43 as follows:

$$\Omega = \prod_i \frac{g_i}{(g_i - N_i)! N_i!} \qquad (15.41)$$

Substitution of 15.41 into 15.36 readily yields

$$\frac{N_i}{g_i - N_i + 1} = \lambda e^{-E_i/kT} \qquad (15.42)$$

Upon neglecting unity compared to g_i and rearranging, one then obtains the Fermi-Dirac distribution law, namely

$$N_i = \frac{g_i}{(1/\lambda)e^{E_i/kT} + 1} \tag{15.43}$$

This is a familiar equation requiring no new interpretation for λ. (See equation 15.40.)

It is now clear that equation 15.36 represents the statistical mechanical distribution law for any system of counting. It is conceivable, of course, that the quotient $\Omega(N - 1, N_i - 1)/\Omega(N, N_i)$ might not be free of all N_j's except N_i. Under such circumstances, a more complicated final formula would result; this could occur if Ω were not a continued product with each factor associated with one level only.

EXERCISE:

Derive the Boltzmann, Bose-Einstein and Fermi-Dirac distribution laws by substituting the proper expressions for their thermodynamic probabilities into equation 15.36 and rearranging the results.

The activity concept can now be summarized for the three specific statistics we have considered through use of one formula:

$$\lambda = \frac{N_i e^{E_i/kT}}{g_i \pm N_i(0)} \tag{15.44}$$

In employing equation 15.44, the zero is used for Boltzmann, the plus sign for Bose-Einstein, and the minus sign for Fermi-Dirac statistics. The significance of the coefficient to the exponential function has already been mentioned for Boltzmann systems. In general, that factor is proportional to the population and inversely proportional to the number of ways another particle can be introduced into the level. For Fermi-Dirac systems, it is clear that the activity becomes infinite when $N_i = g_i$, since no more particles can be accommodated. (If one retained the unity in equation 15.42, one would say the activity is a maximum when $g_i = N_i$ and would become infinite, hence unattainable, for $N_i = g_i + 1$.) For Bose-Einstein systems, on the other hand, an increase in population density of a given level simultaneously increases the number of ways in which another particle can be introduced into that level.

The general distribution formula, equation 15.36, appears to be the most compact formula yet proposed for such matters. One can, of course, minimize the Helmholtz energy for each type of statistics, since stipulating equality of chemical potentials for different levels is really quite equivalent. However, the chemical potential argument, as well as being direct and simple, is advantageous for understanding statistical mechanics from the point of view of microscopic chemical equilibrium.

Problems

1. Using the methods employed to obtain equation 15.23, derive the corresponding set of inequalities for Bose-Einstein statistics, namely:

$$\frac{1 + \dfrac{1}{g_2 + N_2}}{1 + \dfrac{1}{N_2}} \lessgtr \frac{h_1(g_1 + N_1)}{N_1} \cdot \frac{N_2}{h_2(g_2 + N_2)} \lessgtr \frac{1 + \dfrac{1}{N_1}}{1 + \dfrac{1}{g_1 + N_1}}$$

where $h_i = e^{-E_i/kT}$.

Note that the quantity of the left hand side of the above expression is less than unity, whereas that on the right hand side is greater than unity. Letting N_1 and N_2 become large, deduce equation 15.38.

2. In a manner similar to that used to obtain equation 15.23, derive the following set of inequalities for a Fermi-Dirac system:

$$\frac{1}{(1 + 1/N_2)(1 + 1/[g_1 - N_1])} \lessgtr \frac{h_1(g_1 - N_1)}{N_1} \cdot \frac{N_2}{h_2(g_2 - N_2)}$$
$$\lessgtr (1 + 1/N_1)(1 + 1/[g_2 - N_2])$$

where $h_i = e^{-E_i/kT}$.

Hence deduce equation 15.42, except that $g_i + 1$ is replaced by g_i. (This difference comes about because we start with $N_i - 1$ particles instead of N_i.) Note that one or the other of the inequalities of this problem becomes useless if $N_i = g_i$. This really creates no serious problem, however, since g_i is the maximum value for N_i under Fermi-Dirac rules. As long as $N_i \ll g_i$, meaningful conclusions are directly forthcoming and the situation corresponding to $N_i \longrightarrow g_i$ should be regarded as a limiting case.

3. Since radiation can be treated as a photon gas subject to Bose-Einstein statistics, the Planck distribution law is directly obtained from that of Bose-Einstein by setting the chemical potential, μ, equal to zero. This is quite equivalent to setting λ, the generalized activity, equal to unity in equation 15.36. However, the Planck distribution can also be derived by a simple, instructive method corresponding to the first derivation of this chapter.

First of all, set up the expression for the Helmholtz free energy for photons:

$$A = - kT \ln \prod_i \frac{(g_i + N_i)! e^{-N_i h\nu_i/kT}}{g_i! N_i!}$$

To maximize the argument of the logarithm, assume first that you have somehow found the answer, given by $N_1, N_2, N_3, \ldots$, etc. Now add one photon to a particular level; this can be done without removing a photon from any other level, since photons are not conserved. Such addition will render the argument of the logarithm equal to or less than that corresponding to the maximum value. Similarly, you can obtain another inequality by subtracting a photon from the same original level without perturbing any other levels. The two resulting inequalities then enable you to deduce the Planck distribution law, assuming the N_i's are large.

The novelty of this derivation is the addition or subtraction of photons from one and only one level to obtain the condition leading to a minimum Helmholtz free energy.

16

Statistical Thermodynamics of Rubber

Résumé of General Considerations

When dealing with the heat effects and the work involved in deforming a system without producing changes in chemical composition, we have dealt for the most part with gaseous systems. As justification for this emphasis on gases, it was pointed out that the heat and work attending the compression of ordinary solids and liquids, even for large pressure changes, are so small that they can be neglected for most chemical-thermodynamic purposes. There is, however, an important class of non-gases whose deformations are accompanied by appreciable thermal effects; these substances are the high-polymeric plastic materials, and the best example for our immediate purpose is rubber.

The property of rubber that makes it such a useful commodity is its long-range elasticity. Moderate force will stretch a strip of rubber much more than a typical solid, and relatively long extensions can occur more or less reversibly, without breakage or permanent structural changes. To be sure, very long extensions might produce significant internal changes of an irreversible character; nevertheless, we shall assume, in this chapter, that moderate deformations of rubber are reversible. With this in mind, we shall now inquire into the thermodynamics of rubber, first from a macroscopic point of view, and later with particular reference to statistical mechanics.

Let us consider a cylinder of rubber of length L and cross-sectional area A. If a force, f, is applied to the ends of the cylinder, with its sign so chosen that a positive f tends to increase L, then the rubber will, at equilibrium, be elongated by some amount ΔL. It is observed experimentally that upon elongation the cross-sectional area of a piece of rubber will simultaneously diminish by such an amount that the total volume remains nearly constant. In the light of this experimental observation, and for the sake of simplicity, we shall now stipulate that one property of an "ideal rubber" is that it is incompressible, which means that its volume does not change during deformation.* This is also equivalent to saying that no pressure-volume work will attend the elongation of a piece of rubber; other work will nevertheless be involved, and for an infinitesimal change in length, we can write

$$dW = -fdL \qquad (16.1)$$

Evidently the restoring force, f, is a function only of temperature and length, the volume being invariant; this elimination of volume as a variable likewise renders unnecessary any reference to pressure. This means that we can write for the "equation of state" of a rubber strip some expression in the form $f = f(T,L)$. Furthermore, practically all of our familiar basic thermodynamic equations can now be rewritten and applied to rubber by simply substituting $-f$ for p and L for V, as suggested by equation 16.1. The negative sign that accompanies f occurs because the work is negative when a piece of rubber is elongated, in contradistinction to the fact that work is positive when a gas expands.

Insofar as rubber-like elasticity is concerned, all of the principles of thermodynamics previously developed are applicable and many of the functions previously defined can be used directly. The work attending a change in state will be given by an appropriate line integral, using the expression of equation 16.1:

$$W = -\int fdL \qquad (16.2)$$

This line integral will, of course, depend on the path, as does Q, the heat accompanying the process. The concepts of energy, entropy, and of Helmholtz free energy remain unchanged from the earlier definitions applied to isotropic systems. Thus the change in energy, ΔU, must still

---

* This statement also implies that Poisson's ratio for rubber equals $\frac{1}{2}$.

be defined as $Q - W$, and the decrease in Helmholtz free energy, $-\Delta A$, for an isothermal process equals W_{rev}. The enthalpy, however, should now be replaced by an "elastomer enthalpy," here denoted by K and defined by

$$K = U - fL \tag{16.3}$$

This definition takes cognizance of the fact that p and V are no longer useful variables and should be replaced by $-f$ and L respectively. Similarly the Gibbs free energy can be replaced by a quantity J defined by

$$J = A - fL$$
$$= K - TS \tag{16.4}$$

With these definitions, we can now quickly list several thermodynamic equations designed for rubber-like systems.

If Q_L and Q_f are the heats attending constant length and constant force changes in state, then

$$Q_L = \Delta U \tag{16.5}$$

and

$$Q_f = \Delta K \tag{16.6}$$

Moreover the generalized heat capacity can be written

$$C = \left(\frac{\partial U}{\partial T}\right)_L + \left\{\left(\frac{\partial U}{\partial L}\right)_T - f\right\}\frac{dL}{dT} \tag{16.7}$$

or

$$C = \left(\frac{\partial K}{\partial T}\right)_f + \left\{\left(\frac{\partial K}{\partial f}\right)_T + L\right\}\frac{df}{dT} \tag{16.8}$$

Evidently the heat capacities at constant length and constant force will be given by

$$C_L = \left(\frac{\partial U}{\partial T}\right)_L \tag{16.9}$$

and

$$C_f = \left(\frac{\partial K}{\partial T}\right)_f \tag{16.10}$$

In formulating these heat capacities, we are, of course, disregarding any work attending thermal expansion of the total volume, although we consider the effect of temperature on force and length.

EXERCISE:

Show that the difference between the heat capacities at constant force and constant length will be given by

$$C_f - C_L = \left\{ \left(\frac{\partial U}{\partial L} \right)_T - f \right\} \left(\frac{\partial L}{\partial T} \right)_f$$

$$= -\left\{ \left(\frac{\partial K}{\partial f} \right)_T + L \right\} \left(\frac{\partial f}{\partial T} \right)_L$$

The differentials of E, K, A, and J can be set forth as follows:

$$dU = TdS + fdL \tag{16.11}$$

$$dK = TdS - Ldf \tag{16.12}$$

$$dA = -SdT + fdL \tag{16.13}$$

$$dJ = -SdT - Ldf \tag{16.14}$$

From the above equations, it is clear that $(\partial U/\partial S)_L = T$, $(\partial U/\partial L)_S = f$, etc. Moreover, by considering the alternative ways of obtaining mixed partial derivatives of the type $\partial^2 U/\partial L \partial S$, one can derive four counterparts to Maxwell's relations, namely:

$$\left(\frac{\partial T}{\partial L} \right)_S = \left(\frac{\partial f}{\partial S} \right)_L \tag{16.15}$$

$$\left(\frac{\partial T}{\partial f} \right)_S = -\left(\frac{\partial L}{\partial S} \right)_f \tag{16.16}$$

$$\left(\frac{\partial S}{\partial L} \right)_T = -\left(\frac{\partial f}{\partial T} \right)_L \tag{16.17}$$

$$\left(\frac{\partial S}{\partial f} \right)_T = \left(\frac{\partial L}{\partial T} \right)_f \tag{16.18}$$

Finally, we shall conclude our general resume by observing that the dependence of energy on length and the dependence of the "elastomer enthalpy" on force will be given by

$$\left(\frac{\partial U}{\partial L} \right)_T = f - T \left(\frac{\partial f}{\partial T} \right)_L \tag{16.19}$$

and

$$\left(\frac{\partial K}{\partial f} \right)_T = T \left(\frac{\partial L}{\partial T} \right)_f - L \tag{16.20}$$

EXERCISE:

Assuming S to be a function of L and T, substitute the appropriate expression for dS into equation 16.11 and identify $(\partial U/\partial L)_T$. Making use of equation 16.17, prove equation 16.19.

EXERCISE:

Assuming S to be a function of f and T, substitute the appropriate expression for dS into equation 16.12 and identify $(\partial K/\partial f)_T$. Making use of equation 16.18, prove equation 16.20.

Experimentally it is observed that f is proportional to T for moderate extensions of rubber. This means that the equation of state must be of the form $f - T\Phi(L)$ and, in view of equation 16.19, that $(\partial U/\partial L)_T = 0$. We shall now assert that this additional property, namely, that the energy be a function of temperature only, is another characteristic of an ideal rubber.* Upon combining equations 16.19 and 16.17, we conclude that, for an ideal rubber

$$f = -T\left(\frac{\partial S}{\partial L}\right)_T \tag{16.21}$$

This equation implies that the restoring force that tends to bring a piece of stretched ideal rubber back to its original length must be entropic in nature. The work done in stretching a piece of rubber is thermodynamically unlike the work of stretching a coiled metal spring. When a metal spring is subjected to a change in its dimensions, its energy is changed; on the other hand, the energy of a piece of ideal rubber is unchanged when it is isothermally distorted. In other words, when a piece of stretched rubber is allowed to snap back to its original length, the process can be likened to that of the free expansion of an ideal gas into a vacuum, a process involving an increase in entropy without change in energy. Accordingly, in the light of equation 16.21, we should inquire into the dependence of entropy on length if we are to derive an equation of state for a strip of rubber.

Molecular Model of Rubber

To establish the dependence of the entropy of a rubber-like substance on its elongation, we should first examine the nature of the rubber molecules. A single molecule of rubber is a long-chain polymer with a high degree of over-all flexibility. This molecular flexibility results from a considerable degree of free rotation about some of the bonds in the polymer chain. As a result of such rotations, a molecule can assume a tremendous number of different configurations with little change in energy. Thus, a typical rubber-like molecule will give the appearance

---

* This is in direct analogy to one of the properties of an ideal gas. The "elastomer enthalpy," on the other hand, is not independent of f, so the analogy to ideal gases stops there.

of a loose, randomly tangled thread. Although unlikely, it is nevertheless conceivable that a given molecule might by chance exhibit a straightened out configuration, thereby possessing a maximum end-to-end separation. At the other extreme, a molecule might look like a tight little ball of yarn. Most molecules, however, will appear to have random configurations with a considerable amount of free space within their general domains.

Let us now consider the nature of a macroscopic aggregate of uncured rubber. Such an aggregate, which will be amorphous in character, will consist of random molecules intertwined among each other, thus constituting a mass that would maintain some degree of dimensional stability because of the enmeshing of the molecules. The dimensional stability would not be very substantial, however, since an appreciable deformation of such a mass would cause individual molecules to slide by each other, ultimately leading to separation of the over-all mass. The dimensional stability that is realized in commercial rubber products results from chemically bonding the molecules to each other at various points along their chains. The process by which this bonding or cross-linking takes place is known as vulcanization, and the net result of such cross-linking is to make the mass of rubber the equivalent of one giant molecule. An adequately cross-linked piece of rubber can properly be regarded as one molecule because any of its atoms can be reached from any other atom by following an uninterrupted sequence of connected chemical bonds. Between the cross-link points in a mass of vulcanized rubber, the chains will still have highly random configurations, so the contour path from one cross-link to the next may well be many times the shortest geometrical distance between the points in question.

Cross-linking a piece of rubber renders it quite dimensionally stable, so that after it is subjected to moderate deformations, the system will tend to return to its original shape. Extending a piece of vulcanized rubber will, to be sure, affect the statistical dimensions of the molecular chains that go to make up the system, but the original molecules can no longer be easily separated from each other since they are now firmly bound together as a result of the vulcanization process. Our statistical model of rubber will, therefore, consist of a three-dimensional network of polymer chains, in which the ends of each chain are joined to the ends of at least two other chains at cross-link points.

Let us now inquire into the distribution of the lengths of the polymer chains that serve to connect the cross-link points. For the sake of simplicity, we shall first assume that the contour length, or the number of bonds associated with a chain, is the same throughout the mass of rubber.

(Actually, some chain contour lengths will be small and others will be large, as determined by a count of atoms, but these variations will not affect our results.) We shall further assume that the distribution of the end-to-end separations of the molecular chains will be the same as if they were individual molecules, free to assume any possible configurations. This is a reasonable assumption since the introduction of cross-links into a piece of initially uncured rubber need not change the molecular shapes and distributions that existed at the time of vulcanization. Let us therefore focus our attention on a single flexible chain, chosen as that portion of a rubber molecule lying between two cross-link points.

An approximation of the statistical size and shape of a flexible molecule can be obtained by considering the polymer to be like a string of beads in which each successive bead can occupy any position in space, provided only that it is precisely one bond length away from the preceding bead. This simple model fails to take into account any intramolecular interactions; indeed it does not even preclude double occupancy of sites in space. According to this model, the contour of a rubber-like chain would look precisely like the path followed by a particle undergoing Brownian motion, assuming that the distance the particle travels between collisions is constant (and equal to the bond length in the polymer molecule). Such a model corresponds to the so-called "random walk." By further stipulating that one cannot pass through any previous portion of the path, the random walk so modified will closely resemble a random string of beads since the excluded volume effect is thus taken into account. For our present purpose, however, we shall limit our consideration to the unrestricted random walk model.

In using the random walk model to describe the statistical dimensions of a polymer molecule, let us assume that the chain is constrained to a simple cubic lattice with spacings between adjacent points equal to a common bond length. Although the use of such a lattice model might seem oversimplified, the result obtained is substantially the same as one would get for any other lattice, provided that no intramolecular interactions or restrictions concerning double occupancy are imposed. The first bead of the polymer chain will be placed at the origin $(0, 0, 0)$ and the second will be located at one of the sites $(l, 0, 0)$, $(-l, 0, 0)$, $(0, l, 0)$, $(0, -l, 0)$, $(0, 0, l)$ or $(0, 0, -l)$, in which l is the bond length. The third bead would then be placed one bond length away from the second in any of the directions $\pm x$, $\pm y$, or $\pm z$, the fourth would be one bond length from the third, etc. Let us now inquire into the probability that the end of a chain be located at any particular site in the lattice.

Let $p(n, x, y, z)$ be the probability that the nth bond (lying between the nth and $(n + 1)$th beads) terminates at point x, y, z in space. To determine how p changes with increasing n, we shall now establish a difference equation relating $p(n + 1, x, y, z)$ to an appropriate combination of probabilities corresponding to n. Evidently to reach point x, y, z in $n + 1$ steps, the nth bond must have terminated in one of the following six positions: $x \pm l, y, z; x, y \pm l, z; x, y, z \pm l$. The probability of reaching point x, y, z in $n + 1$ steps must therefore equal one-sixth of the sum of the probabilities of reaching each of the six neighboring sites at n steps; the factor one-sixth arises, of course, because each successive step has one-sixth of a chance of being propagated in any particular direction. Therefore

$$p(n + 1, x, y, z) = \tfrac{1}{6}\{p(n, x + l, y, z) + p(n, x - l, y, z)$$
$$+ p(n, x, y + l, z) + p(n, x, y - l, z)$$
$$+ p(n, x, y, z + l) + p(n, x, y, z - l)\} \qquad (16.22)$$

The initial starting conditions are that

$$p(n, 0, 0, 0) = 1 \qquad \text{for } n = 0$$

and

$$p(n, x, y, z) = 0 \qquad \text{for } n = 0 \text{ and } x \neq 0, y \neq 0, \text{ or } z \neq 0 \qquad (16.23)$$

With such starting conditions, the probabilities calculated by use of equation 16.22 are automatically normalized to unity, that is to say

$$\sum_x \sum_y \sum_z p(n, x, y, z) = 1 \qquad (16.24)$$

One can, through successive application of the difference relationship given by equation 16.22, derive numerical values for $p(n, x, y, z)$ for ever increasing values of n. For analytical purposes, strictly numerical procedures of this kind can be awkward; nevertheless the difference equation can be used to derive some important formulas as, for example, that describing the average square end-to-end separation, $\langle r^2 \rangle_n$.

Equation 16.22 can be simplified by reducing it to one-dimensional form. To do so, it is only necessary to define a new quantity $p(n,x)$ equal to the total probability that the n-link chain will have the indicated x component irrespective of y and z. Evidently

$$p(n,x) = \sum_y \sum_z p(n, x, y, z) \qquad (16.25)$$

where the summations are carried over all allowable values of y and z. Of course,

$$\sum_x p(n,x) = 1 \qquad (16.26)$$

From equation 16.22, it follows that

$$p(n + 1, x) = \tfrac{1}{6}\{p(n, x + l) + 4p(n,x) + p(n, x - l)\} \qquad (16.27)$$

Now an average property, such as the average square value of x, will be given by

$$\langle x^2 \rangle_n = \sum_x x^2 p(n,x) \qquad (16.28)$$

To obtain $\langle x^2 \rangle_{n+1}$, it is only necessary to multiply equation 16.27 by x^2 and sum over all values of x. Evidently

$$\langle x^2 \rangle_{n+1} = \tfrac{1}{6}\sum_x x^2 p(n, x + l) + \tfrac{2}{3}\sum_x x^2 p(n,x) + \tfrac{1}{6}\sum_x x^2 p(n, x - l) \qquad (16.29)$$

The second term on the right-hand side of the above equation is precisely $\tfrac{2}{3}\langle x^2 \rangle_n$. The first and third terms, on the other hand, require rearrangements before they can be easily recognized. For the first term let $x = x' - l$ and for the third let $x = x' + l$. Then those two terms, except for the factor $\tfrac{1}{6}$, become

$$\sum_{x'} (x'^2 - 2lx' + l^2)p(n,x') + \sum_{x'} (x'^2 + 2lx' + l^2)p(n,x')$$

Upon consolidation and dropping of primes, these terms become

$$2 \sum_x x^2 p(n,x) + 2l^2 \sum_x p(n,x)$$

Introducing the factor $\tfrac{1}{6}$ and substituting the expression just derived back into equation 16.29, we obtain

$$\langle x^2 \rangle_{n+1} = \sum_x x^2 p(n,x) + \frac{l^2}{3} \sum_x p(n,x)$$

$$= \langle x^2 \rangle_n + \frac{l^2}{3} \qquad (16.30)$$

Since $\langle x^2 \rangle_0 = 0$, successive application of equation 16.30 will yield the result

$$\langle x^2 \rangle_n = nl^2/3 \qquad (16.31)$$

Finally we can assert, through symmetry arguments, that

$$\langle x^2 \rangle_n = \langle y^2 \rangle_n = \langle z^2 \rangle_n = \tfrac{1}{3}\langle r^2 \rangle_n \qquad (16.32)$$

in which

$$r^2 = x^2 + y^2 + z^2 \qquad (16.33)$$

Hence

$$\langle r^2 \rangle_n = nl^2 \qquad (16.34)$$

From equation 16.34 it is clear that the mean square end-to-end separation of an unrestricted flexible chain is proportional to the number of links in the chain. This result, although derived for a simple cubic lattice, is generally true for all lattices provided the sum of the several vector choices available for successive chain additions is zero. In the event that the bond rotation is not free or that the number of immediate vector choices is geometrically limited so as to be unsymmetrical, the coefficient of nl^2 in equation 16.34 may be changed, but the dependence on n will remain the same as long as n is large. On the other hand, if long loops are excluded by reason of limitations against double occupancy of lattice sites, the dependence of $\langle r^2 \rangle_n$ on n can be significantly altered.

Although the average end-to-end separation of coiling-type molecules is an important quantity, for many purposes it is also necessary to know the distribution of lengths. Evidently the distribution function can be obtained by successive application of equation 16.22, subject to the starting conditions, but this does not yield a convenient analytical expression. By an appropriate approximation, however, equation 16.22 can be converted into a differential equation, which can be solved to provide an expression more convenient to work with. To accomplish this conversion, let us express the terms on the right-hand side of equation 16.22 by Taylor's series expansions about x, y, z. Typical terms can, for example, be written as follows:

$$p(n, x \pm l, y, z) = p(n, x, y, z) \pm \left(\frac{\partial p}{\partial x}\right) l + \frac{1}{2}\left(\frac{\partial^2 p}{\partial x^2}\right) l^2 \pm \cdots \quad (16.35)$$

Similar expressions will apply for terms involving increments in y and z. For the left-hand term of equation 16.22 we shall write

$$p(n+1, x, y, z) = p(n, x, y, z) + \left(\frac{\partial p}{\partial n}\right)_{x,y,z} + \cdots \quad (16.36)$$

Upon substituting equations 16.35 and 16.36 into equation 16.22, we obtain the following approximate equation:

$$\frac{\partial p}{\partial n} = \frac{l^2}{6}\left(\frac{\partial^2 p}{\partial x^2} + \frac{\partial^2 p}{\partial y^2} + \frac{\partial^2 p}{\partial z^2}\right) \quad (16.37)$$

This partial differential equation represents only an approximation to the difference equation since only the lowest-order derivatives with non-zero coefficients have been retained. If n is large, under which circumstances the discrete lattice effect should vanish, p must become spherically symmetric, that is to say, a function of n and r only. Converting the Laplacian operator on the right-hand side of equation 16.37 to polar-coordinate form, and suppressing any angular dependence, the equation becomes

$$\frac{\partial p}{\partial n} = \frac{l^2}{6r^2} \frac{\partial}{\partial r} \left(r^2 \frac{\partial p}{\partial r} \right) \qquad (16.38)$$

A complete analysis of possible solutions to equation 16.38 leading to one that satisfies the boundary conditions is quite involved. We shall, accordingly, simply state the solution that satisfies the boundary conditions, leaving verification of its validity as an exercise. Such a solution already normalized is

$$p(n,r) = \left(\frac{3}{2\pi n l^2} \right)^{3/2} e^{-3r^2/2nl^2} \qquad (16.39)$$

The number of molecules with end-to-end separations in the range r to $r + dr$ will, of course, be given by $4\pi r^2 p(n,r)dr$; the normalization referred to earlier requires that

$$\int_0^\infty 4\pi r^2 p(n,r)dr = 1 \qquad (16.40)$$

Using Cartesian coordinates, so that x, y, z represent the components of the end-to-end separations, we can also assert that the probability of an end-to-end separation lying in the range x to $x + dx$, y to $y + dy$, and z to $z + dz$ will be given by $p(n,r)\,dxdydz$ with $r = (x^2 + y^2 + z^2)^{1/2}$.

EXERCISE:

Show that $p(n,r)$ as given by equation 16.39 satisfies the differential equation 16.38.

EXERCISE:

Show that $p(n,r)$ as given by equation 16.39 is normalized in accordance with equation 16.40.

EXERCISE:

Show by evaluating the integral of $4\pi r^4 p(n,r)dr$ that $\langle r^2 \rangle_n = nl^2$.

Theory of Rubber-like Elasticity

We shall now develop a statistical theory of rubber-like elasticity using the Gaussian formula, equation 16.39, to represent the most probable distribution of molecular end-to-end separations. The mass of rubber is assumed to consist of N chains (or molecules) each end of which is joined to the ends of other chains at vulcanization points. We shall assume that the mass of rubber is in the shape of a cylinder of length L and cross-sectional area A, with the length direction parallel to the x axis. Deformations will consist of stretching the rubber in the x direction; the constancy of total volume in the process should be noted.

In its relaxed condition, the system will be statistically isotropic and the end-to-end separations of the chains will be given by equation 16.39, which we shall now rewrite using Cartesian coordinates:

$$p(n, x, y, z) = \left(\frac{3}{2\pi nl^2}\right)^{3/2} e^{-3(x^2 + y^2 + z^2)/2\,nl^2} \qquad (16.41)$$

Now suppose that the rubber is extended by a factor α so that its length becomes equal to αL and its cross-sectional area becomes A/α. Since the y and z directions must diminish by the same relative amounts, each of the external dimensions in those directions will change by the factor $1/\sqrt{\alpha}$. Let us now inquire into what happens to the distribution of end-to-end separations when the macroscopic piece of rubber is stretched. A reasonable assumption, which can be justified by a more detailed statistical-mechanical analysis, is that the Gaussian distribution in the x direction is stretched by a factor α. At the same time the Gaussian distributions in the y and z directions are compressed by the factor $1/\sqrt{\alpha}$. This means that the original three-dimensional isotropy of the system is destroyed, for in the elongated state the x direction becomes unique, although two-dimensional isotropy will still remain in planes perpendicular to the x direction. To derive the molecular-length distribution function for the deformed rubber, we need only rewrite equation 16.41 by replacing x by x/α, y by $\sqrt{\alpha}y$, and z by $\sqrt{\alpha}z$.

$$p(n, x/\alpha, \sqrt{\alpha}y, \sqrt{\alpha}z) = \left(\frac{3}{2\pi nl^2}\right)^{3/2} e^{-3(x^2/\alpha^2 + \alpha y^2 + \alpha z^2)/2nl^2} \qquad (16.42)$$

which remains normalized since $dx\,dy\,dz$ is unchanged by the transformation. In Figure 16-1 there is shown the distribution functions for each of the components of the original unstretched rubber and for rubber that has been stretched by 100 percent ($\alpha = 2$). The problem now simply reduces to calculating the probability of finding a mass of rubber molecules with the deformed distribution and comparing it to the probability of the relaxed state. The entropy change attending the deformation can then be readily calculated from a knowledge of those probabilities.

If the probability that a given chain possesses a particular end-to-end separation is p_i and if N_i molecules actually possess such a separation, then the thermodynamic probability associated with the distribution will be given by

$$\Omega = N! \prod_i \frac{p_i^{N_i}}{N_i!} \qquad (16.43)$$

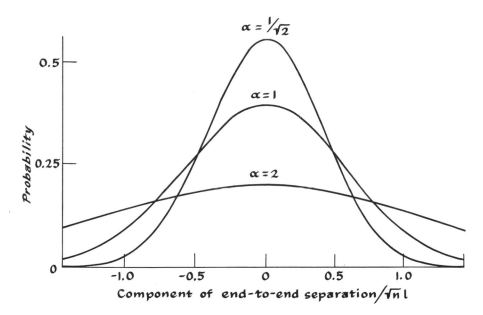

FIGURE 16-1   Graphs showing probability distributions versus components of end-to-
end separations of random rubber-like polymer chains. The curve marked $\alpha = 1$
applies to x, y, and z components alike for unstretched rubber; the curve marked
$\alpha = 2$ applies to x components after rubber is stretched by a factor of 2 in the x
direction; curve marked $\alpha = 1/\sqrt{2}$ applies to y and z components for the same
rubber stretched by a factor of 2 in x direction.

in which N is the total number of chains. Equation 16.43 is quite similar
to equation 11.43 except that the degeneracy is replaced by p_i. It can
be derived by considering an expansion of the type

$$(p_1 + p_2 + p_3 + \ldots)^N = \sum \ldots \sum \frac{N! p_1^{N_1} p_2^{N_2} \ldots}{N_1! N_2! \ldots} \qquad (16.44)$$

in which each term of the multiple summation is precisely the thermodynamic probability associated with the particular distribution represented thereby. Of course it is assumed that the chains are all
distinguishable and that ordinary Boltzmann statistics can be applied.

Taking the logarithm of Ω and multiplying by k, we obtain ΔS, the
change in entropy attending the process of stretching. The result, after
simplification in the light of Stirling's approximation, is given by

$$\Delta S = k \ln \Omega = k \sum_i N_i \ln (N p_i / N_i) \qquad (16.45)$$

Now p_i, the probability that a given chain is characterized by a particular set of values x_i, y_i, and z_i is given by equation 16.41. Moreover, N_i, the actual distribution for deformed rubber will be given by N times $p(n, x/\alpha, \sqrt{\alpha}y, \sqrt{\alpha}z)$, expressed by equation 16.42, which we shall henceforth indicate by $p(\alpha)$. Upon making these substitutions and replacing the summation by integrations, we find that

$$\Delta S = Nk \iiint p(\alpha) \ln \left[p(1)/p(\alpha) \right] dxdydz \qquad (16.46)$$

in which $p(1)$ is, of course, given by equation 16.41 or equation 16.42 with $\alpha = 1$. Further substitution shows that

$$\Delta S = Nk \iiint p(\alpha) \left\{ \frac{3}{2nl^2} \left[\left(\frac{1}{\alpha^2} - 1 \right) x^2 + (\alpha - 1)(y^2 + z^2) \right] \right\} dxdydz$$
$$(16.47)$$

Without actual evaluation of the integrals, it should be clear that the integral

$$\iiint p(\alpha)x^2 dxdydz$$

is precisely equal to the average value of x^2 for the stretched distribution, that is, $\alpha^2 nl^2/3$. Similarly the integrals involving y^2 and z^2 should give the average values of these quantities for the stretched rubber, which is, in fact, compressed in the y and z directions. These average values of y^2 and z^2 are $nl^2/3\alpha$. It follows, therefore, that the entropy change on stretching will be given by

$$\Delta S = -\frac{Nk}{2} \left(\alpha^2 + \frac{2}{\alpha} - 3 \right) \qquad (16.48)$$

Evidently when $\alpha = 1$, $\Delta S = 0$, so we have correctly identified $k \ln \Omega$ as the change in entropy on stretching.

From our earlier discussion, we recall that the restoring force of a deformed piece of ideal rubber is entropic in nature and will be given by equation 16.21. Therefore, since $\alpha = L/L_0$, it follows that

$$f = -\frac{T}{L_0} \left(\frac{\partial S}{\partial \alpha} \right)_T$$
$$= \frac{NkT}{L_0} \left(\alpha - \frac{1}{\alpha^2} \right) \qquad (16.49)$$

This is the "equation of state" for an ideal rubber, an equation that is found to agree fairly well with experiment provided the extensions do not exceed several hundred percent.

EXERCISE:

Show that the heat and work attending the stretching of an ideal piece of rubber is given by $Q = W = -\dfrac{NkT}{2}\left(\alpha^2 + \dfrac{2}{\alpha} - 3\right)$

Problems

1. Suppose that the restoring force, f, of a nonideal strip of rubber is given by a linear function of T of the form $f = \Theta(L) + T\Phi(L)$ in which $\Theta(L)$ and $\Phi(L)$ are each functions of L only. Prove that, if C_L is constant, the energy of the strip of rubber will be given by

$$E = \int \Theta(L)dL + C_L T + \text{constant}$$

2. Show for a strip of rubber that the difference in heat capacities at constant force and constant length will be given by the expression

$$C_f - C_L = -T\left(\frac{\partial f}{\partial T}\right)_L \left(\frac{\partial L}{\partial T}\right)_f$$

3. Show for an ideal rubber strip that $C_f - C_L = f^2/T(\partial f/\partial L)_T$.

4. Show that the differential equation for a reversible adiabatic extension or retraction of an ideal rubber strip can be written, with variables separated, in the form $C_L/T\, dT = \phi(L)\, dL$. Recognizing that $\phi(L)$ increases monotonically with L, show that a reversible adiabatic extension of a rubber strip is accompanied by heating.

5. Show how an ideal rubber strip can be carried through a reversible Carnot cycle, and demonstrate that its efficiency is the same as that for an ideal gas. Assuming that C_L, the heat capacity at constant length of an ideal rubber strip, is independent of temperature, plot f against L for the cycle, making use of equation 16.49.

6. If a block of ideal rubber is subjected to pure shear by compressing the z dimension by a factor α and extending the y dimension by the same factor, while holding the x dimension unchanged, show that the entropy change is given by

$$\Delta S = -\frac{Nk}{2}\left(\alpha^2 + \frac{1}{\alpha^2} - 2\right) = -\frac{Nk}{2}\left(\alpha - \frac{1}{\alpha}\right)^2$$

Hence show that the work done in shearing a piece of rubber is equal to $NkT\,\gamma^2/2$ in which γ, the amount of shear strain, is measured by $\alpha - 1/\alpha$. The shearing force will be given by the derivative (with respect to γ) of the work done in producing the shear. Hence show that an ideal rubber obeys Hooke's law for shear with the modulus of rigidity equal to NkT.

7. The mean square radius of gyration about the center of gravity of a random-coiled polymer chain is $(1/6)\langle r^2 \rangle$ or $(1/6)nl^2$, in which $\langle r^2 \rangle$ is the mean square end-to-end separation. (This does not refer to the radius of gyration about an axis, but about a point.) Assuming the random polymer to be a sphere of uniform density, with a moment of inertia about its center equal to that of a random chain, show that the radius, R, of that sphere is given by $\sqrt{5n/18}\, l$. Taking the volume, v, of such a sphere to be $(4/3)\pi R^3$, derive an expression for the concentration of monomer units, n/v, within the sphere. Assuming n to equal 10,000 and l to be 3 angstroms, calculate the concentration of monomer units in moles per liter and compare with a pure substance like water. (Note how dilute the molecular chains are when taken by themselves. Actually the chains are intermeshed with other chains to bring the over-all density of rubber up to about 1 g cm^{-3}.)

8. Consider the hypothetical polymer consisting of n single bonds alternating with n double bonds:

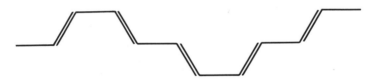

The chain is all trans and the angle between adjacent bonds is θ. The first bond, which is single, starts at the origin and is parallel to the x direction; there is free rotation about the single bonds and no rotation about the double bonds. If the single bonds are of length l_1 and the double bonds of length l_2, what are $\langle x^2 \rangle$, $\langle y^2 \rangle$, and $\langle z^2 \rangle$, where x, y, and z denote end-to-end separations?

17

Nonideal Systems

Chemical Potential and the Definition of Fugacity (Gaseous Activity)

We have already seen that the change in molar Gibbs free energy (or chemical potential) attending an isothermal change in state of an ideal gas is given by the equation

$$\Delta G = \Delta \mu = RT \ln \frac{p_2}{p_1} \tag{17.1}$$

in which p_1 and p_2 are the initial and final pressures. Another statement, equivalent to the one above but obtained by indefinite integration, is

$$\mu = \phi(T) + RT \ln p$$
$$= \mu^\circ + RT \ln p \tag{17.2}$$

in which $\phi(T)$ (or μ°) is a function of temperature only. We saw in Chapter Ten that, if the chemical potentials were expressed in the form of equation 17.2, the reaction potential assumed a particularly tractable form, which led quite naturally to the concept of equilibrium constant.

Unfortunately, nonideal gases cannot be expected to obey equations 17.1 and 17.2, so we are apparently faced with the necessity of revising our formulation of reaction potentials and equilibrium constants when dealing with nonideal gaseous systems. Since such a prospect is not a happy one to contemplate, we shall find it expedient to introduce, as an alternative, a new concept that will permit us to retain the forms that proved to be so convenient for ideal gases. We shall do this by substituting for the pressure a new function so defined that equation 17.2 will be exact. In addition, we shall require that this new function become identified with the pressure as the gas approaches ideality—in other words, as the gas pressure approaches zero. The function so introduced will be called the *fugacity*, f. The first part of its definition will be given by the expression

$$\mu = \phi(T) + RT \ln f \tag{17.3}$$

in which f generally depends upon both temperature and pressure, even though $\phi(T)$ remains a function of temperature only. From the pressure coefficient of the Gibbs energy, we can evidently write that

$$\left(\frac{\partial \mu}{\partial p}\right)_T = RT \left(\frac{\partial \ln f}{\partial p}\right)_T = V \tag{17.4}$$

provided V is the molar volume. This equation suggests that a ratio of fugacities can be determined from the equation of state of the gas by definite integration as follows:

$$\Delta \mu = RT \ln \frac{f_2}{f_1} = \int_{p_1}^{p_2} V dp \tag{17.5}$$

To assign an absolute value to the fugacity, however, we must take recourse to an appropriate reference state. As suggested above, we do this by requiring the fugacity to approach the pressure asymptotically as the pressure approaches zero. This is a natural choice of reference state, for all gases tend to become ideal as their pressures vanish. Hence the fugacity, which is in essence a corrected pressure, can logically be associated with the pressure under the limiting conditions of ideality. We therefore assert that

$$\lim_{p \to 0} \frac{f}{p} = 1 \tag{17.6}$$

This statement, in conjunction with equation 17.5, provides an unambiguous definition of fugacity.* Obviously, fugacity will possess the same dimensions as pressure and accordingly can be expressed in atmospheres, torr, etc.

---

* The idea of *fugacity* is just a special case of a more general concept known as *activity*, which is treated in Chapter Nineteen. In place of *fugacity*, f, it would be quite appropriate to speak

Evaluation of Fugacities from Equations of State

We have seen from equation 17.5 that the ratio of the fugacities of a nonideal gas in two different states at the same temperature can be determined from the actual equation of state; however, to establish numerical values for the fugacity, we must take cognizance of equation 17.6. Let us now proceed by dropping the subscript 2 in equation 17.5 and replacing p_1 and f_1 by p^* and f^*, the starred quantities denoting values in the neighborhood of the reference state. Then

$$\Delta\mu = \mu - \mu^* = RT \ln \frac{f}{f^*} = \int_{p^*}^{p} V dp \qquad (17.7)$$

This equation can be rearranged by appropriate manipulation to give

$$RT \ln f = RT \ln \frac{f^*}{p^*} + RT \ln p + \int_{p^*}^{p} \left(V - \frac{RT}{p} \right) dp \qquad (17.8)$$

Consider now the limiting result obtained from equation 17.8 when p^* approaches zero. Under these circumstances, $f^*/p^* \longrightarrow 1$ by reason of equation 17.6. Hence, after dividing through by RT, we obtain the equation

$$\ln f = \ln p + \int_{0}^{p} \left(\frac{V}{RT} - \frac{1}{p} \right) dp \qquad (17.9)$$

We can use this equation to evaluate the fugacity of any gas from a knowledge of its equation of state. Obviously, if the gas is ideal, the integral will be zero and f will equal p.

Calculation of Fugacities

Consider one mole of a nonideal gas obeying the equation of state

$$pV(1 - \beta p) = RT \qquad (17.10)$$

in which β is a function of T. Solving for V and substituting into equation 17.9, we obtain the equation

$$\ln f = \ln p + \int_{0}^{p} \frac{\beta}{1 - \beta p} dp \qquad (17.11)$$

Upon performing the indicated integration, we see that

$$\ln f = \ln p - \ln (1 - \beta p)$$

---

of an activity, a, using gaseous ideality for reference purposes. The importance of gaseous systems, however, warrants a unique nomenclature that renders unnecessary repeated stipulations about the nature of the reference states.

or that

$$f = \frac{p}{1 - \beta p} \qquad (17.12)$$

If, in place of equation 17.10, we assume the equation of state

$$pV = RT + \alpha p \qquad (17.13)$$

we can show that the fugacity will be given by the expression

$$f = p \exp\left(\frac{\alpha p}{RT}\right) \qquad (17.14)$$

If α is small, it can be further shown that

$$f \approx p(1 + \alpha p/RT) \approx \frac{p}{1 - \alpha p/RT} \qquad (17.15)$$

This result is compatible with equation 17.12 provided $\beta = \alpha/RT$, a conclusion we reached earlier when first considering equations of state (p. 12).

We can put equation 17.12 into another form, one that will not explicitly involve β. We do this by eliminating the factor $(1 - \beta p)$ from equation 17.12 by use of the equation of state. The result so obtained is

$$f = \frac{p^2 V}{RT} \qquad (17.16)$$

This is an equation that we can use to calculate the fugacity directly from p, V, and T. Now, if a gas occupying volume V at temperature T were ideal, it would exert pressure p_{id}, equal to RT/V. Hence equation 17.16 can be rewritten as

$$f = \frac{p^2}{p_{id}} \qquad (17.17)$$

p_{id} being the "ideal pressure." Equation 17.17 thus implies that the actual pressure is a mean proportion between the fugacity and the ideal pressure. This result is not exact, of course, since a particular form of equation of state was assumed for its derivation; nevertheless, it can be regarded as a good approximation for all gases that do not exhibit too large departures from ideality.

EXERCISE:

At 0 °C the second virial coefficient of nitrogen (α of eq. 17.13) is equal to -10.3 ml mol^{-1}. Calculate the fugacity of nitrogen at 0 °C and 2 atm pressure. At what pressure would the fugacity of nitrogen differ from the pressure by 1 percent?

EXERCISE:

At 30 °C and 1 atm pressure the volume of a certain mass of carbon dioxide is 0.5 percent less than the ideal value. Estimate the fugacity of carbon dioxide at 30 °C and 1 atm.

Fugacity of a van der Waals Fluid

When we evaluate the fugacity of a van der Waals fluid, it is not practical to use equation 17.9, for we cannot readily express the volume as a function of p and T. It is better, therefore, to go back to equation 17.7 and take recourse to integration by parts:

$$\mu - \mu^* = RT \ln \frac{f}{f^*} = \int_{p^*}^{p} V dp$$

$$= pV - p^*V^* - \int_{V^*}^{V} p dV \qquad (17.18)$$

In this equation V^* denotes the molar volume in the low-pressure state corresponding to p^*. Since

$$p = \frac{RT}{V - b} - \frac{a}{V^2} \qquad (17.19)$$

it follows upon integration that

$$RT \ln \frac{f}{f^*} = pV - p^*V^* - RT \ln \frac{V - b}{V^* - b} - \frac{a}{V} + \frac{a}{V^*} \qquad (17.20)$$

Hence

$$RT \ln f = RT \ln \frac{f^*}{p^*} + RT \ln p^*(V^* - b) - p^*V^* + \frac{a}{V^*}$$

$$+ pV - RT \ln (V - b) - \frac{a}{V} \qquad (17.21)$$

Letting $p^* \longrightarrow 0$ and recognizing that under such circumstances $V^* \longrightarrow \infty$ and $p^*V^* \longrightarrow RT$, we conclude that

$$\ln f = \ln \frac{RT}{V - b} - \frac{2a}{VRT} + \frac{b}{V - b} \qquad (17.22)$$

To the extent that van der Waals' equation can be used to describe a liquid, we might also employ equation 17.22 to calculate fugacities of liquids. We now turn, however, to the more general question of how to deal with fugacities of liquids (and solids).

Fugacities of Liquids and Solids

Although we introduced the concept of fugacity to provide a convenient means of dealing with nonideal gases, there is no reason why we cannot extend it to cover liquids and solids as well. We can do this by simply assigning to a liquid (or solid) a fugacity numerically equal to that of the vapor in equilibrium with it. Hence the fugacity of a condensed phase will be roughly equal to its vapor pressure, the precise value being obtained from the equation of state of the vapor. This method of assigning a fugacity to a solid or liquid is quite logical, for the Gibbs free-energy change attending the reversible condensation of a vapor is equal to zero at the equilibrium pressure. Thus the fugacity of a substance in any state of aggregation is ultimately referred, for evaluation purposes, to the low-pressure gaseous state.

In formulating a derivative of the fugacity of a substance in a condensed phase, we must take care to make the proper choice of independent variables. Since the pressure applied to a liquid is not necessarily the same as the equilibrium vapor pressure, the pressure derivative of the fugacity of a liquid will be given by the expression

$$RT \left(\frac{\partial \ln f}{\partial p_l} \right)_T = V_l \tag{17.23}$$

in which p_l is the applied pressure and V_l is the molar volume of the liquid. Equation 17.23 is fully compatible with equation 17.4, for we know that at equilibrium, in accordance with equation 8.12,

$$\left(\frac{\partial p_l}{\partial p_v} \right)_T = \frac{V_v}{V_l} \tag{17.24}$$

in which the subscript, v, denotes vapor. Hence, if a liquid has an equilibrium vapor pressure of $p_{v,eq}$ when it is under an applied pressure of $p_{l,eq}$, its fugacity, f_l, at p_l will be

$$\ln f_l = \int_{p_{l,eq}}^{p_l} \frac{V_l}{RT} dp_l + \ln p_{v,eq} + \int_0^{p_{v,eq}} \left(\frac{V_v}{RT} - \frac{1}{p_v} \right) dp_v \tag{17.25}$$

We can use this equation to evaluate fugacities of liquids (or solids) when it is impractical or impossible to extend the integration into what might be a supersaturated region of the vapor.

Temperature Coefficient of Fugacity

Since the fugacity of a substance is, in general, a function of both temperature and pressure, we are naturally led to inquire into the nature of its temperature dependence. The easiest way to get at the temperature coefficient of a fugacity is to consider the substance in relation to its reference state—that is, the low-pressure ideal gas state characterized by μ^* and f^* $(= p^*)$. Thus, upon rearranging equation 17.7, we can write that

$$\ln f = \ln f^* - \frac{\mu^* - \mu}{RT} \tag{17.26}$$

in which expression μ and f are the molar Gibbs energy and fugacity of the substance in any state of aggregation, solid, liquid, or gaseous. Let us now differentiate equation 17.26 partially with respect to T, holding p constant. Since $f^* = p^*$, it is evident that f^* is independent of T; hence, making use of equation 7.49, we obtain the result

$$\left(\frac{\partial \ln f}{\partial T}\right)_p = -\frac{\partial}{\partial T}\left(\frac{\mu^* - \mu}{RT}\right) = -\frac{\partial}{\partial T}\left(\frac{G^* - G}{RT}\right)$$
$$= \frac{H^* - H}{RT^2} \tag{17.27}$$

in which H^* denotes the molar enthalpy in the ideal gaseous state, and H the molar enthalpy of the state under consideration—namely, that of pressure p. If equation 17.27 is applied to a liquid or solid, the enthalpy difference, $H^* - H$, will implicitly include a heat of vaporization or sublimation, provided p is the pressure applied to the liquid or solid.*

EXERCISE:

A solid like ice has such a low vapor pressure that its vapor can be assumed to obey the ideal gas laws. The fugacity of ice can then be set equal to its vapor pressure, p_v. Show, under these circumstances, that equation 17.27 reduces to the equation

$$\left(\frac{\partial \ln p_v}{\partial T}\right)_p = \frac{H^* - H}{RT^2} = \frac{\Delta H_{sub}}{RT^2}$$

in which p is the pressure applied to the ice. Observe how the enthalpy of the reference state, H^*, in conjunction with that of the ice, automatically brings in the heat of sublimation, thereby giving rise to an already familiar equation.

---

* To appreciate further the significance of equation 17.27, see Problem 12 of this chapter.

Fugacities and Standard States

When an ideal gas is under unit pressure (say one atmosphere), its molar Gibbs energy, according to equation 17.2, is equal to $\phi(T)$. Now $\phi(T)$ is also denoted by the symbol μ°, which is called the "standard chemical potential" or the "standard Gibbs free energy" of the gas. Moreover, when an ideal gas is at unit pressure, it is said to be in its "standard state."

Although μ° is equal to μ when p equals unity, it must not be supposed that μ° and μ are identical at that pressure. After all, μ° is a function of temperature only, so its derivative with respect to pressure will equal zero. On the other hand, μ depends on both temperature and pressure, and its derivative with respect to pressure will always equal the molar volume, which will not be zero when p is unity. Hence μ° is endowed with some of the properties of μ at unit pressure but not with all of its properties. Still restricting ourselves to ideal gases, we shall now demonstrate that one of the properties that μ° exhibits in common with μ at unit pressure is its temperature coefficient. Differentiating equation 17.2 partially with respect to T, we obtain

$$\left(\frac{\partial \mu}{\partial T}\right)_p = \frac{d\mu^\circ}{dT} + R \ln p$$

$$= \frac{d\mu^\circ}{dT} + \frac{\mu - \mu^\circ}{T} \tag{17.28}$$

Obviously the temperature coefficient of μ° is precisely the same as that of μ for an ideal gas when the pressure is unity. Moreover,

$$\left(\frac{\partial \mu}{\partial T}\right)_p = -S = \frac{\mu - H}{T} \tag{17.29}$$

Combining equations 17.29 with 17.28, we see that

$$\frac{d\mu^\circ}{dT} = \frac{\mu^\circ - H}{T} \tag{17.30}$$

and that

$$\frac{\partial}{\partial T}\left(\frac{\mu}{T}\right) = -\frac{H}{T^2} = \frac{d}{dT}\left(\frac{\mu^\circ}{T}\right) \tag{17.31}$$

Evidently the temperature coefficients of μ/T and μ°/T are the same for ideal gases at all pressures.

Let us now turn to nonideal gases and introduce the idea of a standard chemical potential by rewriting equation 17.3 as follows:

$$\mu = \mu^\circ + RT \ln f \tag{17.32}$$

Evidently μ will equal $\mu°$ when the fugacity is unity. The resemblance to ideal gases stops there, however, for unlike those of ideal gases, the temperature coefficients of μ and $\mu°$ are not equal when $f = 1$.

Differentiating equation 17.32 partially with respect to T, and taking cognizance of equation 17.27, we obtain the result

$$\left(\frac{\partial \mu}{\partial T}\right)_p = \frac{d\mu°}{dT} + R \ln f + \frac{H^* - H}{T} \tag{17.33}$$

Making use of equation 17.29, the validity of which is not restricted to ideal gases, we can further show, from equation 17.33, that

$$\frac{d\mu°}{dT} = \frac{\mu° - H^*}{T} \tag{17.34}$$

or that

$$\frac{d}{dT}\left(\frac{\mu°}{T}\right) = -\frac{H^*}{T^2} \tag{17.35}$$

Equations 17.34 and 17.35 make it clear that the temperature coefficient of the "standard chemical potential" of a nonideal gas is determined by the enthalphy of the reference state (a low-pressure, ideal gas state) and not by the enthalpy associated with the state for which $f = 1$. Hence we see that, even though $\mu = \mu°$ when $f = 1$, the temperature derivatives of μ and $\mu°$ are not equal under that same condition.

The preceding arguments suggest that the "standard state" of a nonideal gas is a state endowed with a chemical potential equal to $\mu°$ and an enthalpy equal to H^*. Since the former attribute corresponds to $f = 1$ and the latter to $f = 0$, we conclude that the "standard state" is a hypothetical, nonattainable state—a concept introduced solely for convenience.

The foregoing comments regarding the hypothetical character of the standard state may be distressing at first thought, but actually there is an element of simplification involved. Since the enthalpy of a nonideal gas depends upon the pressure, the enthalpy of the gas at unit fugacity will depend upon the units used for measuring the fugacity (atmospheres, torr, etc.). But, since the temperature coefficient of $\mu°$ depends upon the enthalpy at zero pressure, it is evident that the temperature coefficient of $\mu°/T$ will be independent of the units used for measuring fugacity. Moreover, when we turn to mixtures of nonideal gases, we shall find further simplification in the fact that partial molar enthalpies are independent of composition at zero pressure even though they are functions of composition at finite pressures.

EXERCISE:

Show that, if $\overset{\circ}{\mu}_{\text{atm}}$ is the value of the standard chemical potential when f is expressed in atmospheres, and $\overset{\circ}{\mu}_{\text{torr}}$ is the value when f is expressed in torr,

$$\overset{\circ}{\mu}_{\text{atm}} = \overset{\circ}{\mu}_{\text{torr}} + RT \ln 760$$

Show further that

$$\frac{d}{dT}\left(\frac{\overset{\circ}{\mu}_{\text{atm}}}{T}\right) = \frac{d}{dT}\left(\frac{\overset{\circ}{\mu}_{\text{torr}}}{T}\right)$$

This illustrates how the standard enthalpy is independent of the units employed for fugacity.

Standard Enthalpy and Standard Entropy

The preceding discussion of "standard states" of nonideal systems suggests that we define a "standard enthalpy," $H°$, equal to H^*, in accordance with equation 17.35. Hence, by definition,

$$H° = -T^2 \frac{d}{dT}\left(\frac{\mu°}{T}\right) \tag{17.36}$$

and

$$\left(\frac{\partial \ln f}{\partial T}\right)_p = \frac{H° - H}{RT^2} \tag{17.37}$$

Moreover, we can now define a standard entropy, $S°$, by the equation

$$\mu° = H° - TS° \tag{17.38}$$

so that

$$\frac{d\mu°}{dT} = -S° = \frac{\mu° - H°}{T} \tag{17.39}$$

It is important to note that the standard entropy is not the entropy of the gas when $f = 1$ (nor is it the entropy when $f = 0$). The standard entropy is not an entropy associated with any particular realizable state, for it is defined in terms of the molar Gibbs free energy when $f = 1$ and of the enthalpy when $f = 0$. This should make it unmistakably clear that the standard state in the fugacity system of representation is, in general, a hypothetical state. For ideal gases, of course, such a difficulty is not encountered, for the enthalpy is independent of the pressure. Hence, for an ideal gas at unit pressure, the Gibbs free energy, the enthalpy, and the entropy will all equal their standard values.

If liquids or solids are described by means of fugacities, their standard states will be precisely the same as those of their vapors. If the vapor can be regarded as an ideal gas, the standard state associated with a liquid or solid will be defined by the properties of its vapor at unit pressure. The preceding statements regarding condensed phases will be true, of course, only as long as the fugacity, with its zero-pressure reference state, is used for determining the Gibbs energy of the system. We shall later learn about other ways of representing chemical potentials of substances—ways that involve other reference states and still other kinds of standard states.

Fugacity Coefficient

We have seen that the fugacity of a gas is, in essence, a corrected pressure, defined to preserve a convenient form of equation for the partial molar Gibbs free energy. A direct measure of departure from ideality will then be given by the quotient f/p. We shall call this ratio the fugacity coefficient* and give it the symbol γ:

$$\gamma = \frac{f}{p} \tag{17.40}$$

Obviously, $\gamma \longrightarrow 1$ as $p \longrightarrow 0$, in accordance with the choice of reference state. In the light of equation 17.9 we see that

$$\ln \gamma = \int_0^p \left(\frac{V}{RT} - \frac{1}{p} \right) dp \tag{17.41}$$

In particular, if we can assume the validity of the equation of state, $pV(1 - \beta p) = RT$, we obtain, from equation 17.16, the expression†

$$\gamma = \frac{pV}{RT} \tag{17.42}$$

This simple result emphasizes how γ measures departures from ideality. Finally, we can also write that, in general,

$$\left(\frac{\partial \ln \gamma}{\partial T} \right)_p = \frac{H^\circ - H}{RT^2} \tag{17.43}$$

EXERCISE:

Calculate the fugacity coefficient of carbon dioxide at 30 °C and 1 atm, using the data of the second exercise on page 355.

---

* If in place of *fugacity*, we choose to speak of *activity*, the ratio would be called the activity coefficient (see footnote on p. 352).

† The expression pV/RT (for one mole) is also known as the compression factor and is often denoted by Z.

Mixtures of Nonideal Gases

The fugacity concept can readily be extended to gaseous mixtures, although we must exercise some care in specifying the reference states. If f_i denotes the fugacity of the ith component of a mixture, it can be defined by the pair of equations

$$\mu_i = \mu_i^\circ + RT \ln f_i \tag{17.44}$$

$$\lim_{p \to 0} \frac{f_i}{x_i p} = 1 \tag{17.45}$$

in which x_i is the mole fraction of the particular component of interest in the gaseous mixture. It should be observed that it is not sufficient to assert that $f_i/p_i \longrightarrow 1$ as $p_i \longrightarrow 0$. After all, an infinitely dilute solution of a gas in other gases does not behave ideally unless the total pressure approaches zero. Since $p_i = x_i p$ for ideal gases, we see from equation 17.45, that the fugacity does indeed approach a partial pressure, but only as the total pressure approaches zero.

Since $(\partial \mu_i/\partial p)_T = V_i$,

$$RT \ln f_i - RT \ln f_i^* = \int_{p^*}^{p} V_i \, dp \tag{17.46}$$

in which expression f_i^* and p^* refer to low-pressure states. Equation 17.46 can now be manipulated in the same manner as the corresponding equation for a single component; the result is

$$RT \ln f_i = RT \ln \frac{f_i^*}{p^*} + RT \ln p + \int_{p^*}^{p} \left(V_i - \frac{RT}{p} \right) dp \tag{17.47}$$

If we now pass to limit as $p^* \longrightarrow 0$, keeping composition constant and taking cognizance of equation 17.45, we obtain, after division by RT, the result

$$\ln f_i = \ln p x_i + \int_{0}^{p} \left(\frac{V_i}{RT} - \frac{1}{p} \right) dp \tag{17.48}$$

Since the integration indicated in this equation is carried out at constant composition, it is evident that, to obtain f_i, one must know V_i for the composition of interest as a function of pressure. If one is interested in the fugacities corresponding to several compositions, a considerable amount of data is necessary. Since such data are not readily forthcoming, it is often necessary to introduce certain approximations.

The principal difficulty encountered in dealing with fugacities in mixtures is attributable to the interactions between unlike molecules. If the behavior of a gaseous mixture could be regarded as some combination of the behaviors of the separate gases, the problem should be much simplified, at least in principle. Unfortunately, this is not so, for the forces involved include not only those between like molecules, but also those between unlike molecules. It is the latter that cannot be adequately accounted for by any simple model intended to describe the behavior of a mixture in terms of the behaviors of the separated materials.

It sometimes happens, however, that a mixture of gases approximately fulfills at least some of the conditions necessary for ideality without satisfying the complete ideal gas law equations. It is conceivable that the fugacity of a component in such a mixture might be simply related to its fugacity in a pure state. To demonstrate this possibility, let us consider the fugacity, f_{0i}, of the pure ith component when it is at a pressure equal to the total pressure of the mixture under consideration. The fugacity of the pure gas at that total pressure is given by the equation

$$\ln f_{0i} = \ln p + \int_0^p \left(\frac{V_{mi}}{RT} - \frac{1}{p} \right) dp \tag{17.49}$$

in which V_{mi} is the molar volume of the pure substance. If we now subtract equation 17.49 from 17.48, we obtain the equation

$$\ln \frac{f_i}{f_{0i}} = \ln x_i + \int_0^p \left(\frac{V_i - V_{mi}}{RT} \right) dp \tag{17.50}$$

Now it often happens that the partial molar volume of a gas in a mixture at pressure p is very nearly equal to the molar volume of the same gas in a pure state at the same pressure. When this is true, the gaseous mixture is said to obey Amagat's rule, which is nothing more or less than a statement that the volume change attending the isopiestic and isothermal mixing of gases is equal to zero. One should note that this condition, although necessary for ideality, is not sufficient to guarantee ideality and hence leaves some room for departures from the perfect gas laws. If we now assume in accordance with Amagat's rule that the volume terms appearing in the integral of equation 17.50 are equal, we conclude that the fugacity of a component in a mixture is precisely equal to its mole fraction multiplied by the fugacity that the pure component would have at the specified total pressure. In other words, we can write approximately that

$$f_i = x_i f_{0i} \tag{17.51}$$

This equation has been found to agree roughly with experimental observations, but it is by no means exact. Nevertheless, it has found widespread use simply because nothing better has been proposed, short of carrying out laborious experiments on partial molar volumes of various mixtures over a wide range of pressures. It should be noted that equation 17.51 is not the precise counterpart of Dalton's Law. The essential difference is that Dalton's Law provides a relationship between partial pressures and a total pressure, whereas equation 17.51 does not involve the concept of total fugacity of a mixture.

The idea of total fugacity of a mixture can be easily developed, but it has found little utility since it does not introduce any substantial simplifications. If we consider a mixture of gases (like air) to be a single substance with an average molar volume, $\langle V_m \rangle$, its total fugacity can logically be written as

$$\ln f = \ln p + \int_0^p \left(\frac{\langle V_m \rangle}{RT} - \frac{1}{p} \right) dp \tag{17.52}$$

Multiplying this equation through by the total number of moles, n, and then differentiating partially with respect to n_i, the number of moles of any particular component, we obtain the equation

$$\ln f + n \left(\frac{\partial \ln f}{\partial n_i} \right)_{T,p,n_1,n_2,\,\ldots} = \ln p + \int_0^p \left(\frac{V_i}{RT} - \frac{1}{p} \right) dp \tag{17.53}$$

Combining this equation with equation 17.48, we see that the fugacity of a component is related to the total fugacity by the equation

$$\ln f_i = \ln f x_i + n \left(\frac{\partial \ln f}{\partial n_i} \right)_{T,p,n_1,n_2,\,\ldots} \tag{17.54}$$

For a binary mixture, equation 17.54 reduces to the equation

$$\ln f_1 = \ln f x_1 + (1 - x_1) \left(\frac{\partial \ln f}{\partial x_1} \right)_{T,p} \tag{17.55}$$

with a similar equation for f_2. Assuming that the derivative of $\ln f$ can be neglected, an assumption whose validity is by no means justified, we see that a partial fugacity might be considered equal to the mole fraction multiplied by the total fugacity, in direct correspondence to Dalton's Law. Unfortunately, even such a relationship calls for more data about a mixture than are normally forthcoming.

So far as standard states and temperature coefficients of fugacities are concerned, the arguments pertaining to mixtures are not different from those applicable to pure materials. The chemical potential associated with the standard state will correspond to unit fugacity, and the temperature coefficient corresponding to that "standard chemical potential" will depend upon the partial molar enthalpy in the reference state, which is the zero-pressure (ideal) state of the mixture. Since the total enthalpy of a mixture of ideal gases is exactly equal to the sum of the separate enthalpies, the standard partial molar enthalpy, H_i°, of substance i in a mixture will be precisely the same as the molar enthalpy of the pure substance in a zero-pressure state.

EXERCISE:

Prove that for a mixture of gases

$$\left(\frac{f_1}{x_1}\right)^{x_1}\left(\frac{f_2}{x_2}\right)^{x_2}\ldots = f$$

and show that therefore, for an equimolar mixture of gases, the total fugacity, f, cannot equal the sum of the separate fugacities, $f_1 + f_2$, unless $f_1 = f_2$. Since this is generally not true, it is clear that the total fugacity cannot be regarded as the sum of "partial fugacities."

Chemical Reactions Involving Nonideal Systems

Let us consider once again the general chemical reaction

$$aA + bB + \ldots \longrightarrow eE + fF + \ldots$$

in which the various substances can be in any state of aggregation, solid, liquid, or gaseous. Assuming that the chemical potential of each of the substances can be expressed in the form of equation 17.44, we see that the reaction potential will be given by the expression

$$\Delta\tilde{\mu} = \Delta\tilde{\mu}^\circ + RT\ln\frac{f_E^e f_F^f \ldots}{f_A^a f_B^b \ldots} \tag{17.56}$$

in which

$$\Delta\tilde{\mu}^\circ = (e\mu_E^\circ + f\mu_F^\circ + \ldots) - (a\mu_A^\circ + b\mu_B^\circ + \ldots) \tag{17.57}$$

Because of our definition of fugacity, equation 17.56 is now exact for all systems. Moreover, the standard reaction potential, $\Delta\tilde{\mu}^\circ$, is a function of temperature only and can once again be used to define an equilibrium constant, K, by means of the equation

$$\Delta\tilde{\mu}^\circ = -RT\ln K \tag{17.58}$$

Finally, we see that

$$\frac{d\ln K}{dT} = \frac{\Delta\tilde{H}^\circ}{RT^2} \tag{17.59}$$

in which expression $\Delta\tilde{H}^0$, the standard enthalpy of reaction, is

$$\Delta\tilde{H}^\circ = (eH_E^\circ + fH_F^\circ + \ldots) - (aH_A^\circ + bH_B^\circ + \ldots) \tag{17.60}$$

Each of the standard enthalpies, H_i°, is equal to H_i^*, the molar enthalpy in the zero-pressure or ideal gas state. Thus the temperature coefficient of the equilibrium constant, expressed in relation to fugacities, is determined by the enthalpy of the reaction taking place in an infinitely dilute gaseous state. This will be true even if the substances involved are actually liquids or solids.

To utilize equation 17.56, we must, of course, employ individual fugacities instead of partial pressures as we might for ideal gas reactions. When mixtures of gases are encountered, the problem of determining the fugacities in the mixtures naturally arises. Unless more accurate data are forthcoming, the best one can do is to use the approximate rule expressed by equation 17.51, which means that at equilibrium

$$K = \left(\frac{x_E^e x_F^f \cdots}{x_A^a x_B^b \cdots} \right) \left(\frac{f_{0E}^e f_{0F}^f \cdots}{f_{0A}^a f_{0B}^b \cdots} \right) \tag{17.61}$$

The procedure to be followed consists of calculating the fugacities of the pure materials, f_{0i}, at the total pressure of the system and then solving the remaining equation, involving mole fractions, for the extent of reaction.

Problems

1. Prove that, if a gas departs only slightly from ideality, its fugacity will be approximately

$$f = 2p - p_i$$

   if p_{id} is the ideal pressure. (See eq. 17.17.)

2. For low pressures, van der Waals' equation can be written as $pV = RT + (b - a/RT)p$. Set up the expression for $\ln f$ for a low-pressure van der Waals gas. By differentiating $\ln f$ partially with respect to T, deduce an expression for $H° - H$. At what temperature will the enthalpy be independent of the pressure? Compare with the Joule-Thomson inversion temperature.

3. By differentiating equation 17.9 partially with respect to T, show that

$$\left(\frac{\partial \ln f}{\partial T} \right)_p = \int_0^p \frac{1}{RT^2} \left[T \left(\frac{\partial V}{\partial T} \right)_p - V \right] dp$$

   Show further that, in view of equation 6.10,

$$\left(\frac{\partial \ln f}{\partial T}\right)_p = -\int_0^p \frac{1}{RT^2}\left(\frac{\partial H}{\partial p}\right)_T dp$$

Hence, upon integration, derive equation 17.27.

4. (A) A certain gas obeys the equation of state $pV = RT + Ap + Bp^2 + Cp^3 + \ldots$ if A, B, C, ... are functions of temperature. Show that the fugacity of the gas is given by the expression

$$RT \ln f = RT \ln p + Ap + Bp^2/2 + Cp^3/3 + \ldots$$

(B) At 0 °C one mole of carbon monoxide obeys the equation of state

$$pV = 22.414 - 0.01483\,p + 0.000098\,p^2 \ldots$$

if p is expressed in atm and V in liters. Calculate the fugacity and fugacity coefficient of carbon dioxide at 5, 10, and 20 atms and 0 °C.

5. Assuming the volume of a van der Waals liquid to be given by equation 8.42 (namely, $V = b + b^2RT/a$), show that the fugacity (and hence the approximate vapor pressure of a van der Waals liquid) will be given by the expression

$$\ln p = \ln (a/b^2) - a/(bRT)$$

which is the same as equation 8.44.

6. At 0 °C the fugacity of liquid water, under negligible applied pressure, is equal to 4.58 Torr. Assuming that liquid water is incompressible, calculate the fugacity of liquid water at 0 °C when the applied pressure is 100 atm. (Observe that, to calculate the change in fugacity of the liquid, you need not know the equation of state of the vapor.)

7. Prove that

$$\left(\frac{\partial \mu}{\partial T}\right)_f = \frac{\mu - H^\circ}{T} = -S^\circ + R \ln f$$

8. The fugacity coefficient of a liquid is defined to be equal to that of its equilibrium vapor:

$$\gamma_1 = \frac{f_1}{p_v}$$

If the pressure applied to a liquid, p_1, is not necessarily the same as the vapor pressure, p_v, show that

$$\left(\frac{\partial \ln \gamma_1}{\partial p_1}\right)_T = \frac{V_1}{RT}\left(1 - \frac{RT}{p_v V_v}\right)$$

Observe that, although $(\partial \ln f_1/\partial p_1)_T$ is independent of the equation of state of the vapor, the derivative indicated above very definitely depends upon the nature of the vapor. (If the vapor is an ideal gas, γ_1 is independent of the applied pressure, p_1.)

9. At moderate pressures, one mole of a gaseous mixture (as well as a mole of pure gas) can be assumed to obey the equation of state $pV = RT + \alpha p$. The values of α for helium, carbon dioxide, and an equimolar mixture of helium and carbon dioxide are given in the following table:[*]

t	α(He)	α(CO$_2$)	α(Equimolar He-CO$_2$)
30 °C	0.0117 liter mol^{-1}	-0.1193 liter mol^{-1}	-0.0161 liter mol^{-1}
60	0.0115	-0.0955	-0.0087

(A) Calculate the fugacities of pure helium, pure carbon dioxide, and the equimolar mixture at 30 °C and 1 atm.

(B) Suppose that for a mixture of mole fractions x_1 and x_2 the value of α is given by the expression

$$\alpha_{(mixture)} = x_1^2\alpha_1 + 2x_1x_2\alpha_{12} + x_2^2\alpha_2$$

in which α_1 and α_2 refer to the pure materials and α_{12} is a parameter characteristic of the pair of molecules involved (α_{12} is *not* necessarily the value of α for the mixture). Calculate the values of α_{12} for He-CO$_2$ mixtures at 30 and 60 °C.

(C) From the results of part B calculate the compositions for which He-CO$_2$ mixtures would obey Boyle's Law at 30 and 60 °C.

10. Consider the Sackur-Tetrode expression for the entropy and Gibbs free energy of a monatomic ideal gas (eqs. 12.34 and 12.44). If the fugacity, f, were used to replace the pressure, p, in each of these expressions, would the resulting equations be valid for a nonideal monatomic gas?

   Ans.   The free-energy equation would be correct, but the entropy equation would be inexact.

11. An isothermal atmosphere in a uniform gravitational field consists of a nonideal gas. Show that the fugacity at height h must be

$$f(h) = f(0)e^{-Mgh/RT}$$

if $f(0)$ is the fugacity at the earth's surface. Assuming that the gas obeys the equation of state $pV(1 - \beta p) = RT$, derive, through use of the fugacity concept, an expression for p as a function of h. (See Prob. 11, Chap. Seven.)

12. Let us apply equation 17.25 to the situation where the equilibrium vapor pressure of a liquid ($p_{v,eq}$) is equal to the equilibrium applied pressure on the liquid ($p_{1,eq}$). This results in the expression

$$\ln f_1 = \int_{p_{eq}}^{p_1} \frac{V_1}{RT}\,dp_1 + \ln p_{eq} + \int_0^{p_{eq}} \left(\frac{V_v}{RT} - \frac{1}{p_v}\right)dp_v$$

Now, by differentiating $\ln f_1$ partially with respect to T while holding p_1 constant, show that

---

[*] Cottrell and Hamilton, *Trans. Faraday Soc.*, **52**, 156 (1956).

$$\left(\frac{\partial \ln f_1}{\partial T}\right)_{p_1} = -\frac{1}{RT^2}\int_{p_{eq}}^{p_1}\left(\frac{\partial H_1}{\partial p_1}\right)dp_1 - \frac{V_{1,eq}}{RT}\frac{dp_{eq}}{dT} + \frac{d\ln p_{eq}}{dT}$$

$$- \frac{1}{RT^2}\int_0^{p_{eq}}\left(\frac{\partial H_v}{\partial p_v}\right)dp_v + \left(\frac{V_{v,eq}}{RT} - \frac{1}{p_{eq}}\right)\frac{dp_{eq}}{dT}$$

Simplifying the above and making use of the exact Clapeyron equation, show further that

$$\left(\frac{\partial \ln f_1}{\partial T}\right)_{p_1} = \frac{H_{1,eq} - H_1}{RT^2} + \frac{\Delta H_{vap}}{RT^2} + \frac{H_v^* - H_{v,eq}}{RT^2} = \frac{H_v^* - H_1}{RT^2}$$

This result is, of course, the same as equation 17.27; observe, however, the implicit presence of the heat of vaporization in the final result.

13. A certain investigator is given a tube containing the precise equivalent of $1NO_2(g)$, but he is given no information about the chemical identity of the substance. The investigator carefully studies the p-V-T behavior of the gas, which he presumes to be a single chemical species. Since the nitrogen dioxide can dimerize to form nitrogen tetroxide, it appears to the experimenter that the gas departs markedly from ideality.

(A) Assuming that both the NO_2 and the N_2O_4 behave as ideal gases, what does the investigator find for the fugacity of the system when he imagines he has only a single substance?

It is recommended that you begin to solve this problem by starting with $1/2$ mole of N_2O_4, of which a fraction, α, dissociates:

$$N_2O_4 \rightleftharpoons 2NO_2(g)$$

$$\frac{1-\alpha}{2}\text{ moles} \qquad \alpha\text{ moles}$$

If K is the equilibrium constant for the reaction as written, $\alpha = \sqrt{K/(K+4p)}$, from which the apparent equation of state can be readily determined. Substituting the apparent equation of state into the expression for the calculation of fugacities, you will find that

$$f = \frac{2\sqrt{Kp}}{\sqrt{K} + \sqrt{K+4p}} = \frac{2\alpha p}{1+\alpha}$$

Note that the apparent fugacity of the system is precisely equal to the partial pressure of the NO_2. Can you give a logical explanation (without simply repeating the mathematical deduction) why this must be so?

(B) The investigator also studies the temperature coefficient of $\ln f$ to obtain $H^\circ - H$. What will he find? Proceed by differentiating $\ln f$ partially with respect to T, making use of the known relationship for $d\ln K/dT$. Show that

$$H^\circ - H = \frac{(1-\alpha)}{2}\Delta\tilde{H}$$

Explain this result.

14. The fugacity of a certain gas is given by the expression

$$f = p + \alpha p^2$$

in which α is a function of temperature.

(A) Show that the gas must obey the following equation of state

$$\frac{pV}{RT} = 1 + \frac{\alpha p}{1 + \alpha p}$$

(B) Also prove that

$$\left(\frac{\partial H}{\partial p}\right)_T = -\frac{RT^2}{(1 + \alpha p)^2} \cdot \frac{d\alpha}{dT}$$

18

Liquid Solutions

Introduction

Up to the present, our applications of thermodynamics to chemical equilibria have been concerned mostly with gaseous systems. This leaves our treatment far from complete, for a very important part of chemistry is concerned with reactions in solution. Accordingly, we shall now consider, from a thermodynamic point of view, the behavior of solutions. In this chapter we shall be concerned with solutions of nonreacting materials, but later on the discussion will be extended to solutions in which chemical reactions occur.

Our study of solutions will first involve a consideration of certain simple and frequently observed behaviors that are usually designated as "ideal." After that we shall examine systems that depart from such ideality. At appropriate intervals we shall also interject statistical discussions to show how the behavior of solutions can be related to statistical mechanics.

Raoult's Law

Among the most useful quantities that characterize a solution are the equilibrium partial vapor pressures of its components. Since much of solution thermodynamics depends upon vapor pressures, it is natural that we inquire into their significance. Let us consider, for example, a liquid mixture consisting of two components, A and B, with mole fractions x_A and x_B. If the substances A and B are physically and chemically similar to each other, and if neither associates with itself or with the other kind of molecule, it is often observed that the partial vapor pressures can be expressed quite accurately by a simple relationship known as Raoult's Law. If p_A is the vapor pressure of component A and p_{0A} is the vapor pressure of the pure liquid, Raoult's Law assumes the form

$$p_A = p_{0A} x_A \tag{18.1}$$

This equation, if expressed in terms of the relative lowering of vapor pressure, can be rearranged as

$$\frac{p_{0A} - p_A}{p_{0A}} = 1 - x_A \tag{18.2}$$

Similar equations apply to component B or to each of several components if the solution is made up of more than two substances. Raoult's Law does not by any means hold exactly for all combinations of liquids, but it is found to be valid so frequently—for combinations like mixtures of hydrocarbons—that it warrants special recognition in chemical thermodynamics as a kind of ideal behavior that can be regarded as a starting point for the description of all solutions. Accordingly, a solution that conforms to Raoult's Law is said to be an "ideal" solution.

If a binary liquid mixture obeys Raoult's Law, the vapor pressures plotted against the mole fractions will be represented by straight lines, as shown in Figure 18-1. The lines in that figure refer, of course, to pressures observed at a given temperature. As the temperature is changed, the vapor pressures of the pure liquids will necessarily change, and hence the slopes of the partial-pressure lines will be altered and the total-pressure lines correspondingly shifted.

EXERCISE:

Benzene and toluene form solutions that obey Raoult's Law. At 30 °C the vapor pressures of pure benzene and pure toluene are 119 and 37 Torr, respectively. A solution of what composition would have a total vapor pressure of 100 Torr?

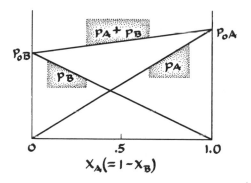

FIGURE 18-1 Vapor pressures as functions of the mole fraction for a binary solution obeying Raoult's Law.

Raoult's Law Expressed in Terms of Fugacities

The discussion of the preceding section was carried out in terms of the partial vapor pressures of solutions. These quantities correspond to what is actually measured experimentally, but for theoretical convenience, to avoid difficulties that might be introduced by gaseous nonideality, we shall now reformulate the basic solution laws in terms of fugacities. Naturally, if the vapor is an ideal gas, this will make no difference, for by definition the fugacity of a liquid component is precisely the same as its fugacity in the equilibrium vapor. However, since we can introduce the fugacity generalization without causing undue complications, we shall now restate Raoult's Law as follows:

$$f_A = f_{0A} x_A \tag{18.3}$$

In this equation f_A is the fugacity of component A and f_{0A} is the fugacity of the corresponding pure substance. Usually the difference between vapor pressure and fugacity can be neglected, but, except as noted below, we simplify little by assuming gaseous ideality. The graphical representation of Raoult's Law expressed in fugacities assumes the form shown in Figure 18-1 so far as the separate fugacities are concerned. The "total" fugacity, of course, does not necessarily equal the sum of the partial fugacities, and it is in this respect that the assumption of gaseous ideality will produce simplification.

EXERCISE:

Substance A in a certain solution obeys Raoult's Law expressed in terms of fugacities. Assuming that the vapor of A is a nonideal gas obeying the equation of state $pV(1 - \beta p) = RT$, show that the equilibrium partial pressure of A above the solution will be

$$p_A = \frac{p_{0A} x_A}{1 - \beta p_{0A} x_B}$$

Heats of Mixing and Volume Changes on Mixing of Ideal Solutions

When two or more liquids form a solution that obeys Raoult's Law, it is found that the enthalpy change and the volume change attending the mixing of the pure liquids are zero. To prove this, let us first inquire into the nature of the Gibbs free energy of mixing. The Gibbs energy of a pure liquid will, by our definition of fugacity, be given by the following equation:

$$\mu_{0i} = \mu_i^\circ + RT \ln f_{0i} \qquad (18.4)$$

In this equation μ_i° is the standard chemical potential or standard Gibbs free energy, the significance of which was discussed in Chapter Seventeen. μ_{0i} represents, as implied, the chemical potential of the pure liquid and f_{0i} the corresponding fugacity. The chemical potential of the ith component of a solution will be given by a similar equation, namely:

$$\mu_i = \mu_i^\circ + RT \ln f_i \qquad (18.5)$$

Moreover, the total Gibbs energy of a solution will be the sum of the chemical potentials, each multiplied by the appropriate number of moles. If n_i is the number of moles of the ith component, the change in Gibbs free energy for the mixing process, ΔG_{mix}, will be

$$\Delta G_{\text{mix}} = \sum_i n_i (\mu_i - \mu_{0i})$$

$$= \sum_i n_i RT \ln \frac{f_i}{f_{0i}} \qquad (18.6)$$

If Raoult's Law holds, the Gibbs energy of mixing simply reduces to the following:

$$\Delta G_{\text{mix}} = \sum_i n_i RT \ln x_i \qquad (18.7)$$

Let us now divide ΔG_{mix} by T and differentiate the quotient partially with respect to T. Since $\Delta G_{\text{mix}}/T$ is independent of T, the enthalpy of mixing, in accordance with equation 7.49, must be zero, for

$$\Delta H_{mix} = -T^2 \frac{\partial}{\partial T}\left(\frac{\Delta G_{mix}}{T}\right) = 0 \tag{18.8}$$

Moreover, by differentiating ΔG_{mix} with respect to p_1, the pressure applied to the liquid solution, we also find that the volume change, ΔV_{mix}, is equal to zero for an ideal solution. Finally, since the enthalpy change of mixing is equal to zero, the Gibbs energy of mixing must simply equal $-T\Delta S_{mix}$. Hence we conclude that the entropy of mixing must be

$$\Delta S_{mix} = -\sum_i n_i R \ln x_i \tag{18.9}$$

This equation for ΔS_{mix} is identical in form to that obtained for the mixing of ideal gases. We shall now turn to a simple statistical interpretation of this conclusion.

EXERCISE:

Twelve grams of benzene are mixed with 25 g of toluene. What is the entropy of mixing in cal deg^{-1}?

Statistical Mechanics of Mixing of Similar Liquids and Derivation of Raoult's Law

Let us consider N_A molecules of kind A and N_B molecules of kind B, which are to be mixed to form a liquid solution. Let us assume that the forces operating between the A molecules are the same as those operating between the B molecules or between mixed molecules A and B. (This assumption will assure us that the energy of mixing will be zero.) To further simplify our model, we shall suppose that the volume occupied by molecule A is the same as that occupied by molecule B.

Let us now imagine that the liquid can be represented by a pseudo lattice—in other words, by an assembly of sites each occupied by one molecule. These sites will not be arranged in perfect crystalline fashion, but there will be a high degree of geometric order between close neighbors; such order or regularity will vanish for pairs of atoms that are far apart. Let us now inquire into the thermodynamic probabilities of the molecules distributed among the various pseudo-lattice sites, considering first the pure liquids and then the solution.

Pure liquid A will consist of N_A sites each occupied by a molecule of A. If the molecules are indistinguishable, then, so far as lattice sites are concerned, the thermodynamic probability, Ω_A, associated with liquid A will simply equal unity. Similarly, the thermodynamic probability, Ω_B, of pure liquid B will also equal unity.* The mixture, on the other hand, will have a different thermodynamic probability. Since there are $N_A + N_B$ total sites available, our problem reduces to calculating the number of ways in which N_A indistinguishable molecules might be distributed among those sites, with the understanding that the remaining sites will be filled with molecules of kind B. Now let us imagine constructing the liquid by successively placing the A molecules one by one in the available sites. The first molecule of A will have $N_A + N_B$ possible locations, leaving a total of $N_A + N_B - 1$ available locations for the second molecule. After the second molecule is placed in the lattice, $N_A + N_B - 2$ sites will remain available for the third, etc. Continuing this kind of computation indefinitely, we find that the ith molecule of A will have $N_A + N_B - i + 1$ sites available. The total number of ways in which N_A *distinguishable* molecules might be placed in the lattice will then be given by the following continued product:

$$(N_A + N_B)(N_A + N_B - 1)(N_A + N_B - 2)\ldots(N_B + 1) = \frac{(N_A + N_B)!}{N_B!}$$

(18.10)

But, since the A molecules are really indistinguishable, the expression above must be divided by $N_A!$ to give the number of distinguishable ways in which N_A molecules can be placed among $N_A + N_B$ sites. Having thus taken care of all the A molecules by placing them in the lattice, we see that there is only one distinguishable way of placing the B molecules in the remaining N_B holes. Hence the thermodynamic probability of the mixture will be

$$\Omega_{\text{mix}} = \frac{(N_A + N_B)!}{N_A! N_B!}$$

(18.11)

According to the model above, the entropy of mixing will therefore be

$$\Delta S_{\text{mix}} = k \ln \frac{\Omega_{\text{mix}}}{\Omega_A \Omega_B} = k \ln \frac{(N_A + N_B)!}{N_A! N_B!}$$

(18.12)

Making use of Stirling's approximation to simplify this equation, we find that

$$\Delta S_{\text{mix}} = -N_A k \ln \frac{N_A}{N_A + N_B} - N_B k \ln \frac{N_B}{N_A + N_B}$$

---

* This is obviously an oversimplification, since a liquid can scarcely be regarded as the equivalent of a crystal at absolute zero. However, no serious error should be introduced as long as we consider only changes in entropy within the framework of a consistent model.

$$= -n_A R \ln x_A - n_B R \ln x_B \qquad (18.13)$$

This result is in full agreement with equation 18.9, which was obtained from the purely thermodynamic consequences of Raoult's Law. Since we assumed that A and B were equivalent in size and in their interaction energies, it is evident that both ΔV_{mix} and ΔH_{mix} will equal zero for this model. Hence the Gibbs energy of mixing must be given by an equation like 18.7. This, in conjunction with equation 18.6, which is generally applicable, completes the derivation of Raoult's Law.

The foregoing statistical demonstration of Raoult's Law is necessarily simple, but it does carry with it the essence of the arguments that are germane to the problem. By appropriate modification of the model— for example, by assuming that the molecules have different sizes—one may obtain a different result, as we shall see later.

Solubility of Solids

If we assume that Raoult's Law holds for a solution obtained by dissolving a solid in some convenient solvent, it is possible to predict the limit of that solubility or, more specifically, the composition of the saturated solution. Consider, for example, the following equilibrium, which involves a solid substance A and its saturated solution:

$$A(c) \rightleftharpoons A(\text{dissolved})$$

When the system is in a state of equilibrium, the chemical potential of pure solid A must equal A's chemical potential in solution; this is equivalent to saying that the fugacity of the solid must equal the fugacity of the same substance in solution. If $f_{0A}(c)$ denotes the fugacity of the pure solid and $f_{0A}(\text{liq})$ that of the corresponding pure liquid at the same temperature, then according to Raoult's Law the following equation must hold at saturation:

$$f_{0A}(c) = f_{0A}(\text{liq}) x_A \qquad (18.14)$$

It is this equation that determines the solubility.

At this point we should offer a comment concerning the significance of $f_{0A}(\text{liq})$, the fugacity of the pure liquid. Since we are dealing with a solid material that is to be dissolved in an appropriate solvent, it is evident that pure liquid A cannot be the stable modification of A under the conditions assumed. Therefore $f_{0A}(\text{liq})$ must represent the fugacity of the super-cooled liquid, which in turn is the fugacity of the vapor that would be in equilibrium with that super-cooled liquid. Such vapor would, of course, be supersaturated in relation to the solid, so that there

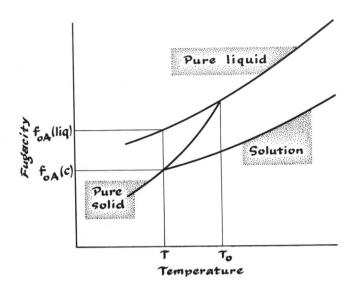

FIGURE 18-2   Fugacity of component A as liquid, as solid, and in solution. The normal
melting point (or freezing point) of the pure material is denoted by T_0, and the
temperature of the solution in equilibrium with the solid is denoted by T. Below
T_0 the curve for the fugacity of pure liquid represents that of *super-cooled* liquid.

would be a spontaneous tendency for it to undergo the reverse of sublimation. The vapor pressure of A in solution must therefore be reduced
at least to the point where equation 18.14 will hold; this provides the
basis for calculating the composition of the saturated solution. The argument involved is best illustrated by Figure 18-2, in which the fugacities
(approximately vapor pressures) of the super-cooled liquid, solid, and
solution are plotted against the temperature. Where the fugacity curve
for the solid meets that of the solution, the system will be in equilibrium.
The solubility at any temperature will, of course, be determined by that
amount of solvent that lowers the fugacity of the pure super-cooled
liquid solute to that of the solid solute at the same temperature. Obviously, as the temperature is raised, the solubility of the solid increases.
At its normal melting point, T_0, the solubility becomes infinite, for then
no solvent need be present for equilibrium to exist between liquid and
solid phases.

It is evident from the foregoing that, to calculate the mole fraction
of solute in a saturated solution, we must by some means or other get at
the fugacity of the pure liquid in the above-mentioned super-cooled
state. To do so, we shall consider the temperature coefficient of fugacity

and its dependence upon enthalpies. Rewriting equation 18.14 in logarithmic form and differentiating partially with respect to T, we find that

$$\left(\frac{\partial \ln f_{0A}(c)}{\partial T}\right)_p = \left(\frac{\partial \ln f_{0A}(\text{liq})}{\partial T}\right)_p + \left(\frac{\partial \ln x_A}{\partial T}\right)_p \tag{18.15}$$

But, in view of equation 17.37, we can also assert that

$$\frac{H_A^\circ - H_{mA}(c)}{RT^2} = \frac{H_A^\circ - H_{mA}(\text{liq})}{RT^2} + \left(\frac{\partial \ln x_A}{\partial T}\right)_p \tag{18.16}$$

In this equation H_A° is the standard molar enthalpy of substance A, and $H_{mA}(c)$ and $H_{mA}(\text{liq})$ are the molar enthalpies of the pure solid and the pure liquid. Rearranging equation 18.16, we see that

$$\left(\frac{\partial \ln x_A}{\partial T}\right)_p = \frac{H_{mA}(\text{liq}) - H_{mA}(c)}{RT^2} = \frac{\Delta H_{\text{fus}}}{RT^2} \tag{18.17}$$

in which expression ΔH_{fus} is the molar heat of fusion—that is, $H_{mA}(\text{liq}) - H_{mA}(c)$. Let us now integrate equation 18.17, assuming ΔH_{fus} to be constant. If we integrate between temperatures T and T', for which the corresponding saturation mole fractions will be x_A and x_A', we find that

$$\ln \frac{x_A}{x_A'} = \frac{\Delta H_{\text{fus}}}{R}\left(\frac{1}{T'} - \frac{1}{T}\right) \tag{18.18}$$

This result gives us the ratio of solubilities at each of two temperatures, those solubilities being expressed in equilibrium mole fractions. To put the solubility on an absolute basis, we now recognize that at T_0, the normal melting point of substance A, the solid can be in equilibrium with its own pure liquid. Hence we can simultaneously set $T' = T_0$ and $x_A' = 1$, thus getting

$$\ln x_A = \frac{\Delta H_{\text{fus}}}{R}\left(\frac{1}{T_0} - \frac{1}{T}\right) \tag{18.19}$$

From this equation we see that the solubility in an ideal solution depends only on the heat of fusion and the normal melting point of the solute. Moreover, for equation 18.19 to be valid, it is not even necessary to assume that Raoult's Law holds for component B, the solvent. It appears that the specific nature of the solvent is immaterial, although, practically speaking, the solvent must bear some resemblance to the solute if Raoult's Law is to hold for the latter.

To complete our discussion of the solubility of a solid in an ideal solution, we should also consider the effect of pressure. To do so, we

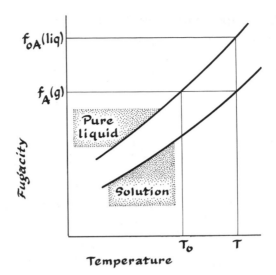

FIGURE 18-3 Fugacity of a pure liquid and
of the same liquid in solution near the normal boiling point, T_0. The temperature of
the solution in equilibrium with gas of fugacity $f_A(g)$ is denoted by T.

need only rewrite equation 18.14 in logarithmic form and then differentiate partially with respect to pressure, which we shall presume to be
the same for both solution and solid. In the light of equation 17.4 we
conclude that the pressure coefficient of the logarithm of the equilibrium mole fraction will be given by the expression

$$\left(\frac{\partial \ln x_A}{\partial p}\right)_T = -\frac{V_{mA}(\text{liq}) - V_{mA}(\text{c})}{RT} = -\frac{\Delta V_{\text{fus}}}{RT} \qquad (18.20)$$

Obviously, if ΔV_{fus}, the change in volume attending the melting process,
is positive, an increase in pressure will diminish the solubility. Conversely, if ΔV_{fus} is negative, an increase in pressure will increase the
solubility.

Solubility of Gases

Just as it was possible to calculate the solubility of a solid in an ideal
solution, so, too, we can calculate the equilibrium mole fraction of a
gas dissolved in such a solution. The equilibrium involved (see Fig. 18-3)
can now be written as follows:

$$A(g) \rightleftarrows A(\text{dissolved})$$

Evidently the fugacities of A in the two states must be equal at equilibrium. Assuming Raoult's Law to hold, we can then assert that

$$f_A(g) = f_{0A}(liq)x_A \tag{18.21}$$

Once more let us take the temperature coefficient of the logarithm of $f_A(g)$, keeping the pressure of the system constant. In this way we obtain the equation

$$\left(\frac{\partial \ln f_A(g)}{\partial T}\right)_p = \left(\frac{\partial \ln f_{0A}(liq)}{\partial T}\right)_p + \left(\frac{\partial \ln x_A}{\partial T}\right)_p \tag{18.22}$$

This equation, in the light of equation 17.37, can be rewritten as follows:

$$\frac{H_A^\circ - H_A(g)}{RT^2} = \frac{H_A^\circ - H_{mA}(liq)}{RT^2} + \left(\frac{\partial \ln x_A}{\partial T}\right)_p \tag{18.23}$$

Rearranging this equation, we can conclude that

$$\left(\frac{\partial \ln x_A}{\partial T}\right)_p = -\frac{H_A(g) - H_{mA}(liq)}{RT^2} = -\frac{\Delta H_{vap}}{RT^2} \tag{18.24}$$

ΔH_{vap} being the molar heat of vaporization of pure liquid A. Once again we can integrate, assuming ΔH_{vap} to be constant. Moreover, since the solution should consist of pure A at its boiling point, one of the limits of integration is fixed in such a way that we obtain the expression

$$\ln x_A = \frac{\Delta H_{vap}}{R}\left(\frac{1}{T} - \frac{1}{T_0}\right) \tag{18.25}$$

in which T_0 is the boiling point, or, in other words, that temperature at which $x_A = 1$. The boiling point to which we refer is that corresponding to an applied pressure, p, which we assume to be the pressure of the gas. It is evident here, as in the case of solids, that the equilibrium solubility depends only upon the solute and not upon the solvent. Also, so far as

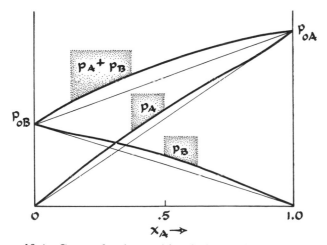

FIGURE 18-4   Curves showing positive deviations from Raoult's Law.

equation 18.25 is concerned, the solvent component need not obey Raoult's Law.

The pressure coefficient of gaseous solubility can likewise be determined very simply. Obviously, if the gas is ideal, its equilibrium mole fraction in solution must, according to Raoult's Law, be proportional to its pressure. If the gas is not ideal but Raoult's Law holds, we find, by the methods employed for solutions of solids, that

$$\left(\frac{\partial \ln x_A}{\partial p}\right)_T = \frac{V_{mA}(g) - V_{mA}(liq)}{RT} = \frac{\Delta V_{vap}}{RT} \tag{18.26}$$

Here again p represents both the applied pressure and the pressure of the gas. The extension to the case where the gas pressure and the pressure on the liquid are different can readily be made, but the result is not ordinarily of much use.

Henry's Law

When actual liquid mixtures depart from Raoult's Law, they do so more or less as represented in Figures 18-4 and 18-5. Figure 18-4 represents a system that exhibits positive deviations from Raoult's Law, so designated because the partial pressures (or fugacities) are in excess of those predicted by the law. Figure 18-5 depicts negative deviations

FIGURE 18-5   Curve for one component showing negative deviations from Raoult's Law. Note the regions in which Henry's Law and Raoult's Law hold.

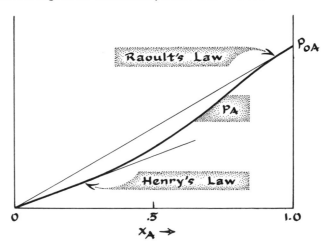

from Raoult's Law, so called because the vapor pressures are below the ideal values. Regardless of the kind of departure, however, careful examination of the experimental curves will disclose certain definite features. First of all, it is observed that, as the composition approaches that of a pure liquid, the partial pressure of the nearly pure liquid component asymptotically obeys Raoult's Law; in other words, the partial pressure curve becomes tangent to the line of Raoult's Law. On the other hand, the vapor pressure of the dilute component, at low enough concentrations, is found to be represented by a straight line with slope different from that expressed by Raoult's Law. For theoretical reasons to be demonstrated shortly, the range of compositions over which Raoult's Law holds for the nearly pure component must coincide with the range over which the dilute component has its vapor pressure expressed as a linear function of the mole fraction.

When the vapor pressure of the dilute component is proportional to the mole fraction, but with a slope different from that required by Raoult's Law, the substance is said to obey Henry's Law, which can be expressed simply as follows:

$$p_B = K_B x_B \qquad (18.27)$$

Stated in terms of fugacity instead of vapor pressure, Henry's Law can be written as

$$f_B = K_B x_B \qquad (18.28)$$

In each instance K_B is called the Henry's Law constant. The general statement of Henry's Law does not, of course, imply as detailed a description of a solution as Raoult's Law does, for the simple reason that the value of K_B is not determined solely by the component under consideration, but depends upon the other components in the mixture as well.

We shall now prove the statement made earlier that, when Raoult's Law holds for one component of a binary mixture, Henry's Law must hold for the other. Let us suppose that component A obeys Raoult's Law over some range of composition starting at $x_A = 1$. Then, in accordance with the Gibbs-Duhem equation, 9.22, we can assert that

$$x_A \frac{\partial \mu_A}{\partial x_A} + x_B \frac{\partial \mu_B}{\partial x_A} = 0 \qquad (18.29)$$

But we also know that

$$\mu_A = \mu_A^0 + RT \ln f_A \qquad (18.30)$$

and that

$$\mu_B = \mu_B^\circ + RT \ln f_B \tag{18.31}$$

Hence

$$x_A \frac{\partial \ln f_A}{\partial x_A} + x_B \frac{\partial \ln f_B}{\partial x_A} = 0 \tag{18.32}$$

Now, if Raoult's Law, equation 18.3, holds for substance A, then

$$\frac{\partial \ln f_A}{\partial x_A} = \frac{1}{x_A} \tag{18.33}$$

Substituting equation 18.33 into equation 18.32 and remembering that $x_A = 1 - x_B$, we obtain, upon integration, the expression

$$\ln f_B = \ln x_B + X(T,p) \tag{18.34}$$

in which $X(T,p)$ is a function of temperature and pressure. In exponential form, equation 18.34 expresses Henry's Law provided we set $X(T,p) = \ln K_B$. The dependence of K_B on the applied pressure can usually be neglected, but the effect of temperature is very marked. To obtain the temperature coefficient of $\ln K_B$, we need only differentiate equation 18.34 partially with respect to T, holding p and x_B constant. Then we see that

$$\left(\frac{\partial \ln K_B}{\partial T} \right)_p = \left(\frac{\partial \ln f_B}{\partial T} \right)_{p,x_B} \tag{18.35}$$

But from equation 17.37 we know that the right-hand side of equation 18.35 depends upon $H_B^\circ - H_B(\text{liq})$, which is essentially the partial molar heat of vaporization of B from solution, or, in other words, the negative of its partial molar heat of solution. (You will recall that H_B° applies to the infinitely dilute vapor.) Hence

$$\left(\frac{\partial \ln K_B}{\partial T} \right)_p = \frac{H_B^\circ - H_B(\text{liq})}{RT^2} = \frac{\Delta H_{\text{vap}}}{RT^2} \tag{18.36}$$

The partial molar heat of vaporization appearing in this equation is not necessarily equal to that of the pure liquid, although the equation has the same form as that for the temperature coefficient of a vapor pressure. Naturally, equation 18.36 can be integrated and used in the same way as the Clausius-Clapeyron equation.

The converse of the theorem leading to equation 18.34 can also be proved. If we assume that Henry's Law holds for the dilute component of a two-component system, Raoult's Law must hold for the other. The proof is similar to that given above, with the added feature that the function of integration is determined by the boundary conditions to equal the logarithm of the fugacity of the pure component.

EXERCISE:

At 20 °C the Henry's Law constant for oxygen dissolved in water is 2.95×10^7 Torr, and that for nitrogen is 5.75×10^7 Torr. If a vessel of water is exposed to the air, what is the equilibrium composition of dissolved air? For this exercise you may assume air to consist of 80 mole percent nitrogen and 20 mole percent oxygen.

Large Deviations from Raoult's Law

Not infrequently it is observed that the deviations from Raoult's Law are so large that the total vapor pressure of a binary mixture plotted as a function of the mole fraction will exhibit a maximum or a minimum. When this occurs, the composition of the mixture corresponding to the maximum or minimum will constitute what is known as a constant-boiling mixture, so named because the equilibrium composition of the vapor is the same as that of the solution. Figures 18-6 and 18-7 illustrate the total vapor pressures from solutions showing such maxima or minima. The compositions of both liquid and vapor corresponding to a given total vapor pressure are shown by the curves marked l and v. Where the curves show a maximum, it is evident that the vapor will have a composition nearer that of the constant-boiling mixture than will the liquid. On the other hand, if the curves show a minimum, the vapor will be less like the constant-boiling mixture than will the liquid.

Negative deviations from Raoult's Law, as shown in Figure 18-7, can be expected when pairs of unlike molecules attract each other more than

FIGURE 18-6   Total vapor pressure curves showing a maximum as a function of composition. The curve marked v refers to the composition of vapor; the curve marked l to that of liquid.

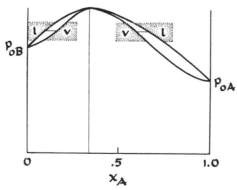

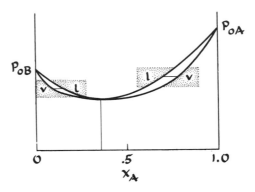

FIGURE 18-7 Total vapor pressure curves showing a minimum as a function of composition. The curve marked *v* refers to the composition of vapor; the curve marked *l* to that of liquid.

pairs of like molecules. Qualitatively, it is reasonable to expect that such attraction will result in a lower total vapor pressure than that predicted by Raoult's Law. Moreover, such a mixture will tend to have a more uniform composition, on a microscopic scale, than might be anticipated from purely statistical calculations alone, such as we employed earlier in deriving Raoult's Law. An extreme case would involve association of one molecule to the other; such association may be attributable to hydrogen bonding or to strong dipole-dipole interaction.

Large positive deviations from Raoult's Law can be expected when each kind of molecule prefers others of its own kind in its environment rather than molecules of the other type. Where positive deviations are observed, one can expect more pronounced microscopic fluctuations in composition that might be predicted from purely random statistical computations.

If the positive deviations from Raoult's Law are great enough (that is to say, if the molecules are sufficiently dissimilar and have sufficiently little attraction for each other), separation into two phases will occur, as illustrated in Figure 18-8. This happens when the Henry's Law constants are so large that the vapor pressures of the dilute components approach, even for small concentrations, the vapor pressures of the corresponding pure liquids. An extreme example of this kind of behavior is given by a mixture of a hydrocarbon like benzene with water. Such a mixture will separate into two layers, a hydrocarbon-rich layer saturated with water and a water-rich layer saturated with hydrocarbon. When the mutual solubility is so very small, it is generally safe to assume that

Raoult's Law holds for the nearly pure component in each phase and, of course, that Henry's Law holds for the other. Since the two separate phases can exist in equilibrium, the fugacity of a component, and hence its partial vapor pressure, must be the same for both phases, since the chemical potentials must be equal. It thus becomes clear that the Henry's Law constants for the dilute components must be very large, since the dilute component must exhibit a vapor pressure almost equal to that of the corresponding pure liquid.

Colligative Properties of Solutions

The lowering of vapor pressure that results from dissolving a foreign substance in a liquid is closely related to a number of other solution properties, which, as a group, are called "colligative properties." In addition to vapor-pressure lowering, this group includes freezing-point lowering, boiling-point elevation, and osmotic pressure, each of which is determined by the vapor-pressure lowering whether that lowering conforms to Raoult's Law or not. The freezing-point lowering is subject to the same kind of thermodynamic considerations that apply to solubility of solids, and the boiling-point elevation is related in a similar

FIGURE 18-8 Partial vapor pressure curves for a system exhibiting such large positive deviations from Raoult's Law that separation into two phases occurs. The region between the broken lines represents the compositions giving rise to separation into two phases. Single-phase regions are outside the broken lines.

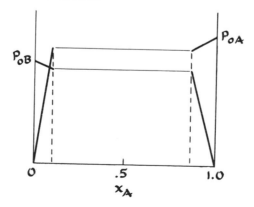

manner to solubility of gases. Osmotic pressure, on the other hand, represents a phenomenon not yet discussed in this chapter.

As mentioned above, the lowering of vapor pressure is the basic phenomenon that determines each of the other three properties mentioned. Accordingly, we shall now take up each of them in turn, pointing out the connection with vapor-pressure lowering and with other subjects already discussed. We shall also simplify the calculations by assuming that Raoult's Law holds for the solvent.

Freezing-point Lowering

When a substance like sugar is dissolved in water, the freezing point of the water is lowered by an amount depending upon the composition of the solution. This phenomenon is an outgrowth of the fact that the fugacity of the water is diminished by the presence of the solute. Assuming that Raoult's Law holds for the solvent, which is reasonable as long as the solution is not concentrated, we can readily establish a relationship between the freezing-point lowering and the composition of the solution. For this purpose we need only apply the equations already developed for solubility of solids in ideal solutions.

The reader will recall that, when we discussed solubility, the essential argument was concerned with finding the composition of solution for which the fugacity of the solute is equal to the fugacity of the corresponding pure solid. Now there is no thermodynamic distinction between solute and solvent; what is important is the equilibrium between phases. Should we reverse the normal use of the words "solute" and "solvent," we could, from a purely thermodynamic point of view, consider the melting of ice into a sugar solution as equivalent to the dissolving of ice in sugar; in other words, the freezing solution is "saturated" with ice. Although this may seem semantically bizarre, it simply reduces to saying that a solution of water will freeze when the chemical potential of water in the liquid phase equals the chemical potential of pure solid ice, which is precisely the argument that formed the basis for calculating solubility. Assuming Raoult's Law to hold for what we shall now call solvent, but not necessarily for solute, we can express the result quantitatively by rewriting equation 18.19:

$$\ln x_A = \ln (1 - x_B) = \frac{\Delta H_{\text{fus}}}{R} \left(\frac{1}{T_0} - \frac{1}{T} \right) \qquad (18.37)$$

In this equation x_A is the mole fraction of the solvent (water) and x_B the mole fraction of the solute (sugar). Once again ΔH_{fus} is the heat of fusion of the solvent, and T_0 is its normal freezing point. If x_B is small, $\ln(1 - x_B) \approx -x_B$, and equation 18.37 can be rewritten in the following approximate form:

$$x_B = -\frac{\Delta H_{fus}}{R}\left(\frac{T - T_0}{T_0 T}\right) \approx \frac{\Delta H_{fus}\Delta T}{RT_0^2} \qquad (18.38)$$

In this equation ΔT equals $T_0 - T$, the lowering of freezing point, which we assume to be small compared with T_0. To simplify further, let us now express the concentration of the solute in moles per thousand grams of solvent, a quantity which we shall designate as the molality, m. With this understanding we see that, for a dilute solution in which m is small,

$$x_B = \frac{m}{m + 1{,}000/M_A} \approx \frac{mM_A}{1{,}000} \qquad (18.39)$$

M_A being the molecular weight of the solvent. Combining this with equation 18.38, we now see that

$$\Delta T = \frac{M_A R T_0^2}{1{,}000\ \Delta H_{fus}}\, m \qquad (18.40)$$

or that

$$\Delta T = K_f m \qquad (18.41)$$

in which expression

$$K_f = \frac{M_A R T_0^2}{1{,}000\ \Delta H_{fus}} \qquad (18.42)$$

The constant K_f is known as the molal freezing-point constant for the particular solvent under consideration. Since K_f depends only on the solvent and not upon the solute, the measurement of freezing-point lowerings can be used to determine molecular weights. For water at one atmosphere pressure, K_f equals 1.86 degrees per unit of molal concentration. Equation 18.41 cannot be used with high accuracy for solutions as concentrated as one molal, but it can be used for dilute solutions, particularly in the following limiting form:

$$\lim_{m \to 0} \frac{\Delta T}{m} = K_f \qquad (18.43)$$

Since the molality of a solute is inversely proportional to its molecular weight, equation 18.43 suggests that a method of limiting freezing-point lowerings can be used for determining molecular weights, just as the method of limiting gas densities can be used for gases. The procedure simply consists of measuring freezing-point lowerings for various values of the concentration, c, measured in grams per 1,000 grams of solvent, then plotting $\Delta T/c$ against c, and finally extrapolating the data to $c = 0$.

Since $m = c/M_B$, provided M_B is the molecular weight of the solute, we see that

$$M_B = K_f \lim_{c \to 0} \frac{c}{\Delta T} \qquad (18.44)$$

EXERCISE:

Calculate K_f for water, using the value 79.7 cal for the specific heat of fusion.

Boiling-point Elevation

Just as the solubility of a solid is closely related to freezing-point lowering, so, too, the solubility of a gas is directly connected with a phenomenon known as boiling-point elevation. When a solute such as sugar is dissolved in a solvent like water, it is found that the boiling point—that is, the temperature at which the vapor pressure equals one atmosphere—is raised. This elevation of boiling point is a direct result of the fact that the solution, at any temperature, has a lower vapor pressure than the pure solvent. Since boiling points involve equilibrium vapor pressures, the thermodynamic relationship between boiling-point elevation and gas solubility should be clear, for a boiling solution can be regarded as "saturated" with its vapor. To use the results of our calculations of gas solubility, we must once again interchange the words "solute" and "solvent" and think of the vapor as dissolved in the solute. The net result can be expressed in the form of equation 18.25, which we rewrite as follows:

$$\ln x_A = \ln (1 - x_B) = \frac{\Delta H_{vap}}{R} \left(\frac{1}{T} - \frac{1}{T_0} \right) \qquad (18.45)$$

Like the equation for freezing-point lowering, the boiling-point equation can also be modified to a form valid for dilute solutions. By approximating to the logarithm of $1 - x_B$ with $-x_B$ and proceeding as before, we see that

$$x_B = \frac{\Delta H_{vap} (T - T_0)}{R T T_0} \approx \frac{\Delta H_{vap} \Delta T}{R T_0^2} \qquad (18.46)$$

ΔT being the boiling-point elevation. Finally, if we express the concentration of solute in terms of the molality, m, or moles per thousand grams of solvent, we find that

$$\Delta T = \frac{M_A R T_0^2}{1,000 \, \Delta H_{vap}} m = K_b m \qquad (18.47)$$

The constant K_b is known as the molal boiling-point constant, which is a quantity characteristic of the solvent. For water this equals 0.52 degrees per unit of molal concentration. The mathematical limitations on the use of equation 18.47 for boiling-point elevation are precisely the same as those on the approximate equation for freezing-point lowering. The equation is most reliable for dilute solutions and is used most effectively through extrapolation to infinitely dilute solutions.

EXERCISE:

Calculate K_b for water, using the value 540 cal for the specific heat of vaporization.

Osmotic Pressure

Closely related to freezing-point lowering and boiling-point elevation is the solution property known as osmotic pressure. We have seen that the freezing point of a solution is precisely that temperature at which the fugacity of the solvent component equals that of the corresponding pure solid. On the other hand, the boiling point of a solution is just that temperature at which the vapor pressure of the solvent component equals one atmosphere, which is the vapor pressure of pure solvent at its normal boiling point. Not unrelated is the concept of osmotic pressure, which arises from making the fugacity of the solvent component of a solution equal to that of pure solvent by isothermally increasing the applied pressure on the solution.

Now an increase of the pressure applied to a solution will, of course, increase the chemical potentials of the substances in the solution; in particular it will increase the chemical potential of the solvent, together with its fugacity and vapor pressure. In the light of this fact, the osmotic pressure of a solution is defined as the excess pressure that must be applied to the solution to bring the chemical potential of its solvent up to that of pure solvent at the same temperature. Although one generally assumes a nonvolatile solute, it is not necessary to make this assumption so far as the following development is concerned.

If V_A is the partial molar volume of solvent in solution, p_1 the applied pressure, and μ_A and V_A the chemical potential and fugacity of the solvent, then

$$d\mu_A = V_A\, dp_1 = RT\, d\ln f_A \qquad (18.48)$$

When the applied pressure is one atmosphere, the fugacity of the solvent in the solution of interest will have some value, f_A. Let us now ask how much extra pressure (osmotic pressure) must be applied to the solution to bring the fugacity of the solvent up to f_{0A}, which the pure solvent would exhibit under an applied pressure of one atmosphere. Assuming the solution to be incompressible, and hence V_A to be independent of pressure, we see, upon integration of equation 18.48, that the osmotic pressure, Π ($= \Delta p_1$, will be given by the expression

$$\Pi = \frac{RT}{V_A} \ln \frac{f_{0A}}{f_A} \tag{18.49}$$

Whether the solution is ideal or not, equation 18.49 can be used to calculate the osmotic pressure; for this purpose the fugacities can usually be replaced by vapor pressures. However, if the solvent obeys Raoult's Law, then $f_A = f_{0A} x_A$, and

$$\Pi = -\frac{RT}{V_A} \ln x_A$$

$$= -\frac{RT}{V_A} \ln (1 - x_B) \tag{18.50}$$

If the mole fraction of solute, x_B, is small, the osmotic pressure will be approximately

$$\Pi = \frac{RT}{V_A} x_B \tag{18.51}$$

For a dilute solution, however, x_B is nearly equal to n_B/n_A, the ratio of the number of moles of solute to solvent. Moreover, since the total volume, V, will be nearly equal to $n_A V_A$, we see further that for a very dilute solution

$$\Pi = \frac{n_B RT}{V} = \frac{c\,RT}{M_B} \tag{18.52}$$

provided c is the concentration of solute in grams per unit volume. Equation 18.52, which is identical in form to the ideal gas law equation, is known as the van't Hoff equation. It is not very accurate except for dilute solutions, but it is useful for analyzing data that are properly extrapolated to infinite dilution.

Osmotic pressures can be measured directly with an apparatus in which the solution of interest is separated from the pure solvent by a semi-permeable membrane that permits passage of solvent but not of solute. The experimental procedure consists of allowing the solvent to pass through the membrane into the solution to build up a hydrostatic head, which, at equilibrium, will equal the osmotic pressure. Osmotic pressures are large even for solutions of moderate concentration. If the van't Hoff equation were valid at a concentration of 0.1 molar (which it is not), the osmotic pressure of such a solution at zero degrees celsius would be 2.24 atmospheres. Since it is not easy to obtain membranes strong enough to withstand the very large pressures involved, osmotic pressures are not often measured for solutions of low molecular weight solutes. Although it is not practical to measure osmotic pressures for such solutions, the method becomes quite useful for high polymers, which, because of their large molecular weights, can be used to make solutions of low mole fractions even at appreciable weight concentrations. The actual lowering of vapor pressure will be exceedingly small for high polymers, but the osmotic pressure will be measurable and can be used as a means of determining molecular weights. In practice the osmotic pressure is measured for each of several weight concentrations, and the quotient, Π/c, is plotted against c. These results, extrapolated to zero concentration, give the limiting value of Π/c, which can be used for calculating the molecular weight of the solute by the following equation:

$$M_B = RT \lim_{c \to 0} (c/\Pi) \qquad (18.53)$$

Polymer solutions exhibit marked deviations from ideality even at low mole fractions. Accordingly, it is very important to extrapolate the results to zero concentration if one wishes to arrive at a meaningful measurement of molecular weight.

Problems

1. Consider a dilute solution of a nonvolatile solute, B, in a volatile solvent, A. Assuming Trouton's rule to hold for the solvent (that is, $\Delta H_{vap} = 21\ T_0$, if T_0 is the normal boiling point), show that the boiling point, T_b, of the solution is related to the mole fraction of solvent by the following equation:

$$-\ln x_A = 10.5 \left(1 - \frac{T_0}{T_b}\right)$$

Show further that, if $\ln x_A$ is small,

$$T_b x_A^{.095} = T_0$$

2. Several liquid substances are mixed together to form an ideal solution. For a fixed total number of moles of the substances (that is, for $n_1 + n_2 + n_3 \ldots = $ constant), show that the entropy of mixing is a maximum when $n_1 = n_2 = n_3 = \ldots$.

3. The temperature and applied pressure of a solution saturated with a solid are simultaneously changed in such a way that the composition remains unchanged and the solution remains saturated. If x_A is the mole fraction of solute, show, assuming Raoult's Law to hold, that the temperature and applied pressure must be related in such a way that

$$\left(\frac{\partial p_1}{\partial T}\right)_{x_A} = \frac{\Delta H_{fus}}{T \Delta V_A}$$

Show that a similar equation applies to a saturated solution of a gas in a liquid.

4. Naphthalene, which melts at 80 °C, forms an ideal solution in benzene. If the heat of fusion of naphthalene is 35.6 cal g^{-1}, calculate the solubility of naphthalene in benzene at 25 °C. At what temperature would the solubility of naphthalene be 85 g per mole of benzene?

5. The solubility of a certain gas in a liquid obeys Henry's Law, but the value of K is so large that the solutions are exceedingly dilute even at moderately high pressures. Prove that the volume of the dissolved gas, when measured as a gas under the equilibrium pressure, is independent of the applied pressure. Assume the gas to be ideal. (The equilibrium volume of a gas dissolved in a unit volume of liquid is called the solubility coefficient.)

6. At 0 °C the solubility coefficient (see Problem 5) of carbon dioxide in water is 1.71. Suppose that a 2-liter vessel originally contains 1 liter of pure water and 1 liter of pure carbon dioxide gas at 0 °C and 1 atm. If we neglect the vapor pressure of water and any change in the volume of the aqueous phase, what will be the equilibrium pressure of the carbon dioxide after it has been permitted to dissolve in the water at 0 °C?

7. The Henry's Law constants for hydrogen dissolved in water are 4.42×10^7 at 0 °C, 5.20×10^7 at 20 °C, and 5.78×10^7 at 40 °C. Calculate the heat of solution of hydrogen in water in the temperature ranges 0–20 °C and 20–40 °C.

8. Consider a pair of immiscible liquids, like water and carbon tetrachloride, in which a third substance, like iodine, can be dissolved. Assuming that iodine obeys Henry's Law for each of the two immiscible solutions, show that, when the two solutions are placed in contact with each other, the iodine will distribute itself between the two layers in accordance with the following equilibrium relationship:

$$\frac{x_{I_2} \text{ (in CCl}_4)}{x_{I_2} \text{ (in H}_2\text{O)}} = \frac{K_{I_2} \text{ (in H}_2\text{O)}}{K_{I_2} \text{ (in CCl}_4)} = \text{constant}$$

Assuming further that the solutions of I_2 are both infinitely dilute, show that

$$\frac{C_{I_2} \text{ (in CCl}_4)}{C_{I_2} \text{ (in H}_2\text{O)}} = K$$

if C denotes a concentration (in moles per liter, for example) and K is a distribution coefficient.

9. At 25 °C the distribution coefficient, K, for iodine dissolved in carbon tetrachloride and water is 85. (This K appears in the last equation of Prob. 8.)

(A) If 0.25 g of iodine is dissolved in 1 liter of water at 25 °C and the solution is subsequently equilibrated with 1 liter of carbon tetrachloride, how much iodine will be left in the water?

(B) Suppose that the liter of aqueous solution is shaken up with one-half liter of carbon tetrachloride and the two layers are separated, and that the aqueous solution is then shaken again with another half liter of carbon tetrachloride. How much iodine remains in the water after the two-stage extraction?

(c) If the aqueous solution were shaken up successively with $1/s$ liter of carbon tetrachloride and the procedure repeated s times so that, in all, one liter of carbon tetrachloride was used, how much iodine would remain in the water? What is the limit of this expression as $s \longrightarrow \infty$?

Ans. to limit. $C_{\text{final}} = C_0 e^{-K}$

or $m = 0.25 e^{-85}$ g

10. A solid, such as iodine, is distributed between two immiscible liquids, such as water and carbon tetrachloride. Prove that, if one phase is saturated with iodine, the other phase must also be saturated. (No particular form of solution law need be assumed for this proof. If the proposition stated above were not true, it would be possible to construct a perpetual-motion machine. How?)

11. Consider a dilute solution of solute B in solvent A, which freezes to give a dilute solid solution instead of pure solid A. Let us suppose that Raoult's Law holds with respect to component A for both the liquid solution and the solid solution; for the solid we can therefore write

$$f_A(c) = f_{0A}(c) x_A(c)$$

(A) Assuming the heat of fusion, ΔH_{fus}, of A to be a constant, show that

$$\ln \frac{x_A(\text{liq})}{x_A(c)} = \frac{\Delta H_{\text{fus}}}{R} \left(\frac{1}{T_0} - \frac{1}{T} \right)$$

if T_0 is the normal freezing point of pure A.

(B) Show that, if $x_B(\text{liq})$ and $x_B(c)$ are both small compared with unity,

$$x_B(\text{liq}) - x_B(c) \approx \frac{\Delta H_{\text{fus}}}{R T_0^2} \Delta T$$

if ΔT is the freezing-point lowering. (Observe that, if the solid is richer in B than the liquid is, ΔT is negative; in other words, there is a freezing-point rise.)

12. Consider the apparatus in the accompanying illustration. It contains a vertical tube of solution, some pure solvent at the bottom, and some solvent vapor, all in the earth's gravitational field. The tube of solution is separated at its bottom from the pure solvent by a semi-permeable membrane, M, which permits the transfer of solvent molecules but is impermeable to the solute, which is also assumed to be nonvolatile. The system pictured will reach equilibrium when the following conditions are fulfilled: The fugacity, f, of the solution at height h must be less than that of the pure solvent, f_0, by whatever amount the fugacity of a gas, in an isothermal atmosphere, diminishes by reason of the effect of the earth's gravitational field. Simultaneously, the extra hydrostatic pressure on the solution in the immediate neighborhood of the semi-permeable membrane must equal its osmotic pressure so that osmotic equilibrium is attained.

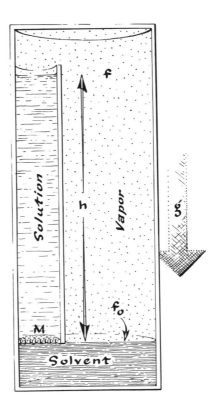

(A) If the equilibrium height of the solution is h, show that the osmotic pressure will be

$$\Pi = \rho g h$$

if ρ is the density of the solution. Assuming the density of the solution to be the same as that of the pure solvent, show that the osmotic pressure will be approximately

$$\Pi = \frac{Mgh}{V_m}$$

if V_m is the molar volume of pure solvent and M its molecular weight.

(B) Show also that the height, h, can be determined from the fugacity equation for an isothermal atmosphere, namely:

$$f = f_0 e^{-Mgh/RT}$$

or

$$h = \frac{RT}{Mg} \ln \frac{f_0}{f}$$

(C) By eliminating the height, h, between the equations in A and B, show that the osmotic pressure is

$$\Pi = \frac{RT}{V_m} \ln \frac{f_0}{f}$$

(D) Show that, if the vapor is an ideal gas and the decrease in vapor pressure is small, the osmotic pressure can be related to the vapor pressure through the following approximate equation:

$$\Pi = \frac{RT}{V_m} \frac{p_0 - p}{p_0}$$

Observe that this derivation does not require assuming that the solution is ideal.

13. Prove, for an exceedingly dilute solution, that the osmotic rise, h (measured at temperature T), the freezing-point lowering, ΔT, and the relative lowering of vapor pressure, $(p_0 - p)/p_0$, are related as follows:

$$\frac{p_0 - p}{p_0} = \frac{M \Delta T}{1,000 \, K_f} = \frac{Mgh}{RT}$$

(In these equations M is the molecular weight of the *solvent*.)

14. (A) A solution of a polymer of high molecular weight dissolved in benzene exhibits an osmotic rise of 1.15 cm at 25 °C. If $K_f = 5.12$ for benzene, calculate the freezing-point lowering of the solution and the vapor-pressure lowering. $p_0 = 94$ Torr at 25 °C. What do these results mean concerning the practicability of measuring freezing-point lowerings and vapor-pressure lowerings of high polymer solutions?

(B) The polymer solution described above contains 3.15 g of polymer per liter of solution. A solution half as concentrated shows at 25 °C an osmotic rise of 0.51 cm. Estimate the molecular weight of the polymer.

15. A certain aqueous solution of sucrose freezes at -0.15 °C. What is its osmotic pressure at 25 °C? How high a column of water would correspond to this pressure? What does this suggest concerning the practicability of measuring osmotic pressures for solutions of substances of low molecular weights?

16. Suppose you allow the vapor from a solution of benzoic acid in benzene to come into equilibrium with the vapor from a solution of phenanthrene in benzene by permitting the benzene vapor to distill from one solution to the other, in a closed apparatus, until the vapor pressures of the two solutions are equal. (The benzoic acid and phenanthrene can be regarded as nonvolatile.) At 32.68 °C, 18.20 ml of solution, containing 0.3845 g of phenanthrene, has the same benzene vapor pressure as 18.24 ml of solution, containing 0.4915 g of benzoic acid. Assuming that the solvent obeys Raoult's Law for both solutions, calculate the average molecular weight of the benzoic acid, assuming further that phenanthrene exists as single molecules, $C_{14}H_{10}$. At 32.68 °C the densities of the solutions in g ml^{-1} are expressed by the following equations, in which w is the weight percentage of solute:

benzoic acid in benzene: $\rho = 0.8653 + 0.001735\,w$
phenanthrene in benzene: $\rho = 0.8653 + 0.001920\,w$

The calculated molecular weight of benzoic acid suggests that it is largely associated into dimers. From the average molecular weight obtained above calculate the degree of dissociation, α, of benzoic acid dimer into monomer:

$$(C_6H_5COOH)_2 \rightleftharpoons 2C_6H_5COOH$$
$$1 - \alpha \qquad\qquad\quad 2\alpha$$

From the degree of dissociation calculate a value for the equilibrium constant

$$\frac{4\alpha^2 c}{1 - \alpha} = K$$

in which c denotes the total concentration in moles of dimer per liter of solution, assuming no dissociation.

Ans. $K = 2.2 \times 10^{-3}$

17. At 25 °C the solubility of rhombic sulfur in carbon tetrachloride is 0.84 g per 100 g of solvent. At the same temperature the Gibbs energy change attending the transition of rhombic to monoclinic sulfur is 23 cal mol^{-1}. What is the solubility of monoclinic sulfur in carbon tetrachloride at 25 °C? (Assume that the solutions are infinitely dilute and obey Henry's Law.) If one had a saturated solution of monoclinic sulfur at 25 °C, what relationship would that solution bear to rhombic sulfur?

18. (A) A pair of binary liquids obeys Raoult's Law. If $p_{0A} = 500$ Torr, and $p_{0B} = 250$ Torr, plot the total pressure as a function of the mole fraction of the equilibrium vapor. Compare these results on one graph with the usual plot of pressure against composition of liquid.

(B) Prove that the difference $(x'_A - x_A)$ between the mole fraction of the vapor, x'_A, and that of the liquid, x_A, is a maximum when

$$x_B/x_A = \sqrt{p_{0A}/p_{0B}}$$

(C) Show, as a consequence, that the composition of the equilibrium vapor that differs most from that of the liquid is

$$x'_A = \frac{(p_{0A})^{1/2}}{(p_{0A})^{1/2} + (p_{0B})^{1/2}}$$

19

Activities

Definition of Activity

The treatment of fugacities in Chapter Seventeen was designed primarily to describe the behavior of nonideal gases and to deal with equilibria involving such gases. To be sure, the fugacity concept was also extended to solids and liquids, and was applied to them in connection with vapor pressures, but its application to nongaseous systems often becomes awkward, if not impractical. If we should be interested, for example, in a reaction involving sodium chloride in aqueous solution, we would find the fugacity of the sodium chloride so small and so hard to measure that we would be compelled to seek some other means of describing the contribution of the sodium chloride to a reaction potential. With that end in mind we shall now introduce a new concept, which will really be a generalization of fugacity, of such character that it can readily be applied to nongaseous systems. This more general function will be named *activity*.

The reader will recall that, in Chapter Fifteen, we introduced a quantity, λ, called the absolute activity, which was directly related to the chemical potential of a single substance through equation 15.30.

The same kind of definition can be used for the absolute activity, λ_i, of any component i in a mixture, as follows:

$$\mu_i = RT \ln \lambda_i \qquad (19.1)$$

μ_i being, of course, the partial molar Gibbs energy or chemical potential. The absolute activity of a substance is of theoretical importance, but it is impractical for numerical calculations since it cannot be determined by ordinary thermodynamic measurements; in other words, we do not have available absolute values of the chemical potential. Nevertheless, we can measure isothermal changes in chemical potentials and hence ratios of absolute activities. This provides the basis for establishing numerous specific kinds of activities, a subject that we shall now systematically develop.

Just as we did for fugacity, we shall employ two equations to define an activity, the first of which is identical in form with one of those used for fugacity. Thus we shall relate the chemical potential of a substance to its activity, a_i, by means of the equation

$$\mu_i = \phi_i(T) + RT \ln a_i \qquad (19.2)$$

in which $\phi_i(T)$ is a function of T only; as before, $\phi_i(T)$ will be called a "standard chemical potential." From this portion of our definition of activity it is obvious that, as with fugacity, the ratio of the activities of a substance in two different states—at the same temperature, of course— will depend only on the difference in chemical potential. It is from here on that we shall encounter the difference between an activity, which is quite general, and fugacity, which is more specific. The difference between activity and fugacity lies in the choice of the reference states and in the use of functions other than pressure for reference purposes. It should be clear that, before a numerical value can be assigned to an activity, some kind of reference function must be assumed (analogous to p for fugacity) and some particular state must be chosen in which the activity approaches or equals the value of the reference function.

For illustrative purposes, let us consider a liquid solution and assume for the second part of our definition of activity that

$$\lim_{c_i \to 0} \frac{a_i}{c_i} = 1 \qquad (19.3)$$

c_i being the concentration of the component of interest. Under these circumstances we say that the activity approaches the concentration as the solution becomes infinitely dilute; the concentration is said to be the "reference function" and the infinitely dilute solution the "reference state." In the same way, the molality, m_i, could equally well be used.

As another example, consider a binary liquid mixture for which x_i is the mole fraction of one of the components. Under certain circumstances it is convenient to complete the definition of activity by means of the equation

$$\lim_{x_i \to 1} \frac{a_i}{x_i} = 1 \tag{19.4}$$

In this example the pure liquid is the reference state, and the activity of the pure liquid is equal to unity.

The foregoing illustrations can now be generalized as follows: Let us assume for reference purposes a function, $g(\zeta)$, in which ζ is a concentration, pressure, mole fraction, or some similar variable that might be used in conjunction with temperature† for describing a system, gaseous, liquid, or solid. We shall now define our generalized activity by stipulating that

$$\lim_{\zeta \to \zeta^*} \frac{a_i}{g(\zeta)} = 1 \tag{19.5}$$

ζ^* being the value of ζ in the reference state, which is that state in which the activity equals the value of the reference function. ζ^* may correspond to infinite dilution or to a finite concentration; no restriction need be imposed at this time regarding the precise value of ζ^*. Equation 19.5, in conjunction with equation 19.2, completes our general definition. The only thing that remains to be done is to find some means of establishing the change in chemical potential that attends the change in going from the reference state to the state of interest. Once this is done, a precise value for the activity can be determined.

That the fugacity is a special kind of activity should now be clear, for the fugacity has all the attributes of an activity; specifically, it involves the pressure as the reference function and zero pressure as the reference state. Obviously, a ratio of fugacities for any two states at the same temperature must be identical with the ratio of activities associated with the same two states regardless of what function the activity is referred to. This follows from equations 17.44 and 19.2 since

$$\mu_i' - \mu_i = RT \ln \frac{f_i'}{f_i} = RT \ln \frac{a_i'}{a_i} \tag{19.6}$$

in which the primed quantities denote states different from those with-

---

† The function $g(\zeta)$ may be replaced by the function $g(\zeta, T)$; such a change will be inconsequential except when we deal with temperature coefficients of the activity.

out the primes.† Hence, if we know the fugacity of a substance in each of two states, we can determine a ratio of activities for those states. If one of those states is the reference state for the activity, we can evidently determine the activity for the other state without difficulty. A similar argument can be applied to switching from one activity representation to another, without recourse to fugacity.

Activity Coefficients

Just as we found it convenient to define fugacity coefficient (eq. 17.40), we shall also define activity coefficient as the ratio of the activity to its reference function:

$$\gamma = \frac{a}{g(\zeta)} \tag{19.7}$$

It follows from equation 19.5 that the limit of γ as ζ approaches ζ^* is unity; in other words, the activity coefficient equals unity in the reference state.

The reference function, $g(\zeta)$, is usually chosen as some function that appears in a statement of an idealized physical law; the reference state, corresponding to ζ^*, is then chosen as a state in which the law is valid, at least asymptotically. For example, if we think of Raoult's Law as our ideal, it is convenient to adopt the mole fraction as the reference function and the pure liquid as the reference state, as illustrated by equation 19.4. Under these circumstances, γ measures the departure from ideality. If $g(\zeta)$ is chosen arbitrarily without regard to possible conformity to an idealized behavior, γ simply measures the departure of a from $g(\zeta)$, and no special significance can be attached to it.‡ Later in this chapter we shall illustrate some practical choices of reference functions for describing solutions and shall show how the activity coefficient measures nonideality.

---

† Obviously, equation 19.6 can also be applied to ratios of absolute activities, λ_i'/λ_i, in the light of equation 19.1.

‡ Consider, for example, an ideal gas, for which $f = p$, and let us choose our reference function as p^2 and our reference state as $p_0 (\neq 0)$. Then $\lim_{p \to p_0} a/p^2 = 1$. If a_0 is the activity in the reference state, $a_0 = p_0^2$. But, in the light of equation 19.6, we can also assert that $a/a_0 = p/p_0$ and that $a = p_0 p$ and $\gamma = p_0/p$. Obviously, since the gas is ideal, a must be proportional to p; this means that with respect to the chosen reference function—namely, p^2—the activity coefficient γ must equal p_0/p. Although this example is quite artificial, it illustrates how an activity coefficient corrects for departures of an activity from an assumed functional form.

Temperature Coefficient of Activity; Standard States

To obtain the temperature coefficient of an activity, we shall proceed in a manner analogous to that employed in connection with fugacities. Consider two states at the same temperature, one of which approaches the reference state as a limit. Using an asterisk to denote the reference state (or approach thereto), we can assert that

$$\mu_i - \mu_i^* = RT \ln a_i - RT \ln a_i^*$$
$$= RT \ln a_i - RT \ln g(\zeta^*) \qquad (19.8)$$

Dividing through by RT and differentiating partially with respect to T while holding p constant, we obtain

$$-\frac{H_i}{RT^2} + \frac{H_i^*}{RT^2} = \left(\frac{\partial \ln a_i}{\partial T}\right)_p - \left(\frac{\partial \ln g(\zeta^*)}{\partial T}\right)_p \qquad (19.9)$$

H_i^* being the value of the partial molar enthalpy in the reference state, regardless of what that state might be. Now let us suppose that ζ does not depend on T; this situation might be realized, for example, if ζ were a function of p, x_i, etc. Under such circumstances, equation 19.9 reduces to

$$\left(\frac{\partial \ln a_i}{\partial T}\right)_p = \frac{H_i^* - H_i}{RT^2} \qquad (19.10)$$

Let us now introduce the concept of "standard state" by rewriting equation 19.2 in the form

$$\mu_i = \mu_i^\circ + RT \ln a_i \qquad (19.11)$$

in which μ_i° is the "standard chemical potential." Evidently μ_i° will equal the value of the partial molar Gibbs energy when the activity is equal to unity; its temperature coefficient, however, will not necessarily equal the temperature coefficient of the actual partial molar Gibbs energy when $a_i = 1$. For by dividing equation 19.11 through by RT and differentiating, we see, upon comparing with equation 19.9, that

$$\frac{d}{dT}\left(\frac{\mu_i^\circ}{RT}\right) = -\frac{H_i^*}{RT^2} - \left[\frac{\partial \ln g(\zeta^*)}{\partial T}\right]_p \qquad (19.12)$$

Once again, if ζ does not depend upon T,

$$\frac{d}{dT}\left(\frac{\mu_i^\circ}{RT}\right) = -\frac{H_i^*}{RT^2} \qquad (19.13)$$

Hence we conclude that the temperature coefficient of the "standard chemical potential" is determined by the enthalpy of the reference state; this is equivalent to saying that $H_i^\circ = H_i^*$. Therefore the "standard state" in the activity system of representation, like that of the fugacity system, is generally a hypothetical state, since H_i^* is not necessarily the partial molar enthalpy when $a_i = 1$. However, if the formulation is such that the activity is equal to unity in the reference state, the standard state becomes identical with the reference state and is not hypothetical. For example, if the reference function is the mole fraction of a particular substance and the reference state is the pure substance, the pure substance will also represent the standard state.

If $g(\zeta)$ depends upon T, even implicitly, then, of course, the more complicated expressions 19.9 and 19.12 must be used in place of equations 19.10 and 19.13.

For the numerical data given in Appendix D (as well as certain tables in Chapters Three and Ten), the reference function for ions in solution, such as $Na^+(aq)$, is the molality, m. Accordingly, their standard states are hypothetical. On the other hand, the standard states for liquid or solid elements and compounds are the pure substances. Finally, the standard states for gases are hypothetical states, each corresponding to a fugacity of 1 atm.

Application of the Activity Concept to Solutions

We shall now illustrate the activity concept by applying it to binary solutions that do not obey Raoult's Law. Consider, for example, a pair of liquids exhibiting negative deviations, as shown in Figure 19-1. (To avoid making the diagram unduly complicated, we have indicated the fugacity of only one of the components, A.) If we think of Raoult's Law as the ideal, departures from which are to be measured, it is convenient to adopt as our reference function for activity the mole fraction of the component of interest and to set the pure liquid [$x(liq) = 1$] as the reference state. In accordance with equation 19.6 it follows, for any pair of concentrations, that

$$\frac{a_A'}{a_A} = \frac{f_A'}{f_A} \tag{19.14}$$

Since, by our choice of reference state, the activity of the pure liquid equals unity, the activity at any other concentration will simply equal the fugacity at that concentration divided by the fugacity of the pure liquid. In other words,

$$a_A = f_A/f_{0A} \tag{19.15}$$

if f_{0A} is the fugacity of the pure liquid.

By definition, the activity coefficient, γ_A, must equal the activity, a_A, divided by the reference function, x_A, so that

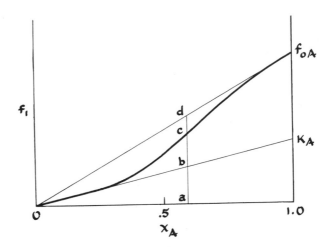

FIGURE 19-1   Fugacity as a function of composition for one component of a solution. If Raoult's Law is used for reference, the activity coefficient, γ_A, is equal to ac/ad. If Henry's Law is used for reference, $\gamma_A = ac/ab$.

$$\gamma_A = \frac{f_A}{f_{0A}x_A} \qquad\qquad (19.16)$$

It is evident, therefore, that the activity coefficient is simply equal to the ratio of the actual fugacity to what the fugacity would be if Raoult's Law held. (Graphically, γ_A equals ac/ad in Fig. 19-1.) Obviously, if Raoult's Law holds, γ_A will equal unity; thus it is clear that γ_A measures a departure from ideality. In the region where Henry's Law holds, the actual fugacity will be $K_A x_A$, and the activity coefficient will then be constant:

$$\gamma_A = \frac{K_A x_A}{f_{0A}x_A} = \frac{K_A}{f_{0A}} \qquad\qquad (19.17)$$

This conclusion is, of course, based on the premise that the reference function is the mole fraction and that the reference (or standard) state is still the pure liquid.

As an alternative, it is sometimes convenient to think of Henry's Law as the ideal, departures from which are to be measured. The Henry's Law portion of the curve is then extrapolated and an abnormality is measured by the ratio of the actual fugacity to what the fugacity would be if Henry's Law held. The reference function is again taken to be the mole fraction, x_A, but now the reference state will be that of infinite dilution. With this understanding, we can assert that

$$\frac{a_A}{a_A^*} = \frac{f_A}{f_A^*} = \frac{f_A}{K_1 x_A^*} \qquad (19.18)$$

The asterisks in this expression denote infinitely dilute states in which Henry's Law is valid. Hence we see that

$$a_A = \frac{f_A}{K_A} \cdot \frac{a_A^*}{x_A^*} \qquad (19.19)$$

But, since our reference function has been chosen as x_A, it follows that $a_A^* = x_A^*$; therefore

$$a_A = \frac{f_A}{K_A} \qquad (19.20)$$

and

$$\gamma_A = \frac{a_A}{x_A} = \frac{f_A}{K_A x_A} \qquad (19.21)$$

In other words, the activity coefficient is equal to the actual fugacity divided by what the fugacity would be if Henry's Law held. (Graphically, γ_A equals ac/ab in Fig. 19-1.) The "standard state" is now a hypothetical one in which the fugacity, $f°$, equals the Henry's Law constant, K_A, and the standard enthalpy, $H_A°$, equals H_A^*, the partial molar enthalpy in infinitely dilute solution. It should be evident now that, when the Henry's Law end of the fugacity curve is used for reference purposes, a solution exhibiting negative deviations from Raoult's Law will have an activity coefficient equal to or greater than unity. On the other hand, if the Raoult's Law portion of the curve is used for reference, the activity coefficient will be equal to or less than unity. If positive deviations from Raoult's Law occur, the equations are identical in form, but the sizes of the activity coefficients, compared with unity, are the reverse of those stated above.

Statistical Mechanics of Nonideal Solutions

In Chapter Eighteen we saw that a very simple model sufficed to provide a statistical-mechanical basis for Raoult's Law. In that model it was assumed that a liquid could be represented by a quasi lattice, and that the two (or more) kinds of molecules involved were equivalent in size and in their intermolecular interaction energies. We shall now modify that model slightly to provide an approximate description of nonideal systems. Although this new model will appear to be highly artificial, it will nevertheless prove instructive and helpful in our thinking.

Let us assume once again that our mixture consists of N_A molecules of kind A and N_B molecules of kind B. We imagine that the molecules of A are simple spheres, each of which can occupy one lattice site in our liquid model. The B molecules, on the other hand, consist of two spheres, each of size A, joined together to form a dumbbell. The distance between the spheres that make up a B molecule is precisely equal to the distance between lattice sites, so that a B molecule occupies two sites in the liquid pseudo lattice. (See Fig. 19-2.) We further suppose that each site in the lattice is immediately surrounded by z other sites as nearest neighbors; hence, if one end of a B molecule is placed in a particular site, the other end might conceivably be placed in any of z other sites. For the present we assume that the molecular energies are independent of molecular environments, so that the heat of mixing will be zero. Our statistical-mechanical problem therefore reduces to finding the thermodynamic probability associated with placing N_A molecules of type A and N_B molecules of type B in a lattice containing $N_A + 2N_B$ sites.

To obtain the thermodynamic probability of the mixture, we need only count the number of distinguishable ways in which the B molecules can be placed in the available sites; the A molecules can then be placed in the remaining sites. For the sake of simplicity we assume that the two ends of the B molecules are distinguishable from each other. As we shall see later, this assumption will make no difference so far as the entropy of

FIGURE 19-2 Model of liquid made up of spherical molecules, A, and double-sized molecules, B. Each B molecule occupies the equivalent of two A sites.

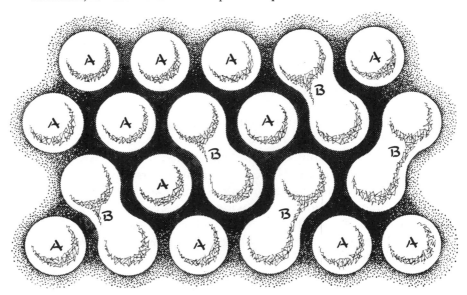

mixing A with B is concerned, although it would affect the absolute entropy of B. We are now in a position to proceed with our derivation, which will consist of placing each of the B molecules, figuratively speaking, in the lattice.

A particular end of the first B molecule can evidently be placed in any of $N_A + 2N_B$ sites, after which the other end will have available z possibilities. Hence the contribution of the first molecule toward the thermodynamic probability will equal $(N_A + 2N_B)z$. One end of the second B molecule can now be placed in any of $N_A + 2N_B - 2$ sites. The other end of the second molecule, however, will not always have z choices, but on the average will have $z(1 - p_2)$ if p_2 is the probability that the first molecule has blocked out one or two of the z potential sites for the second half of the second molecule. But the probability, p_2, that the sites occupied by the first molecule coincide with any of the sites around the first end of the second is just $2/(N_A + 2N_B - 1)$. Hence the average number of sites available for the second end of the second molecule will be

$$z[1 - 2/(N_A + 2N_B - 1)] \quad \text{or} \quad z(N_A + 2N_B - 3)/(N_A + 2N_B - 1).$$

Now the first end of the third molecule will have $N_A + 2N_B - 4$ sites to fit into, after which its second end can enter any of $z(1 - p_3)$ sites if p_3 is the probability that the first two molecules have blocked out part of the surrounding sites. Evidently p_3 will equal $4/(N_A + 2N_B - 1)$; so the desired factor will be

$$z[1 - 4/(N_A + 2N_B - 1)] \quad \text{or} \quad z(N_A + 2N_B - 5)/(N_A + 2N_B - 1)$$

Continuing the process indefinitely, we find that the first end of the ith molecule of B can be placed in any of $N_A + 2N_B - 2(i - 1)$ sites and its second end into

$$z[1 - 2(i - 1)/(N_A + 2N_B - 1)]$$

or

$$z(N_A + 2N_B - 2i + 1)/(N_A + 2N_B - 1)$$

sites. Taking the continued product of terms of this kind for all of the B molecules, we find, as the principal contribution to the thermodynamic probability, the following expression:

$$\prod_{i=1}^{N_B} \frac{[N_A + 2N_B - 2i + 2]z(N_A + 2N_B - 2i + 1)}{(N_A + 2N_B - 1)}$$
$$= \left(\frac{z}{N_A + 2N_B - 1}\right)^{N_B} \frac{(N_A + 2N_B)!}{N_A!}$$

However, since the different B molecules are really indistinguishable from each other, the expression should be divided by $N_B!$ to give the thermodynamic probability of the mixture. If the two ends of the B molecules were indistinguishable, we should also divide by 2^{N_B}, but our ultimate answer of interest will not be affected thereby. So far as the A molecules are concerned, it should be clear that after all the B molecules are placed in the liquid, the A molecules can enter the remaining N_A sites in only one indistinguishable way. Hence we write for the thermodynamic probability of the mixture the expression

$$\Omega_{AB} = \left(\frac{z}{N_A + 2N_B - 1}\right)^{N_B} \frac{(N_A + 2N_B)!}{N_B!N_A!} \tag{19.22}$$

To calculate the entropy of mixing, however, we really need $\Omega_{AB}/\Omega_A\Omega_B$, in which expression Ω_A and Ω_B are the thermodynamic probabilities of the qure liquids. Ω_A can evidently be set equal to unity, but to obtain Ω_B we should set N_A equal to zero in the expression for Ω_{AB}; in this way we allow for the various orientations of the B molecules in the pure liquid state. Hence

$$\Omega_B = \left(\frac{z}{2N_B - 1}\right)^{N_B} \frac{(2N_B)!}{N_B!} \tag{19.23}$$

Therefore

$$\frac{\Omega_{AB}}{\Omega_A\Omega_B} = \left(\frac{2N_B - 1}{N_A + 2N_B - 1}\right)^{N_B} \frac{(N_A + 2N_B)!}{N_A!(2N_B)!} \tag{19.24}$$

Neglecting unity, compared with $2N_B$, we can rewrite this equation as

$$\frac{\Omega_{AB}}{\Omega_A\Omega_B} = \left(\frac{2N_B}{N_A + 2N_B}\right)^{N_B} \frac{(N_A + 2N_B)!}{N_A!(2N_B)!} \tag{19.25}$$

The entropy change on mixing will then equal

$$\Delta S_{mix} = k \ln \frac{\Omega_{AB}}{\Omega_A\Omega_B} \tag{19.26}$$

Upon substituting equation 19.25 into equation 19.26 and using Stirling's approximation, we find that

$$\Delta S_{mix} = -N_A k \ln \frac{N_A}{N_A + 2N_B} - N_B k \ln \frac{2N_B}{N_A + 2N_B} \tag{19.27}$$

But $N_A/(N_A + 2N_B)$ and $2N_B/(N_A + 2N_B)$ are precisely the *volume* fractions, ϕ_A and ϕ_B, of the two components. Therefore

$$\Delta S_{mix} = -n_A R \ln \phi_A - n_B R \ln \phi_B \tag{19.28}$$

This equation resembles that for the entropy of mixing of ideal solutions (18.13) except that the mole fractions are replaced by volume fractions.

Although our model was quite naive, equation 19.28 represents a marked improvement in the description of nonideal systems. For high polymers, further corrections might be made, but we shall not deal with them here.

Under the assumption made concerning zero heat of mixing, it is clear that the Gibbs energy of mixing equals $-T\Delta S_{mix}$. Therefore, if we let the pure liquids serve as standard states, we see that the total Gibbs energy of a solution equals the Gibbs energies of the separate pure liquids plus the Gibbs energy of mixing:

$$G = n_A\mu_A^\circ + n_B\mu_B^\circ + n_A RT \ln \phi_A + n_B RT \ln \phi_B \qquad (19.29)$$

We can now obtain the chemical potentials by partial differentiation:

$$\mu_A = \left(\frac{\partial G}{\partial n_A}\right)_{T,p} = \mu_A^\circ + RT \ln \phi_A + \frac{RT\phi_B}{2} \qquad (19.30)$$

$$\mu_B = \left(\frac{\partial G}{\partial n_B}\right)_{T,p} = \mu_B^\circ + RT \ln \phi_B - RT\,\phi_A \qquad (19.31)$$

We can also obtain the activity of either component by comparing equations 19.30 and 19.31 with equation 19.11. Thus we obtain, for component A,

$$\ln a_A = \ln \phi_A + \frac{\phi_B}{2} \qquad (19.32)$$

or

$$a_A = \phi_A e^{\phi_B/2} \qquad (19.33)$$

Converting from volume fractions to mole fractions, we obtain

$$a_A = \frac{x_A}{1 + x_B}\, e^{x_B/(1+x_B)} \qquad (19.34)$$

from which expression we conclude that the activity coefficient will be

$$\gamma_A = \frac{a_A}{x_A} = \frac{e^{x_B/(1+x_B)}}{1 + x_B}$$

$$= 1 - \frac{x_B^2}{2} + \frac{2x_B^3}{3} + \qquad (19.35)$$

It is evident now that the solution will exhibit negative deviations from Raoult's Law, as shown in Figure 19-3.

If the heat of mixing of the two liquids is not zero, we can modify the preceding treatment by adding to the Gibbs energy of mixing a term like that given by equation 9.31:

$$\Delta H_{mix} = \text{constant } V\, \phi_A\phi_B \qquad (19.36)$$

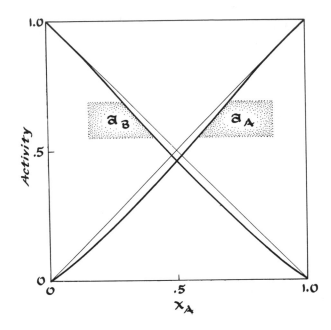

FIGURE 19-3 Activities of components A and B of the liquid mixture pictured in
    Figure 19-2.

By adding ΔH_{mix} to $-T\Delta S_{\text{mix}}$, we find that the partial molar Gibbs
energies are augmented by the amounts given by equations 9.33. Actually
this procedure is rather arbitrary, for the statistical mechanics involved in
our derivation would not be strictly valid if the energies of interaction
between the different molecules were not identical. Since a heat of
mixing different from zero implies different intermolecular interactions,
the entropy of mixing would also be affected, for the degree of randomness in the solution would be altered. However, since a complete treatment can become exceedingly complicated, we shall simply combine
a heat-of-mixing term with the separately calculated entropy term to
get a corrected Gibbs free energy of mixing.

EXERCISE:

Show by direct substitution that μ_A and μ_B of equations 19.30 and 19.31
satisfy the Gibbs-Duhem equation,

$$x_A \frac{\partial \mu_A}{\partial x_A} + x_B \frac{\partial \mu_B}{\partial x_A} = 0$$

Measurement of Solvent Activity from Freezing-point Lowering

In Chapter Eighteen we saw that the freezing-point lowering of a solution is determined by the fugacity of its solvent in relation to the fugacities of the pure solvent and the pure solid. Moreover, by assuming that the solution obeys Raoult's Law, we were able to determine the freezing-point lowering from the composition of the solution. On the other hand, if Raoult's Law does not hold, we can use the measured freezing-point lowering to calculate the actual activity of the solvent and its departure from Raoult's Law, as we shall see presently.

The basis of the calculation is the use of equation 19.10 in conjunction with the proper choice of reference state. For this purpose it is most expedient to let the pure liquid represent the standard state; in other words, we shall set the activity of the pure liquid equal to unity at all temperatures. Now the activity of the solid solvent at the normal freezing point, T_0, must be exactly equal to that of the pure liquid solvent—namely, unity; and the activity of the solid at any other temperature, T, must equal the activity of the solvent in solutions freezing at T. Hence, if we could establish an equation relating the activity of a solid to temperature, we should have an expression for the activity of a solvent as a function of the freezing point of the solution. Letting $a(c)$ denote the activity of a pure solid, we see, in accordance with our choice of reference state, that

$$\left(\frac{\partial \ln a(c)}{\partial T}\right)_p = \frac{H^{\circ}(\text{liq}) - H_m(c)}{RT^2} = \frac{\Delta H_{\text{fus}}}{RT^2} \tag{19.37}$$

if ΔH_{fus} is the heat of fusion of solid to pure liquid. Integrating this equation from T to T_0, for which the activities will be $a(c)$ and 1, we find that

$$-\ln a(c) = \int_T^{T_0} \frac{\Delta H_{\text{fus}}}{RT^2} dT \tag{19.38}$$

When we use this equation, we must remember that $a(c)$ is the activity of the solid, and hence of the freezing solution, at temperature T and not at T_0. If ΔH_{fus} were constant, equation 19.38 could be integrated directly. For a more precise computation, however, it is necessary to take into account the dependence of heat of fusion upon temperature. Assuming that ΔC_p, the difference in heat capacities, is constant, we can evidently write that

$$\Delta H_{\text{fus}} = \Delta H(T_0) + \Delta C_p (T - T_0) \tag{19.39}$$

if $\Delta H(T_0)$ is the heat of fusion at temperature T_0. To simplify the calculations, we shall now introduce a new variable, θ, defined as the lowering of the freezing point, $T_0 - T$. With this understanding equation 19.38 becomes

$$-\ln a(\text{c}) = \int_0^\theta \frac{\Delta H(T_0) - \Delta C_p \theta}{R(T_0 - \theta)^2} \, d\theta \qquad (19.40)$$

By expanding the integrand as a power series in θ and retaining only the first two terms, we obtain

$$-\ln a(\text{c}) = \int_0^\theta \left\{ \frac{\Delta H(T_0)}{RT_0^2} + \left(\frac{2\Delta H(T_0)}{RT_0^3} - \frac{\Delta C_p}{RT_0^2} \right) \theta + \ldots \right\} d\theta$$

$$= \frac{\Delta H(T_0)}{RT_0^2} \theta + \left(\frac{\Delta H(T_0)}{RT_0^3} - \frac{\Delta C_p}{2RT_0^2} \right) \theta^2 + \ldots \qquad (19.41)$$

Assuming that water is the solvent, and inserting the proper numerical values for the constants involved, we find, upon changing to common logarithms, that

$$-\lg a(\text{c}) = 4.209 \times 10^{-3}\, \theta + 0.215 \times 10^{-5}\, \theta^2 + \ldots \qquad (19.42)$$

This equation can be used directly for calculating the activity of water in solutions freezing at $-\theta$ °C. If θ should be greater than 10 °C, it would be desirable to include higher-power terms in the equation.

Having obtained the activity of the solvent in a solution at its freezing point, we naturally ask how this result might be converted to an activity for the same solution at another temperature. We can accomplish this by applying equation 19.10 to the solvent and then integrating from the freezing point of the solution to the temperature of interest. If the partial molar enthalpy of the solvent in the solution is the same as the molar enthalpy of the pure solvent, the activity will be independent of temperature. If we desire the activity of the solvent in a solution at the normal freezing point of the pure solvent, it would be correct to replace the heat of fusion of the pure solvent by its partial molar heat of fusion, in equation 19.38. The proof of this is left to the student as an exercise.

Measurement of Solvent Activity from Boiling-point Elevation

By this time it should be clear that most of the solution properties that can be determined by measurement of freezing-point lowerings could, at least in principle, likewise be determined by measurement of boiling-point elevations. As a specific illustration, we shall now consider the determination of solvent activity from the measurement of the rise in boiling point, using a procedure closely resembling that of the preceding section. First, we shall adopt the pure liquid as representing the standard state of unit activity. We shall then ask how the activity of the vapor, kept at constant pressure, changes with temperature. Since the activity of the solvent in a solution must equal that of its equilibrium vapor, a knowledge of the activity of the vapor as a function of temperature should give us precisely the activity of the solvent in boiling solutions.

To proceed with the derivation of the necessary equations, we shall apply equation 19.10 to the vapor, using the afore-mentioned choice of standard state. Letting $a(v)$ be the vapor activity, we see that

$$\left(\frac{\partial \ln a(v)}{\partial T}\right)_p = \frac{H°(\text{liq}) - H_m(v)}{RT^2} \tag{19.43}$$

Integrating this equation from T_0, the normal boiling point, to any other temperature, T, we find, taking cognizance of the fact that $a(v)$ is equal to unity at T_0, that

$$\ln a(v) = -\int_{T_0}^{T} \frac{\Delta H_{\text{vap}}}{RT^2} dT \tag{19.44}$$

Once again, assuming that the heat of vaporization is constant, we can carry out the integration without any difficulty. It is more accurate, however, to go to the next approximation by assuming that ΔC_p is constant, in which case we write that

$$\ln a(v) = -\int_{T_0}^{T} \frac{\Delta H(T_0) + \Delta C_p(T - T_0)}{RT^2} dT \tag{19.45}$$

$\Delta H(T_0)$ being the heat of vaporization at T_0. Once again let us introduce a new variable, $\theta = T - T_0$, the boiling-point elevation. Making this substitution and proceeding as before, we arrive at the following result:

$$-\ln a(v) = \frac{\Delta H(T_0)}{RT_0^2}\theta - \left(\frac{\Delta H(T_0)}{RT_0^3} - \frac{\Delta C_p}{2RT_0^2}\right)\theta^2 + \cdots \tag{19.46}$$

Since the use and interpretation of this equation are precisely the same as for equation 19.41, we shall not elaborate on them.

Determination of Solvent Activity from Osmotic Pressure

It is also possible to measure solvent activities simply by measuring the osmotic pressures of solutions. According to equation 18.49, osmotic pressure is related to the fugacities of pure liquid and solution in the following way:

$$\Pi = \frac{RT}{V_A} \ln \frac{f_{0A}}{f_A} \tag{19.47}$$

If we now assign unit activity to the pure liquid, it is evident that $a_A = f_A/f_{0A}$, if a_A is the solvent activity, so we can rewrite equation 19.47 in the following form:

$$-\ln a_A = \frac{\Pi V_A}{RT} \tag{19.48}$$

Hence it follows that the logarithm of the solvent activity is directly proportional to the osmotic pressure. The measurement of osmotic pressures is particularly useful for high-polymer solutions, since the freezing-point lowerings and boiling-point elevations are generally too small to be measured directly.

Determination of Activity of Solute from Activity of Solvent

In Chapter Eighteen we proved, by use of the Gibbs-Duhem equation, that, if Raoult's Law holds for the solvent component of a solution, Henry's Law must hold for the solute. Let us now suppose that Raoult's Law does not hold, but that we do know, from experimental measurements, the actual activity of the solvent as a function of composition. We can again use the Gibbs-Duhem equation, this time to determine the actual activity of the solute and how its behavior departs from Henry's Law. The detailed development is most conveniently carried out in the following manner: If a_A is the activity of the solvent and a_B that of the solute, then, upon rewriting the Gibbs-Duhem equation, 18.29, we obtain the expression

$$x_A \frac{\partial \ln a_A}{\partial x_B} + x_B \frac{\partial \ln a_B}{\partial x_B} = 0 \tag{19.49}$$

in which $x_A = 1 - x_B$. In addition we have the easily proved identity

$$x_A \frac{\partial \ln x_A}{\partial x_B} + x_B \frac{\partial \ln x_B}{\partial x_B} = 0 \tag{19.50}$$

Subtracting this equation term by term from equation 19.49, we obtain

$$x_A \frac{\partial \ln (a_A/x_A)}{\partial x_B} + x_B \frac{\partial \ln (a_B/x_B)}{\partial x_B} = 0 \tag{19.51}$$

Rearranging this equation, we see that

$$\frac{\partial \ln (a_B/x_B)}{\partial x_B} = -\left(\frac{1 - x_B}{x_B}\right) \frac{\partial \ln (a_A/x_A)}{\partial x_B} \tag{19.52}$$

Let us now integrate this equation from $x_B = 0$ to some other value of x_B. This will give us

$$\ln \left(a_B/x_B\right) - \lim_{x_B^* \to 0} \ln \left(a_B^*/x_B^*\right) = -\int_0^{x_B} \left(\frac{1 - x_B}{x_B}\right) \frac{\partial \ln \left(a_A/x_A\right)}{\partial x_B} dx_B \qquad (19.53)$$

If we now choose x_B as the reference function for a_B and let the infinitely dilute state be the reference state, then $\lim \left(a_B^*/x_B^*\right) = 1$; hence we can write that

$$\ln \left(a_B/x_B\right) = \ln \gamma_B = -\int_0^{x_B} \left(\frac{1 - x_B}{x_B}\right) \frac{\partial \ln \left(a_A/x_A\right)}{\partial x_B} dx_B \qquad (19.54)$$

Since it is presumed that we know a_A as a function of composition, evaluation of the integral appearing in equation 19.54 should enable us to determine a_B/x_B, or the activity coefficient γ_B. If a_A is known in analytical form, the integral may be susceptible to direct evaluation; more often a_A is known for a series of points, in which case recourse must be taken to approximate numerical or graphical methods. The presence of x_B in the denominator of the integrand may make it appear that the integral will not converge; actually it will converge as long as the solvent obeys Raoult's Law asymptotically. If Raoult's Law does not hold in the limit as one approaches the pure solvent, Henry's Law will similarly not hold for the solute, and the limit of the quotient a_B/x_B may not even exist as x_B approaches zero.

In practice the most convenient method of using equation 19.54 involves plotting $(1 - x_B)/x_B$ against $\ln \left(a_A/x_A\right)$. The area under the curve so obtained is then measured over the range of concentrations represented by the interval of integration, thus giving directly the logarithm of (a_B/x_B). This procedure is quite practical when a_A is known at each of several compositions through measurement of freezing-point lowerings, boiling-point elevations, osmotic pressures, or vapor pressures (fugacities), and the method will work even if the solute has an exceedingly low vapor pressure, so low that it would be difficult or impossible to measure the absolute partial pressure directly.

EXERCISE:

Suppose that activity of the solute in a dilute nonideal solution is $a_B = x_B(1 + x_B)^2$. Show that the activity of the solvent, if the pure liquid is the standard state, is $a_A = x_A(1 - x_B^2)$.

Conversion of Activity Coefficients from One Reference System to Another

We have just seen how the activity coefficient of a solute can be determined from the solvent activities. The method described, however, gives the activity coefficient when the reference function is the mole fraction of solute and the reference state is infinite dilution. It is often desirable to use a concentration in place of the mole fraction as the reference function for solute. Such a concentration might be the molality, m, which equals moles per 1,000 grams of solvent, or the molarity, c, which equals moles per liter of solution. In either event infinite dilution is still usually chosen for reference purposes. Since the ratio of the activities in a pair of states is independent of the reference functions employed (see equation 19.6), we can assert that

$$\frac{a}{a^*} = \frac{\gamma_x x}{\gamma_x^* x^*} = \frac{\gamma_m m}{\gamma_m^* m^*} = \frac{\gamma_c c}{\gamma_c^* c^*} \tag{19.55}$$

in which the asterisks denote approach to infinite dilution and the subscripts on the γ's denote the reference function. (The subscript A, designating the solute, is omitted for the sake of simplicity.) Now the mole fraction of solute will be related to the molality by the equation

$$x = \frac{m}{m + 1,000/M_A} \tag{19.56}$$

if M_A is the molecular weight of the solvent. Similarly, the molality is related to the molarity by the expression

$$m = \frac{1,000\,c}{1,000\rho - cM_B} \tag{19.57}$$

in which ρ is the density of the solution and M_B the molecular weight of the solute. From equations 19.56 and 19.57 we see, upon passing to a state of infinite dilution, that

$$\lim_{m^* \to 0} \frac{x^*}{m^*} = \frac{M_A}{1,000} \tag{19.58}$$

and that

$$\lim_{c^* \to 0} \frac{m^*}{c^*} = \frac{1}{\rho_{0A}}$$

in which ρ_{0A} is the density of the pure solvent. Moreover, we know that at infinite dilution all of the γ^*'s approach unity; therefore, upon substitution and rearrangement, we find that

$$\frac{\gamma_x}{\gamma_m} = 1 + 0.001\,mM_A \tag{19.59}$$

and that

$$\frac{\gamma_m}{\gamma_c} = \frac{\rho - 0.001\,cM_B}{\rho_{0A}}$$

$$= \frac{\rho/\rho_{0A}}{1 + 0.001\,mM_B} \tag{19.60}$$

Transformations of this type are often made, especially for solutions of electrolytes.

EXERCISE:

At 25 °C the density of a solution of phenanthrene in benzene is given by the expression

$$\rho = 0.8736 + 0.00187\, w \text{ g ml}^{-1}$$

in which w is the weight percentage of phenanthrene. Calculate the ratios γ_x/γ_m and γ_m/γ_c for phenanthrene in benzene when $w = 2.0$ and 4.0 percent.

Use of Activities in Formulation of Reaction Potentials

Let us consider again the general chemical reaction that can be represented as

$$aA + bB + \ldots \rightleftarrows eE + fF + \ldots$$

and let us recognize that the various substances involved might exist in any state of aggregation—gaseous, liquid, or solid—and either pure or in solution. Our formulation of the reaction potential, $\Delta\bar{\mu}$, given by equations 10.29 and 17.56 should then be modified by the use of generalized activities, whenever expedient, in place of partial pressures or fugacities. After choosing suitable reference functions and reference states, we can write the chemical potentials of the substances in terms of activities in the following familiar form:

$$\mu_i = \mu_i^{\circ} + RT \ln a_i \tag{19.61}$$

The reaction potential (eq. 10.5) can then be expressed exactly by the equation

$$\Delta\bar{\mu} = \Delta\bar{\mu}^0 + RT \ln \frac{a_E^e a_F^f \ldots}{a_A^a a_B^b \ldots} \tag{19.62}$$

in which

$$\Delta\bar{\mu}^{\circ} = (e\mu_E^{\circ} + f\mu_F^{\circ} + \ldots) - (a\mu_A^{\circ} + b\mu_B^{\circ} + \ldots) \tag{19.63}$$

This formulation is identical in form to those used earlier in equations 10.29 and 17.56, but it has the advantage that it can be applied in practice even to substances whose vapor pressures are so low as to be quite immeasurable. The standard reaction potential, $\Delta\bar{\mu}^{\circ}$, is once again related to an equilibrium constant, K, as follows:

$$-\Delta\bar{\mu}^{\circ} = RT \ln K \tag{19.64}$$

Dividing through by RT and differentiating with respect to the temperature, we see that

$$-\frac{d}{dT}\left(\frac{\Delta\bar{\mu}^\circ}{RT}\right) = \frac{d\ln K}{dT} = \frac{\Delta\tilde{H}^\circ}{RT^2} \qquad (19.65)$$

in which expression

$$\Delta\tilde{H}^\circ = (eH_E^\circ + fH_F^\circ + \ldots) - (aH_A^\circ + bH_B^\circ + \ldots) \qquad (19.66)$$

Thus we have shown that the temperature coefficient of the equilibrium constant, expressed in terms of activities, is determined by the standard enthalpy change for the reaction, which in turn is obtained from the enthalpies of the various materials in their *reference* states.

To illustrate the use of activities in chemical reactions, let us consider the reaction

$$\text{C(c)} + \text{H}_2\text{O(g)} \rightleftarrows \text{H}_2\text{(g)} + \text{CO(g)}$$

The most convenient expression for the reaction potential in this example would involve fugacities of the gases and an activity for solid carbon if we let the pure solid represent the standard state. Hence we should write that, at equilibrium,

$$\frac{f_{\text{H}_2}f_{\text{CO}}}{a_\text{C}f_{\text{H}_2\text{O}}} = K \qquad (19.67)$$

However, the activity of the carbon, which is assumed to be pure, will be constant and equal to unity. We can therefore omit a_C from equation 19.67 to get

$$\frac{f_{\text{H}_2}f_{\text{CO}}}{f_{\text{H}_2\text{O}}} = K \qquad (19.68)$$

However, in writing the standard heat of reaction, $\Delta\tilde{H}^\circ$, we must take cognizance of the enthalpy of the carbon even though its activity is constant. This important point should not be overlooked when we deal with reactions involving condensed phases of constant activity.

EXERCISE:

Explain the underlying reason for the statement of the text immediately following equation 19.68. After all, it might be argued that, since a_C is a constant equal to unity at all temperatures, the temperature coefficient of K should be independent of the enthalpy of the carbon. (Suggestion: The temperature coefficient of $f_{\text{H}_2}f_{\text{CO}}/f_{\text{H}_2\text{O}}$ at constant *composition* is indeed independent of the enthalpy of carbon. However, equation 19.68 holds only at equilibrium, and the equilibrium composition changes with temperature.) Consider this argument in connection with the reaction

$$\text{CaCO}_3\text{(c)} \rightleftarrows \text{CaO(c)} + \text{CO}_2\text{(g)}$$

At constant pressure, the temperature coefficient of f_{CO_2} depends only on the properties of CO_2. At equilibrium, however, the temperature coefficient of f_{CO_2} depends also on the properties of the solids.

Problems

1. Using the model of the text that leads to equation 19.29, show that the Henry's Law constants for A dissolved in B and for B dissolved in A are

$$K_A = \frac{e^{1/2}}{2} f_{0A}$$

$$K_B = 2e^{-1} f_{0B}$$

if f_{0A} and f_{0B} denote the fugacities of pure A and B. (To obtain K_A, for example, one must evaluate the limit of f_A/x_A as $x_A \longrightarrow 0$.)

2. A certain solution obeys Raoult's Law, so that we can write $f_i = x_i f_{0i}$ for a particular component of interest. Suppose that we choose the expression x_i^2 as the reference function for activity, using the pure liquid for the reference state. Show that the activity coefficient, γ_i, must equal $1/x_i$.

3. At 35.17 °C the equilibrium partial pressures from solutions of chloroform and acetone have values summarized in the following table:

100 x_{CHCl_3}	p_{CHCl_3}	$p_{(CH_3)_2CO}$
0.0	0.0 Torr	344.5 Torr
5.88	9.2	323.2
12.32	20.4	299.3
18.53	31.9	275.4
29.10	55.4	230.3
42.32	88.9	174.3
51.43	117.8	135.0
58.12	139.9	108.5
66.35	170.2	79.0
79.97	224.4	37.5
91.75	267.1	13.0
100.0	293.1	0

(A) Plot the partial pressures and total pressures as functions of the mole fraction.

(B) Calculate the activities and activity coefficients of the acetone, using Raoult's Law and the pure liquid for reference purposes.

(C) Assuming Henry's Law and infinite dilution for reference purposes, calculate the activities and activity coefficients of the chloroform.

4. At 60 °C the vapor pressure of aniline is 5.7 Torr, and that of water is 149 Torr. When aniline and water are mixed at 60 °C, two layers are formed, the compositions of which are 4.40 and 93.4 percent aniline by weight.

(A) Assuming that Raoult's Law holds for the more nearly pure component in each phase and that Henry's Law holds for the dilute component, calculate the values of the two Henry's Law constants.

(B) Derive an expression for the activity coefficient of each of the components for each phase, using first Raoult's Law for reference purposes and then Henry's Law.

5. At 25 °C the distribution coefficient, expressed in terms of concentrations, of hydrogen peroxide between ether and water is 9.1 in favor of the ether $(K = C_{ether}/C_{water})$.

(A) Suppose that a large volume of water containing 0.1 mole of hydrogen peroxide is extracted with an equal volume of ether. Assuming the volumes to be so large that the solutions can be considered infinitely dilute (and ideal), calculate the change in chemical potential of the peroxide. If you neglect the changes in the chemical potentials of the solvents, what is the total Gibbs energy change in calories?

(B) Suppose that the aqueous solution is successively extracted with two portions of ether, each of half the volume of the water. What is the total Gibbs energy change of the peroxide?

(C) If the aqueous solution is successively extracted with s portions of ether, each of volume $1/s$ times that of the water, what is the *total* Gibbs energy change? What is the limit of the expression as $s \longrightarrow \infty$? (For this problem you may assume that water and ether are completely immiscible.)

$$\text{Ans.} \quad (c) \lim_{s \to \infty} \Delta G = -nRT(1 - e^{-K})$$

6. Suppose that we choose, as the reference function for the activity of a nonideal gas, the concentration, $c_i = n_i/V$. Let us further choose the state of zero pressure for reference purposes, so that

$$\lim_{p \to 0} \frac{a_i}{c_i} = 1$$

(A) Prove that

$$\left(\frac{\partial \ln a_i}{\partial T} \right)_p = \frac{U_i^* - H_i}{RT^2}$$

in which expression U_i^* is the partial molar *energy* of component i in the low-pressure gas.

(B) Also show that

$$\left(\frac{\partial \ln \gamma_i}{\partial T} \right)_p = \frac{U_i^* - H_i}{RT^2} + \frac{1}{V} \left(\frac{\partial V}{\partial T} \right)_p$$

in which expression V is the total volume of the gas at p and T.

(C) Hence show for an ideal gas that the above-defined activity depends upon the temperature although the activity coefficient does not.

(D) Prove also that

$$\left(\frac{\partial \ln a_i}{\partial T}\right)_V = \left(\frac{\partial \ln \gamma_i}{\partial T}\right)_V = \frac{U_i^* - U_i}{RT^2}$$

(E) Finally show that μ°, the standard chemical potential, has more attributes of a Helmholtz energy than of a Gibbs energy since

$$\frac{d}{dT}\left(\frac{\mu_i^\circ}{T}\right) = -\frac{U_i^*}{T^2}$$

7. Suppose that the equilibrium constant for the gaseous reaction

$$aA + bB + \ldots \rightleftharpoons eE + fF + \ldots$$

is expressed in terms of activities referred to concentrations, as described in Problem 6.

(A) Prove that

$$\frac{d \ln K_c}{dT} = \frac{\Delta \tilde{U}^*}{RT^2}$$

if

$$\Delta \tilde{U}^* = (eU_E^* + fU_F^* + \ldots) - (aU_A^* + bU_B^* + \ldots)$$

(B) Show also that

$$\Delta \tilde{U}^* = \Delta \tilde{H}^\circ - \Delta \nu\, RT$$

if $\Delta \tilde{H}^0$ is the "standard heat of reaction" that appears in the temperature coefficient of the usual equilibrium constant, K_p, referred to pressures—namely,

$$\frac{d \ln K_p}{dT} = \frac{\Delta \tilde{H}^\circ}{RT^2}$$

8. The activity of a solute in a dilute solution is referred to its concentration ($c_i = n_i/V$) with the understanding that

$$\lim_{c_i \to 0} \frac{a_i}{c_i} = 1$$

(A) Prove that

$$\left(\frac{\partial \ln a_i}{\partial T}\right)_p = \frac{H_i^* - H_i}{RT^2} - \alpha_{\text{solvent}}$$

if α_{solvent} is the coefficient of thermal expansion of the pure solvent.

(B) Show also that

$$\left(\frac{\partial \ln \gamma_i}{\partial T}\right)_p = \frac{H_i^* - H_i}{RT^2} + \Delta \alpha$$

if $\Delta \alpha$ equals $\alpha_{\text{solution}} - \alpha_{\text{solvent}}$, the difference between the coefficients of thermal expansion of the solution and the pure solvent.

9. The freezing-point lowerings of aqueous solutions of sucrose are summarized in the following table:

m (conc. in mol kg^{-1} H$_2$O)	θ °C
0.005	0.00930
.05	.0935
.10	.188
.20	.380
.40	.776
.50	.980
.70	1.40
1.00	2.06
1.20	2.52
1.50	3.26
2.0	4.6

Calculate the activity and activity coefficient of the water for each of the solutions.

10. The lowerings of the vapor pressure of aqueous solutions of sucrose at 0 and 30 °C are summarized in the following table:

m (conc. in mol kg^{-1} H$_2$O)	100 $(p_0 - p)/p_0$	
	at 0 °C	at 30 °C
0.2	.372	
0.5	.945	
1.0	1.97	1.94
3.0		6.75
3.5	8.37	
4.0		9.44
4.5	11.25	
5.0	12.75	12.30

Calculate activities and activity coefficients for the water, and compare the 0 °C results with those of Problem 9.

11. Derivation of formulas for calculation of the activity of a solute from freezing-point lowerings:

(A) Equation 19.54 shows how the activity of a solute can be calculated from that of the solvent. If $\ln a_A$ is known from the freezing-point lowering, as suggested by equation 19.41, the computation of $\ln \gamma_B$, using the mole fraction for reference purposes, can be carried out as follows: First, write

$$\ln a_A = -\int_0^\theta (A + B\theta + \ldots)d\theta$$

in which expression

$$A = \frac{\Delta H(T_0)}{RT_0^2}, \quad B = \left(\frac{2\,\Delta H(T_0)}{RT_0^3} - \frac{\Delta C_p}{RT_0^2}\right), \quad \text{etc.}$$

Next, show that

$$\frac{\partial \ln (a_A/x_A)}{\partial x_B} = -(A + B\theta + \ldots)\frac{\partial \theta}{\partial x_B} + \frac{1}{x_A}$$

and that

$$-\frac{(1 - x_B)}{x_B} \cdot \frac{\partial \ln (a_A/x_A)}{\partial x_B} = \frac{(A + B\theta + \ldots)}{x_B}\frac{\partial \theta}{\partial x_B}$$

$$- (A + B\theta + \ldots)\frac{\partial \theta}{\partial x_B} - \frac{1}{x_B}$$

But, since

$$\frac{A}{x_B}\frac{\partial \theta}{\partial x_B} - \frac{1}{x_B} = \frac{\partial}{\partial x_B}\left(\frac{A\theta}{x_B} - 1\right) + \frac{A\theta/x_B - 1}{x_B}$$

you can see that

$$-\frac{(1 - x_B)}{x_B}\frac{\partial \ln (a_A/x_A)}{\partial x_B} = \frac{\partial}{\partial x_B}\left(\frac{A\theta}{x_B} - 1\right) + \frac{A\theta/x_B - 1}{x_B}$$

$$+ \frac{\partial \ln a_A}{\partial x_B} + \left(\frac{B\theta + \ldots}{x_B}\right)\frac{\partial \theta}{\partial x_B}$$

Show now, by equation 19.54, that

$$\ln \gamma_B = \ln a_A + \int_0^{x_B} \frac{\partial}{\partial x_B}\left(\frac{A\theta}{x_B} - 1\right) dx_B + \int_0^{x_B} \left(\frac{A\theta/x_B - 1}{x_B}\right) dx_B$$

$$+ \int_0^{\theta} \left(\frac{B\theta + \ldots}{x_B}\right) d\theta$$

But, from the assumption of limiting ideality at infinite dilution, observe, from equation 18.38, that

$$\lim_{x_B \to 0} \frac{A\theta}{x_B} = 1$$

Therefore

$$\ln \gamma_B = \ln a_A + \left(\frac{A\theta}{x_B} - 1\right) + \int_0^{x_B} \left(\frac{A\theta/x_B - 1}{x_B}\right) dx_B$$

$$+ \int_0^{\theta} \left(\frac{B\theta + \ldots}{x_B}\right) d\theta$$

The two integrals can be determined graphically, or otherwise, from experimental data.

(B) The activity coefficient can also be calculated if the molality, m, is used as the reference function. Replacing mole fractions by molality, m, with the understanding that $x_B = m/(m + 1{,}000/M_A)$, show, upon appropriate identification of the various terms, that

$$\frac{\partial \ln a_B}{\partial m} = -\frac{1{,}000}{M_A m}\frac{\partial \ln a_A}{\partial m}$$

$$= \frac{1}{K_f m}(1 + B\theta/A + \ldots)\frac{\partial \theta}{\partial m}$$

and that

$$\frac{\partial \ln (a_B/m)}{\partial m} = \frac{\partial \ln \gamma_B}{\partial m} = \frac{1}{K_f m}\frac{\partial \theta}{\partial m} - \frac{1}{m} + \frac{B\theta + \ldots}{AK_f m}\frac{\partial \theta}{\partial m}$$

Rearranging the last equation above, observe that

$$\frac{\partial \ln \gamma_B}{\partial m} = \frac{\partial}{\partial m}\left(\frac{\theta}{K_f m} - 1\right) + \left(\frac{\theta}{K_f m} - 1\right)\left(\frac{1}{m}\right) + \frac{B\theta + \ldots}{AK_f m}\frac{\partial \theta}{\partial m}$$

Recognizing that

$$\lim_{m \to 0}\left(\frac{\theta}{K_f m} - 1\right) = 0$$

show, upon integration, that

$$\ln \gamma_B = \frac{\theta}{K_f m} - 1 + \int_0^m \frac{(\theta/K_f m - 1)}{m}\,dm + \int_0^\theta \left(\frac{B\theta + \ldots}{AK_f m}\right)d\theta$$

(c) Using data from Problem 9, calculate values for the activities of sucrose in water in the neighborhood of 0 °C.

12. Suppose that equation 19.28 correctly expresses the entropy of mixing even when the molar volume of B is not exactly twice the molar volume of A. Assuming that the heat of mixing and the volume change of mixing are zero, derive expressions for a_A and a_B if the molar volume of B is precisely s times that of A.

$$\text{Ans.} \quad \ln a_A = \ln \phi_A + \frac{(s-1)}{s}\phi_B$$

$$\ln a_B = \ln \phi_B - (s-1)\phi_A$$

13. Consider a reaction that might occur either in the vapor phase or in the liquid phase—for example,

$$\text{ester} + \text{water} \rightleftarrows \text{acid} + \text{alcohol}$$

Suppose that the liquid-phase reaction takes place in the presence of a large excess of water so that the acid, alcohol, and ester are all in dilute solution, each obeying Henry's Law. Assume that the water, on the other hand, obeys Raoult's Law.

(A) Show that the equilibrium constant for the vapor reaction, $K(v)$, expressed in terms of fugacities, is related to that for the liquid reaction, $K(liq)$, expressed in terms of activities referred to appropriate limiting mole fractions, by the expression

$$K(v) = K(liq) \frac{K_{acid} K_{alcohol}}{K_{ester} f_{0\,water}}$$

in which K_{acid} etc. denote Henry's Law constants and f_{0water} is the fugacity of pure water.

(B) Show that $\Delta\tilde{H}°(v)$, the standard heat of reaction in the vapor, is related to $\Delta\tilde{H}°(liq)$, that for the liquid reaction, by the equation

$$\Delta\tilde{H}°(v) = \Delta\tilde{H}°(liq) + \Delta H_{vap}(acid) + \Delta H_{vap}(alc)$$
$$- \Delta H_{vap}(ester) - \Delta H_{vap}(water)$$

in which ΔH_{vap} denotes the partial molar heat of vaporization or the negative of the heat of solution of the substance indicated, referred, of course, to infinite dilution in all cases except that of the water.

14. Prove by methods similar to those used in Problem 11 that the activity coefficient of a solute can be related to the osmotic pressure of the solution as follows:

(A) Using the mole fraction of the solute as the reference function for activity, show that

$$\ln \gamma_B = \ln (a_B/x_B) = -(1 - x_B) X - x_B - \int_0^{x_B} X\, d \ln x_B$$

if

$$X = 1 - \frac{\Pi V_A}{x_B R T}$$

(B) Using the molality as reference function for activity, show that

$$\ln \gamma_B = \ln (a_B/m) = -X - \int_0^m X\, d \ln m$$

provided

$$X = 1 - \frac{1{,}000\ \Pi V_A}{M_A m\, R T}$$

(Assume for both parts, A and B, that V_A the partial molar volume of the solvent, is independent of concentration.)

20

Solutions of Electrolytes; Electrolytic Cells

Activities of Ions

A large part of chemistry is concerned with solutions of ionic materials, such as aqueous solutions of acids, bases, and salts. For this reason a great many physical-chemical measurements have been made on ions in solution, and much has been done to develop theories of their behavior. We shall, therefore, devote some attention to this subject, emphasizing those aspects that are peculiar to ions, but drawing freely upon the thermodynamic principles formulated in the earlier chapters. In carrying out this purpose, we shall initially concern ourselves with solutions of strong electrolytes—that is, with solutions of materials that ionize completely. We can then extend the treatment to weak electrolytes without difficulty by simply introducing equilibrium considerations to cover partial ionization.

Consider, for example, an electrolyte of the form $C_{\nu_+}A_{\nu_-}$, which in solution might undergo complete ionization according to the following chemical equation:

$$C_{\nu_+}A_{\nu_-} \longrightarrow \nu_+ C^{+z_+} + \nu_- A^{-z_-}$$

Since the solution must be electrically neutral, we can stipulate that

$$\nu_+ z_+ = \nu_- z_-$$ (20.1)

Let us now imagine that each of the ions has its own chemical potential and its own activity; we can then write for the separate positive and negative ions the following equations:

$$\mu_+ = \mu_+^\circ + RT \ln a_+$$

$$\mu_- = \mu_-^\circ + RT \ln a_-$$ (20.2)

The partial molar Gibbs energy of the over-all electrolyte will be the sum of the ionic chemical potentials, each multiplied by the appropriate number of ions derived from the undissociated "molecule." In other words, we can write for the electrolyte, which will be denoted by subscript B, the equation

$$\mu_B = \mu_B^\circ + RT \ln a_B$$

$$= \nu_+ \mu_+^\circ + \nu_- \mu_-^\circ + \nu_+ RT \ln a_+ + \nu_- RT \ln a_-$$ (20.3)

It now becomes convenient to define what will be known as the mean ion activity, $a_\pm$, which is a quantity that will satisfy the following equation:*

$$a_\pm^\nu = a_+^{\nu_+} a_-^{\nu_-}$$ (20.4)

In this equation, $\nu = \nu_+ + \nu_-$, the total number of ions formed from the electrolyte. This definition now permits us to write the overall chemical potential in the simple form

$$\mu_B = \mu_B^\circ + \nu RT \ln a_\pm$$ (20.5)

in which

$$\mu_B^\circ = \nu_+ \mu_+^\circ + \nu_- \mu_-^\circ$$ (20.6)

When we define the activity of a particular ion, it is convenient to use m, the molal concentration, or c, the molar concentration, as the reference function. We then choose the state of infinite dilution as the reference state, such dilution applying to all ions in solution and not just to the particular ion under consideration. This stipulation with respect to all ions is important because it is the ionic environment that is mostly responsible for departures from ideality and not just the characteristics of the particular ion under consideration. With such a choice of reference function and reference state, we can now introduce activity coefficients for the ions as follows:

$$a_+ = \gamma_+ \nu_+ m$$

$$a_- = \gamma_- \nu_- m$$ (20.7)

---

* This mean ion activity is a geometric mean.

In these equations, $\nu_+ m$ and $\nu_- m$ equal the concentrations of the two kinds of ions. Similar equations can be written, of course, with c in place of m, and a conversion from γ_m to γ_c can always be made by use of equation 19.60. In keeping with our definition of mean activity, we can also refer to a mean activity coefficient, $\gamma_\pm$, which, in general, will be defined by the expression

$$\gamma_\pm^\nu = \gamma_+^{\nu_+} \gamma_-^{\nu_-} \tag{20.8}$$

The mean activity of a simple uni-univalent salt like sodium chloride is simply the square root of the product of the separate activities, and the mean activity coefficient is likewise the square root of the product of the separate activity coefficients. In general, the activity of the electrolyte will be

$$a_B = a_\pm^\nu = \gamma_\pm^\nu \nu_+^{\nu_+} \nu_-^{\nu_-} m^\nu \tag{20.9}$$

Determination of Ion Activities

It is experimentally observed that ions in solution are by no means ideal in their behavior. We might expect, a priori, that the freezing-point lowering of a dilute solution of sodium chloride would be just twice that of an undissociated material at the same molal concentration. Actually, for reasons that will appear later, this is not so. Before formulating theories of ion activity, however, we should inquire into how ion activities might be determined experimentally.

Since ions in solution always come in combinations that are electrically neutral, it turns out to be impossible to measure the activity of a single ion by conventional means. The reason for this is the same as the reason for the impossibility of obtaining single ionic enthalpies, which we mentioned in Chapter Three. The mean activity, on the other hand, is a quantity that can be measured by procedures equivalent to those discussed in Chapter Nineteen. Let us consider once again a solution of a salt $C_{\nu_+} A_{\nu_-}$. At infinite dilution its freezing-point lowering, boiling-point elevation, osmotic pressure, etc. should be exactly ν times what one might expect for an undissociated solute. Proceeding from the colligative properties of the solution, we can evaluate the activity of the solute by use of the Gibbs-Duhem equation. In essence, this involves the kind of treatment described in Chapter Nineteen; but, since some modifications are necessary when the method is applied to electrolytes, we shall describe the procedure in some detail, using freezing-point lowering for illustrative purposes.

If subscript A denotes solvent, we see, upon rewriting equation 19.41, that

$$-\ln a_{\rm A} = \int_0^\theta \frac{\Delta H(T_0)}{RT_0^2}\left[1 + \left(\frac{2}{T_0} - \frac{\Delta C_p}{\Delta H(T_0)}\right)\theta + \ldots\right]d\theta \qquad (20.10)$$

But the Gibbs-Duhem equation, 19.49, can be rearranged as follows:

$$\frac{\partial \ln a_{\rm B}}{\partial n_{\rm B}} = -\frac{n_{\rm A}}{n_{\rm B}}\frac{\partial \ln a_{\rm A}}{\partial n_{\rm B}} \qquad (20.11)$$

Let us now set $n_{\rm B}$, the number of moles of solute, equal to m, the molal concentration of salt. In accord with this change in variable, we must simultaneously require that $n_{\rm A}$ equal $1{,}000/M_{\rm A}$, since the molal concentration is defined as the number of moles of solute per 1,000 grams of solvent. We can therefore write that

$$\frac{\partial \ln a_{\rm B}}{\partial m} = -\frac{1{,}000}{M_{\rm A}m}\frac{\partial \ln a_{\rm A}}{\partial m} \qquad (20.12)$$

If we now differentiate equation 20.10 partially with respect to m, we obtain the equation

$$\frac{\partial \ln a_{\rm A}}{\partial m} = -\frac{\Delta H(T_0)}{RT_0^2}\{1 + \alpha\theta + \ldots\}\frac{\partial \theta}{\partial m} \qquad (20.13)$$

in which α is the coefficient of θ in the power-series expansion appearing in the integrand of equation 20.10. Combining equation 20.13 with equation 20.12, we find that

$$\frac{\partial \ln a_{\rm B}}{\partial m} = \frac{1}{K_f m}[1 + \alpha\theta + \ldots]\frac{\partial \theta}{\partial m} \qquad (20.14)$$

in which expression K_f denotes the molal freezing-point lowering constant defined earlier by equation 18.42. Let us now replace $a_{\rm B}$ by its equivalent as expressed by equation 20.9. This gives us

$$\frac{\partial \ln a_{\rm B}}{\partial m} = \nu\frac{\partial \ln a_\pm}{\partial m} = \nu\frac{\partial \ln \gamma_\pm}{\partial m} + \frac{\nu}{m} \qquad (20.15)$$

Solving for the dependence of the mean-activity coefficient on concentration, we obtain from equations 20.14 and 20.15 the following:

$$\frac{\partial \ln \gamma_\pm}{\partial m} = \frac{1}{\nu K_f m}[1 + \alpha\theta + \ldots]\frac{\partial \theta}{\partial m} - \frac{1}{m} \qquad (20.16)$$

If, to simplify the integration of this equation, we make use of the easily verified identity*

$$\frac{1}{\nu K_f m}\frac{\partial \theta}{\partial m} - \frac{1}{m} = -\frac{\partial}{\partial m}\left(1 - \frac{\theta}{\nu K_f m}\right) - \frac{1}{m}\left(1 - \frac{\theta}{\nu K_f m}\right) \qquad (20.17)$$

---

* It may be helpful to the reader to define a new quantity, $j = 1 - \theta/(\nu K_f m)$, so that the right-hand side of equation 19.17 will be $-(\partial j/\partial m) - j/m$. This is often done, but the final answer, of course, is not different from that of the text.

we find that

$$\frac{\partial \ln \gamma_{\pm}}{\partial m} = -\frac{\partial}{\partial m}\left(1 - \frac{\theta}{\nu K_f m}\right) - \left(1 - \frac{\theta}{\nu K_f m}\right)\frac{1}{m} + \left\{\frac{\alpha\theta + \ldots}{\nu K_f m}\right\}\frac{\partial\theta}{\partial m}$$

(20.18)

Since the system approaches ideality as $m \longrightarrow 0$, we can also assert that in the limit $\gamma_{\pm} \longrightarrow 1$ and that $\theta \longrightarrow \nu K_f m$. Hence the limit, as $m \longrightarrow 0$, of $1 - \theta/\nu K_f m$ must also be zero. Integrating equation 20.18 from $m = 0$, $\theta = 0$ to m, θ, and taking cognizance of limiting behavior at infinite dilution, we see that

$$\ln \gamma_{\pm} = -\left(1 - \frac{\theta}{\nu K_f m}\right) - \int_0^m \left(1 - \frac{\theta}{\nu K_f m}\right)\frac{dm}{m} + \int_0^\theta \frac{\alpha\theta + \ldots}{\nu K_f m}d\theta$$

(20.19)

The first of the terms on the right-hand side of this equation can be measured directly. The values of the integrals, however, are best obtained by graphical or numerical computations based on experimental results. One can evaluate the first integral by plotting $1 - \theta/\nu K_f m$ against $\ln m$ and measuring the area under the curve. (It will be noted that $1 - \theta/\nu K_f m$ approaches zero as m approaches zero.) Alternatively, one can plot $(1 - \theta/\nu K_f m)/m^{1/2}$ against $m^{1/2}$; the area under such a curve will equal one-half of the integral. The second integral appearing in equation 20.19 can be neglected for solutions less concentrated than 0.1 molal, although recourse can always be taken to graphical integration if necessary.

At this time we should comment on the temperature for which equation 20.19 gives the correct value for $\gamma_{\pm}$. First we observe that equation 20.10 correctly gives the activity of the solvent at the freezing point of the solution, $T_0 - \theta$. However, the computation of the activity of the solute by use of the Gibbs-Duhem equation involves integration over a range of concentrations from zero up to that of the solution of interest; to be strictly accurate, this procedure should be carried out for a fixed temperature. It is evident, therefore, that the procedure followed above is not exact since the values of $\ln a_A$ used for the computation cover the range of temperatures from T_0 to $T_0 - \theta$. Hence a value of $\gamma_{\pm}$ computed by use of equation 20.19 must correspond to some intermediate temperature, T, such that $T_0 - \theta < T < T_0$. If one used the partial molar heat of fusion of the solvent in place of the molar heat of fusion of solid to pure liquid, one could obtain the precise value of $\gamma_{\pm}$ at T_0. However, this requires a knowledge of the dependence of the heat of solution on the concentration.

In principle, one could use equation 19.10 to adjust all the values of $\ln a_A$ to any temperature T (say 25 °C), and then use the Gibbs-Duhem equation to calculate $\ln \gamma_\pm$ at T. Alternatively, one might adjust values of $\ln \gamma_\pm$ computed at or in the neighborhood of T_0; this requires a knowledge of the partial molar enthalpies of the solute at infinite dilution and at molality m. Since this involves more computation than appears warranted here, we shall not go into the matter further.

The methods described above can be extended to measurements of boiling-point elevation, but, of course, the same complications and limitations regarding temperature will arise. On the other hand, the measurement of osmotic pressures can be carried out for each of several concentrations at the same temperature. From such measurements one can, by use of the Gibbs-Duhem equation, get at solute activities for the particular temperature of the experiments.

Debye-Hückel Theory of Ion Activities

When dealing with nonelectrolytes, we saw that a simple model involving molecules of different sizes provides a basis for calculating the entropy of mixing of nonideal solutions. For solutions of ions, however, a different approach must be made, for the deviations from ideality are attributable principally to electrical effects rather than to differences in size. Since unlike charges attract each other and like charges repel, it is reasonable to expect that in a solution of an electrolyte each positive ion, on the average, will tend to be surrounded by a cloud of negative ions, and vice versa. With this idea in mind, Debye and Hückel formulated a theory for the activities of ions in dilute solutions by making use of electrostatic considerations and the Boltzmann Distribution Law. This theory, as we shall here describe it, is valid only for very dilute solutions, although numerous attempts have been made to extend its range of applicability with some success.

A basic premise of the Debye-Hückel theory is that deviations from ideality can be attributed solely to the presence of the ionic charges. According to that theory, it is assumed that, if a solute were to dissociate completely into uncharged fragments, the solution would still obey the ideal laws—that is, Raoult's Law for the solvent and Henry's Law for each kind of solute fragment. With this understanding, the Debye-Hückel theory can be reduced, in essence, to the calculation of the Gibbs free-energy change attending the hypothetical process of electrically charging the fragments obtained from the dissociation of a nonionic solute. Such a charging process corresponds to converting a nonionic solute into an ionic one, and hence an ideal solution into one that is nonideal. The total Gibbs energy of a solution of an electrolyte can then be regarded as the sum of two terms, $G(\mathrm{id})$, the ideal Gibbs energy, and $G(\mathrm{el})$, the electrical (or extra) Gibbs energy associated with charging the particles. Therefore, if we could calculate the electrical Gibbs free energy, we should be able to calculate the activity coefficients, since the electrical Gibbs energy measures departures from ideality. Reduced to quantitative form, the activity of the electrolyte will be related to the chemical potentials as follows:

$$\mu_{\mathrm{B}} = \mu_{\mathrm{B}}^{\circ} + RT \ln a_{\mathrm{B}} = \mu_{\mathrm{B}}(\mathrm{id}) + \mu_{\mathrm{B}}(\mathrm{el}) \qquad (20.20)$$

But the activity, a_{B}, will be the product of an appropriate reference function, say $g(c)$, and $\gamma_{\pm}^{\nu}$, provided $\gamma_{\pm}$ is the mean-activity coefficient and ν is the total number of ions obtained from the dissociation. Therefore

$$\mu_{\mathrm{B}}^{\circ} + RT \ln g(c) + \nu RT \ln \gamma_{\pm} = \mu_{\mathrm{B}}(\mathrm{id}) + \mu_{\mathrm{B}}(\mathrm{el}) \qquad (20.21)$$

But, since the ideal chemical potential must be precisely the value of the chemical potential when $\gamma_{\pm} = 1$,

$$\ln \gamma_{\pm} = \frac{\mu_{\mathrm{B}}(\mathrm{el})}{\nu RT} \qquad (20.22)$$

Now, to calculate the electrical free energy of a solution, a step that appears to be necessary if we are to obtain $\gamma_{\pm}$, we must first find the electrical potential energy of each of the ions. To do so, Debye started out with Poisson's equation, a basic equation of electrostatics:

$$\frac{\partial^2 \psi}{\partial x^2} + \frac{\partial^2 \psi}{\partial y^2} + \frac{\partial^2 \psi}{\partial z^2} = -\frac{4\pi \rho}{D} \qquad (20.23)$$

In this equation ψ represents the electrical potential, ρ the electrical charge density, and D the dielectric constant of the medium. Next Debye assumed that the charge density is determined by the Boltzmann distribution. Let us suppose once again that the electrolyte $C_{\nu_+}A_{\nu_-}$ dissociates into ν_+ positive ions of charge z_+e (e being the charge on the electron) and ν_- negative ions of charge $-z_-e$. Then the potential energy of a positive ion in a region of potential ψ will be $z_+e\psi$; similarly, the potential energy of a negative ion will be $-z_-e\psi$. Let us now suppose that the average numbers of ions per unit volume are N_+ and N_-, which will also be their concentrations in regions of zero electrical potential. Then, according to the Boltzmann Distribution Law, the number of positive ions per unit volume in a region of potential ψ will be $N_+ \exp(-z_+e\psi/kT)$, and the positive ion contribution to the charge density will be $N_+z_+e \exp(-z_+e\psi/kT)$. The contribution of the negative ions to the charge density will be $-N_-z_-e \exp(z_-e\psi/kT)$. Hence the total charge density will be

$$\rho = N_+ z_+ e \exp\left(-z_+ e\psi/kT\right) - N_- z_- e \exp\left(z_- e\psi/kT\right) \qquad (20.24)$$

If we now assume that the potential energies of the ions are both small compared with kT, we can expand the exponential functions and, to a good degree of approximation, retain only constant terms and terms linear in ψ. Thus we obtain the expression

$$\rho = (N_+ z_+ e - N_- z_- e) - \left(\frac{N_+ z_+^2 e^2 + N_- z_-^2 e^2}{kT}\right)\psi + \ldots \qquad (20.25)$$

But since $N_+ z_+ = N_- z_-$, because of electrical neutrality, the constant term in equation 20.25 will vanish. Moreover, since the total concentration of "molecules" is N_B per unit volume, we can write that $N_+ = v_+ N_B$ and that $N_- = v_- N_B$ to get the expression

$$\rho = -\frac{N_B(v_+ z_+^2 + v_- z_-^2)e^2\psi}{kT} \qquad (20.26)$$

Substituting this equation into equation 20.23, we obtain the equation

$$\frac{\partial^2\psi}{\partial x^2} + \frac{\partial^2\psi}{\partial y^2} + \frac{\partial^2\psi}{\partial z^2} = \kappa^2\psi \qquad (20.27)$$

in which

$$\kappa^2 = \frac{4\pi N_B(v_+ z_+^2 + v_- z_-^2)e^2}{DkT} \qquad (20.28)$$

Our problem now is to find a solution of equation 20.27 that will satisfy the boundary conditions for a given physical situation. Suppose that we place at the origin of our coordinate system one ion, either positive or negative. All the other ions can then be considered as distributed about the central ion in a spherically symmetrical ionic cloud. In other words, the potential, ψ, at any point x,y,z, will depend only on the distance from the origin—that is, upon the radius vector, $r = \sqrt{x^2 + y^2 + z^2}$. Hence, if we transform equation 20.27 from rectangular coordinates (x,y,z) to polar coordinates (r,θ,ϕ), we can suppress any derivatives with respect to θ and ϕ, leaving the following ordinary differential equation:

$$\frac{1}{r^2}\frac{d}{dr}\left(r^2\frac{d\psi}{dr}\right) = \kappa^2\psi \qquad (20.29)$$

We can readily obtain a solution to this differential equation if we first replace ψ by a new variable, R, such that $\psi = R/r$. Upon making this change, we obtain the following result:

$$\frac{d^2R}{dr^2} = \kappa^2 R \qquad (20.30)$$

This is a familiar differential equation, for which the solution is well known to be given by the expression

$$R = C_1 e^{-\kappa r} + C_2 e^{\kappa r} \qquad (20.31)$$

in which C_1 and C_2 are arbitrary constants. Since the potential, ψ, cannot become infinite as r increases without limit, C_2 must be zero. Hence we can say that

$$\psi = \frac{C_1 e^{-\kappa r}}{r} \qquad (20.32)$$

Now let us further suppose that the concentration of ions is small enough so that κr is small compared with unity. We can then write, approximately, that

$$\psi = \frac{C_1}{r}(1 - \kappa r) = \frac{C_1}{r} - C_1 \kappa \qquad (20.33)$$

The first term on the right-hand side of this equation must represent the potential attributable to the central ion, for, as r becomes small, the potential must be dominated by that ion. If we imagine a positive central ion of charge $z_+ e$, its contribution to the potential will be precisely $z_+ e / Dr$. The constant C_1 must therefore equal $z_+ e / D$, from which we conclude that

$$\psi = \frac{z_+ e}{Dr} - \frac{z_+ e \kappa}{D} \qquad (20.34)$$

(If our reference ion were chosen to be negative, we should simply replace $z_+ e$ by $-z_- e$.)

Equation 20.34 can be interpreted as follows: The first term is, of course, the contribution of the central ion to the potential, whereas the second term is that of the environment. It appears, therefore, that the ionic cloud surrounding a central ion acts as if it were a spherical shell of electricity of radius $1/\kappa$ and of charge equal to that of the central ion but of opposite sign. Thus every positive ion, on the average, will find itself subject to an environmental potential equal to $-z_+ e \kappa / D$ and every negative ion to a potential of $z_- e \kappa / D$.

EXERCISE:

Calculate in centimeters the radius $(1/\kappa)$ associated with the hypothetical sphere of ions surrounding a central ion of each of the following salts:

Molar Concentration, c	Salt
0.10	KCl
0.01	KCl
0.05	$MgCl_2$
0.005	$MgCl_2$
0.05	$MgSO_4$
0.005	$MgSO_4$

We are now in a position to calculate the electrical Gibbs energy of a solution of an electrolyte. To do so, let us imagine the process of simultaneously charging all the ions from zero to their final charges. The electrical work done on the system (or the negative of that appearing in the surroundings) will then equal the electrical free energy, $G(\text{el})$, as previously mentioned. We can most easily imagine this charging process if we replace each electronic charge, e, by fe (f being a variable representing the fraction of the ultimate charge to be attained, subject to the range of values $0 \leqslant f \leqslant 1$). In view of equation 20.28, the effective value of κ during the charging process will be $f\kappa$; the environmental potential, which ends up at $-z_+e\kappa/D$ (or $z_-e\kappa/D$), will accordingly be $-f^2z_+e\kappa/D$ (or $f^2z_-e\kappa/D$). If an infinitesimal charge of magnitude z_+edf is brought into a region of potential $-f^2z_+e\kappa/D$, the work done on the system will be $-(z_+^2e^2\kappa/D)f^2df$. For a negative ion the corresponding quantity is $-(z_-^2e^2\kappa/D)f^2df$. Multiplying each potential by the total number of the appropriate kind of ion and adding, we see that

$$dG(\text{el}) = -\frac{N_Be^2\kappa V(\nu_+z_+^2 + \nu_-z_-^2)f^2df}{D} \tag{20.35}$$

V being the volume. Integrating from $f = 0$ to $f = 1$, we obtain the expression

$$G(\text{el}) = -\frac{N_Be^2\kappa V(\nu_+z_+^2 + \nu_-z_-^2)}{3D} \tag{20.36}$$

We can now obtain the electrical contribution to the chemical potential, $\mu_B(\text{el})$, by differentiating partially with respect to n_B, the total number of moles of solute. Evidently $n_B = N_BV/N_0$, if N_0 is Avogadro's number. Hence, remembering that κ depends on N_B, we see that

$$\mu_B(\text{el}) = \frac{N_0}{V}\frac{\partial G}{\partial N_B} = \frac{N_0}{V}\left(\frac{G(\text{el})}{N_B} + \frac{G(\text{el})}{\kappa}\frac{\partial\kappa}{\partial N_B}\right) \tag{20.37}$$

But since we see, from equation 20.28, that $\partial\kappa/\partial N_B = \kappa/(2N_B)$, it is apparent that

$$\mu_B(\text{el}) = \frac{3N_0G(\text{el})}{2VN_B}$$

$$= -\frac{N_0e^2\kappa(\nu_+z_+^2 + \nu_-z_-^2)}{2D} \tag{20.38}$$

Combining this equation with equation 20.22, we see that

$$\ln\gamma_\pm = -\frac{N_0e^2\kappa(\nu_+z_+^2 + \nu_-z_-^2)}{2\nu DRT}$$

$$= -\frac{e^2\kappa(\nu_+z_+^2 + \nu_-z_-^2)}{2\nu DkT} \tag{20.39}$$

and, since $\nu_+z_+ = \nu_-z_-$, we see also that $\nu_+z_+^2 + \nu_-z_-^2 = \nu z_+z_-$. Therefore

$$\ln \gamma_\pm = -\frac{z_+z_-e^2\kappa}{2DkT} \qquad (20.40)$$

This equation agrees reasonably well with experiment for very dilute solutions. It is also possible, from the theory, to calculate the activity coefficients of single ions, even though these cannot be determined experimentally. Thus it turns out, theoretically, that

$$\ln \gamma_+ = -\frac{z_+^2e^2\kappa}{2DkT} \qquad (20.41)$$

with a similar equation for $\ln \gamma_-$.

In carrying out the preceding treatment of ion activities, we assumed that we were dealing with a solution of a single electrolyte. This assumption is not necessary, for the theory can be applied to mixtures of electrolytes, provided the quantity κ is modified accordingly. For any mixture of electrolytes let us redefine κ by the expression

$$\kappa^2 = \frac{4\pi e^2}{DkT} \sum_i N_iz_i^2 \qquad (20.42)$$

in which N_i is the number of ions of type i per unit volume and z_i the number of charges on the ion. Converted from molecular to molar quantities, equation 20.42 becomes

$$\kappa^2 = \frac{4\pi e^2 N_0^2}{1{,}000 \, DRT} \sum_i c_iz_i^2 \qquad (20.43)$$

c_i being the concentration in moles per liter and N_0 Avogadros number. (The factor 1,000 converts to moles per cubic centimeter.) Let us now define the *ionic strength*, I_c, as follows:*

$$I_c = \tfrac{1}{2} \sum_i c_iz_i^2 \qquad (20.44)$$

This definition is made so that the ionic strength of a solution of a uni-univalent electrolyte like sodium chloride will simply equal its concentration. By introducing the ionic strength into equation 20.43, we obtain the expression

---

* The ionic strength can also be defined using molalities instead of concentrations, so that $I_m = \tfrac{1}{2}\sum_i m_iz_i^2$. For infinitely dilute solutions, the ionic strength defined in terms of concentrations (equation 20.44) is ρ_0 times the ionic strength defined using molalities, if ρ_0 is the density of solvent. For aqueous solutions, this makes little difference, but for precise calculations and for calculations involving nonaqueous systems, special care should be exercised to reconcile different sources of data.

$$\kappa^2 = \frac{8\pi e^2 N_0^2 I_c}{1,000\ DRT} \tag{20.45}$$

Notwithstanding this generalization, the expressions for $\ln \gamma_{\pm}$ and $\ln \gamma_i$ are unchanged from equations 20.40 and 20.41. Putting in the numerical values of the physical constants that appear in these formulas, we find that at 0 °C

$$\kappa = 0.3244 \times 10^8 \sqrt{I_c}$$

and

$$\lg \gamma_{\pm} = -0.4896\ z_+ z_- \sqrt{I_c} \tag{20.46}$$

At 25 °C the equations become

$$\kappa = 0.3286 \times 10^8 \sqrt{I_c}$$

$$\lg \gamma_{\pm} = -0.5092\ z_+ z_- \sqrt{I_c} \tag{20.47}$$

Activity Coefficients from Measurements of Electromotive Force

One of the best ways of getting thermodynamic information about ions in solution is through the measurement of electromotive forces. We shall, accordingly, turn to a consideration of this topic, starting first with so-called concentration cells, which are cells whose electromotive forces are attributable solely to differences in concentration. Consider, for example, the electrolytic cell

$$\text{Pt,}H_2(g)\ |\ H^+Cl^-\ |\ H^+Cl^-\ |\ H_2(g)\text{,Pt}$$
$$p \qquad\quad c_1 \qquad\quad c_2 \qquad\quad p$$

which consists of two hydrogen electrodes, each at pressure p, and two aqueous solutions of hydrochloric acid of concentrations c_1 and c_2. Now, if one faraday of electricity should pass through the cell from left to right, the following changes in state would take place:

(A) $\tfrac{1}{2}H_2(g) \longrightarrow H^+ + e^-$
$\quad\ p \qquad\qquad\quad c_1$

(B) $t_+ H^+ \longrightarrow t_+ H^+$
$\quad\ c_1 \qquad\qquad c_2$

(C) $t_- Cl^- \longrightarrow t_- Cl^-$
$\quad\ c_2 \qquad\qquad c_1$

(D) $H^+ + e^- \longrightarrow \tfrac{1}{2}H_2(g)$
$\quad\ c_2 \qquad\qquad\quad p$

Processes A and D involve the conversion of hydrogen gas into hydrogen ions and electrons, or the reverse thereof, whereas processes B and C are concerned with the transference of ions across the boundary separating the two solutions of hydrochloric acid. The symbols t_+ and t_- represent the transference numbers of the ions, here assumed to be independent of concentration. (Since a transference number is the fraction of the total current carried by the ion under consideration, it is necessary that $t_+ + t_- = 1$.) The over-all change in state accompanying the passage of one faraday of electricity through the cell will be given as the sum of processes A through D. Simply adding the chemical equations and rearranging the result, one obtains the expression

$$t_-H^+Cl^- \longrightarrow t_-H^+Cl^-$$
$$\quad c_2 \qquad\qquad\qquad c_1$$

In other words, the net thermodynamic change per faraday is equivalent to transferring t_- moles of hydrogen chloride from a solution of concentration c_2 to a solution of concentration c_1. The Gibbs free-energy change, or, more properly, the reaction potential, attending this change in state will be

$$\Delta\tilde{\mu} = t_-RT \ln \frac{a_1}{a_2} \tag{20.48}$$

if a_1 and a_2 are the activities of the hydrogen chloride in the two solutions. The electrical work, which will be zEF if z is the number of faradays, E the electromotive force, and F the value of the faraday, must equal the decrease in Gibbs energy, $-z\Delta\tilde{\mu}$. Hence, since $z = 1$, we see that

$$E = \frac{t_-RT}{F} \ln \frac{a_2}{a_1} \tag{20.49}$$

Now the activity ratio, a_2/a_1, must equal the fugacity ratio, f_2/f_1, which will be very nearly the same as the ratio of the partial pressures, p_2 and p_1, of hydrogen chloride above the two solutions. Hence, assuming that HCl is an ideal gas, we see that

$$E = \frac{t_-RT}{F} \ln \frac{p_2}{p_1} \tag{20.50}$$

This treatment, while perfectly correct, still says nothing about ion activities. However, if we choose the concentrations of the ions for reference purposes, we can write that

$$a = \gamma_{\pm}^2 c^2 \tag{20.51}$$

Hence, from equation 20.49, we see that

$$E = \frac{t_-RT}{F} \ln \frac{\gamma_{\pm 2}^2 c_2^2}{\gamma_{\pm 1}^2 c_1^2}$$

or that

$$E = \frac{2t_-RT}{F} \ln \frac{c_2}{c_1} + \frac{2t_-RT}{F} \ln \frac{\gamma_{\pm 2}}{\gamma_{\pm 1}} \tag{20.52}$$

Now let us replace c_1 and $\gamma_{\pm 1}$ by c^* and $\gamma_{\pm}^*$, the asterisks indicating an approach to infinite dilution. Upon taking cognizance of the fact that $\lim_{c^* \to 0} \gamma_{\pm}^* = 1$, we obtain, after dropping subscript "2", the following:

$$\ln \gamma_{\pm} = -\ln c + \lim_{c^* \to 0} \left(\frac{EF}{2t_-RT} + \ln c^* \right) \tag{20.53}$$

By graphical or other means it is possible to determine the limit appearing in this equation and thus to obtain a measure of the mean-activity coefficient of the hydrogen chloride in a solution of concentration c. (To make the computation more precise, we should make a correction for the change in transference number with concentration.) It should be apparent from this example, as with the freezing-point method, that, to obtain an activity, one must ultimately refer to the reference state for activity, which, in this case, is a state of infinite dilution.

Another type of concentration cell, which is really a double cell, is the following:

$$\text{Pt,H}_2(g) \mid \text{H}^+\text{Cl}^-,\text{AgCl(c)} \mid \text{Ag(c)} \mid \text{AgCl(c)H}^+\text{Cl}^- \mid \text{H}_2(g),\text{Pt}$$
$$p \qquad c_1 \qquad\qquad\qquad\qquad\qquad c_2 \qquad p$$

When one faraday passes through this cell from left to right, the following net thermodynamic change in state occurs:

$$\text{H}^+\text{Cl}^- \longrightarrow \text{H}^+\text{Cl}^-$$
$$c_2 \qquad\qquad c_1$$

In this case the transference number of the hydrochloric acid does not appear in the equation describing the change in state. Here the electromotive force is

$$E = \frac{RT}{F} \ln \frac{a_2}{a_1} = \frac{2RT}{F} \ln \frac{c_2 \gamma_{\pm 2}}{c_1 \gamma_{\pm 1}} \tag{20.54}$$

Manipulating this equation by previously employed methods, we can show that

$$\ln \gamma_{\pm} = -\ln c + \lim_{c^* \to 0} \left(\frac{EF}{2RT} + \ln c^* \right) \tag{20.55}$$

The foregoing treatment can equally well be carried out using the molality, m, in place of volume concentration, c. In accordance with equation 19.60, this leads to a different value for $\gamma_{\pm}$, but the difference is not large for aqueous solutions. Since molality is generally used as the reference function for ions in solution, we shall in the ensuing sections adhere to that practice.

Standard Electromotive Forces

Let us now turn to a cell that does not simply involve a change in concentration—that is, one whose operation is accompanied by a chemical change in state. Consider, for example, one-half of the cell described above, namely:

$$\text{Pt,H}_2(g) \mid \text{H}^+\text{Cl}^-, \text{AgCl}(c) \mid \text{Ag}(c)$$
$$\quad\; p \qquad\quad m$$

in which the concentration of H^+Cl^- is expressed in terms of the molality, m. When one faraday of electricity passes through that cell from left to right, the net change in state will be represented by

$$\tfrac{1}{2}\text{H}_2(g) + \text{AgCl}(c) \longrightarrow \text{H}^+\text{Cl}^- + \text{Ag}(c)$$
$$\quad p \qquad\qquad\qquad\quad m$$

The rate of change of Gibbs free energy, or reaction potential, will be given by the expression

$$\Delta\tilde{\mu} = \Delta\tilde{\mu}^0 + RT \ln \frac{a_{\text{HCl}}a_{\text{Ag}}}{a_{\text{H}_2}^{\frac{1}{2}}a_{\text{AgCl}}} \tag{20.56}$$

in which $\Delta\tilde{\mu}^\circ$ is the standard reaction potential. If we choose the pure materials as standard states for the solids, represent the activity of hydrogen by its fugacity, and use infinitely dilute molalities as reference functions for the ions of hydrochloric acid, we can rewrite the equation as

$$\Delta\tilde{\mu} = \Delta\tilde{\mu}^0 + RT \ln \frac{m^2\gamma_\pm^2}{f_{\text{H}_2}^{\frac{1}{2}}} \tag{20.57}$$

Since $-\Delta\tilde{\mu} = EF$, we can further write that

$$E = E^\circ - \frac{RT}{F} \ln \frac{m^2\gamma_\pm^2}{f_{\text{H}_2}^{\frac{1}{2}}} \tag{20.58}$$

In this expression E° is equal to $-\Delta\tilde{\mu}^\circ/F$ and is called the "standard electromotive force" of the cell. (Alternatively, one could substitute the fugacity of hydrogen chloride in place of its activity; this would involve a different standard state for HCl and hence a different E° for the cell.) This treatment can now be extended to any kind of cell that, when operated isothermally, produces a chemical change in state for which a reaction potential can be written. E° will obviously be related to the equilibrium constant for the reaction, since

$$E^\circ = - \frac{\Delta\tilde{\mu}^\circ}{zF} = \frac{RT}{zF} \ln K \tag{20.59}$$

The temperature coefficient of the standard electromotive force will be given by the expression

$$\frac{dE^\circ}{dT} = \frac{d}{dT}\left(-\frac{\Delta\tilde{\mu}^\circ}{zF}\right) = -\frac{1}{zF}\left(\frac{\Delta\tilde{\mu}^\circ - \Delta\tilde{H}^\circ}{T}\right) = \frac{E^\circ}{T} + \frac{\Delta\tilde{H}^\circ}{zFT} \qquad (20.60)$$

in which $\Delta\tilde{H}^\circ$ is the standard enthalpy of reaction. The temperature coefficient of the actual electromotive force, $(\partial E/\partial T)_p$, will be given by a similar expression involving E as well as $\Delta\tilde{H}$, the rate of change of enthalpy per unit of reaction under the actual conditions.

Since the standard electromotive force is an important quantity, being directly related to the reaction potential, it is worth our while to inquire into means for its determination. Assuming that the necessary Gibbs energy data are not available from other sources, we can determine E° by direct measurement of electromotive forces at various concentrations, which in turn are subjected to the following type of analysis. Using the example discussed immediately above, we see, upon rearrangement of equation 20.58, that

$$E + \frac{RT}{F}\ln\frac{m^2}{f_{H_2}^{\frac{1}{2}}} = E^\circ - \frac{2RT}{F}\ln\gamma_\pm \qquad (20.61)$$

Since $\gamma_\pm \longrightarrow 1$ as $m \longrightarrow 0$, E° is simply the limiting value at infinite dilution of $E + (RT/F)\ln(m^2/f_{H_2}^{\frac{1}{2}})$, in which expression f_{H_2} can usually be replaced by p_{H_2}. Various graphical procedures can be devised for plotting the experimental results in convenient form, but a particular method is suggested by the Debye-Hückel limiting law. Applying equation 20.46 to an electrolyte like hydrochloric acid, we see that for low concentrations, $\ln\gamma_\pm = A\sqrt{m}$ where A is a constant. This suggests that, if the function $E + (RT/F)\ln(m^2/f_{H_2}^{\frac{1}{2}})$ is plotted against $\sqrt{m}$, the curve should become linear at low concentrations; extrapolation of zero concentration will then yield E° as the ordinate intercept. This technique can be modified to fit the actual problem at hand, but the salient features arc illustrated above.

EXERCISE:

The value of the faraday, F, is 96,487 coulombs. Evaluate $(R/F)\ln 10$ in V deg^{-1} and $(RT/F)\ln 10$ for 25 °C in volts. (These factors, which equal 0.00019841 V deg^{-1} and 0.059157 V, are regularly used in calculations involving electromotive force.) Hence show that, at 25 °C, a typical electromotive-force expression will be

$$E = E^\circ + \frac{0.05916}{z}\lg\left\{\prod_i a_i\right\}$$

in which $\prod_i a_i$ denotes an appropriate product (and quotient) of activities.

EXERCISE:

(A) If E is the electromotive force of a reversible cell operating at pressure p and temperature T, prove that

$$\left(\frac{\partial E}{\partial p}\right)_T = -\frac{\Delta \tilde{V}}{zF}$$

if $\Delta \tilde{V}$ is the change in volume attending the cell reaction.

(B) Assuming that hydrogen is an ideal gas, and neglecting the volumes of liquids and solids, derive an expression for $(\partial E/\partial p)_T$ and evaluate it in volts per atm for 25 °C and 1 atm pressure for the following cell:

$$\text{Pt,H}_2(\text{g}) \mid \text{H}^+\text{Cl}^-,\text{AgCl(c)} \mid \text{Ag(c)}$$
$$p m$$

Ans. $\left(\dfrac{\partial E}{\partial p}\right)_T = \dfrac{RT}{2zFp}$

Electrode Potentials and "Half-cells"

It should be clear from the foregoing that the electromotive force of a cell is determined by over-all changes in state that do not involve the net appearance or disappearance of electrons. This is true because every electron that appears at one electrode has its counterpart disappearing at the other. It is often convenient, nevertheless, to divide a cell arbitrarily into two hypothetical "half-cells," each of which will involve the appearance or disappearance of electrons. Every cell can then be regarded as a combination of half-cells, and the behavior of the total cell can be described by appropriate coupling of the characteristics of the half-cells. For the cell described in the preceding section, the two half-cells are conventionally represented by

$$\text{H}^+ \mid \text{H}_2(\text{g}),\text{Pt}$$
$$m p$$

and

$$\text{Cl}^-,\text{AgCl(c)} \mid \text{Ag(c)}$$
$$m$$

The corresponding "half-reactions" are

$$\text{H}^+ + e^- \longrightarrow \tfrac{1}{2}\text{H}_2(\text{g})$$

and

$$\text{AgCl(c)} + e^- \longrightarrow \text{Cl}^- + \text{Ag(c)}$$

The first of these half-cells is called the hydrogen electrode and the second is called the silver silver-chloride electrode. By turning the hydrogen electrode around and placing it to the left of the other, we construct the original electrolytic cell. Moreover, by subtracting the half-cell reaction for the hydrogen electrode from that of the one involving silver, we obtain the total cell reaction, which in turn represents the passage of positive electricity from left to right within the cell.

Of course, many other electrodes exist or can be imagined, and innumerable pairs of them can be combined to make up electrolytic cells. In general, a complete cell consists of one half-cell on the right joined to the reverse of another half-cell on the left. The corresponding cell reaction will be the algebraic sum of the half-reactions, provided we assign a negative sign to the reaction characterizing that portion of the cell that is written in reverse. Evidently, if a whole cell is turned around, the overall cell reaction will be reversed, and the electromotive force will change sign.

Pursuing the half-cell analogy further, we shall now break up the electromotive force of a cell into two parts by assigning to each electrode an electromotive force, their difference being equal to the electromotive force of the cell. For the half-cells above, we assert in accordance with their chemical reactions that

$$E_{H^+/H_2} = E^{\circ}_{H^+/H_2} - \frac{RT}{F} \ln (f_{H_2}^{\frac{1}{2}}/a_{H^+}) \tag{20.62}$$

and

$$E_{Cl^-,AgCl/Ag} = E^{\circ}_{Cl^-,AgCl/Ag} - \frac{RT}{F} \ln a_{Cl^-} \tag{20.63}$$

In these two equations the electron activity is omitted because that quantity is not obtainable by any ordinary method. In practice, when we consider pairs of electrodes in juxtaposition, the electron activities cancel out. We have also omitted a_{AgCl} and a_{Ag} since we have chosen the pure solids to represent their standard states, just as we did in connection with equations 20.57 and 20.58. Now, by subtracting equation 20.62 from equation 20.63, we obtain a result resembling equation 20.58 provided

$$E_{Cl^-,AgCl/Ag} - E_{H^+/H_2} = E \tag{20.64}$$

with E applicable to the entire cell. We can also assert that

$$E^{\circ}_{Cl^-,AgCl/Ag} - E^{\circ}_{H^+/H_2} = E^{\circ} \tag{20.65}$$

Now it is not possible by ordinary thermodynamic means to obtain a single electrode potential, since we do not have methods for getting at electron activities. However, if we arbitrarily assign a value to the hydrogen electrode, we can determine the electromotive forces of other electrodes in relation to that of hydrogen. Just as it was convenient to assign a value of zero to the enthalpy of hydrogen ions at infinite dilution, so, too, it is convenient to set $E^{\circ}_{H^+/H_2}$ equal to zero; this is equivalent to assigning zero to the standard Gibbs energy, $\Delta\tilde{\mu}^{\circ}_{H^+/H_2}$, of the hydrogen-electrode reaction:

$$H^+ + e^- \longrightarrow \tfrac{1}{2}H_2(g)$$

The foregoing discussion is fully compatible with an international convention regarding the definition of an electrode potential. An electrode potential is defined as the electromotive force of a cell in which the electrode of interest is on the right-hand side with a standard hydrogen electrode on the left-hand side. It follows, therefore, that the electromotive force of any cell is the electrode potential of the right-hand electrode less the electrode potential of that on the left. This definition of a half-cell potential is, of course, a pragmatic one, and the data of Table 20-1 are presented accordingly.

Since a standard hydrogen electrode is said to have an electrode potential equal to zero at all temperatures, we see that the temperature coefficient of $E^{\circ}_{H^+/H_2}$ or of $\Delta\tilde{\mu}^{\circ}_{H^+/H_2}$ will also be zero. Hence the standard entropy change associated with the operation of the hydrogen-electrode reaction, $\Delta S^{\circ}_{H^+/H_2}$, must likewise be zero.

We now go one step further by assigning a value of zero to the standard entropy of hydrogen ions in solution. Because of the convention concerning standard entropy changes, this is equivalent to setting the standard electron entropy equal to that of one-half mole of hydrogen gas. The arbitrariness of such assignments of value should not distress us, for, in any practical applications, combinations of electrode reactions will automatically compensate for the conventions. We shall say, for example, that the standard electromotive force or standard electrode potential of the silver electrode $(Ag^+ + e^- \longrightarrow Ag(c))$ is 0.799 V at 25 °C; this really means that its standard value is 0.799 V greater than that of the hydrogen electrode. Similarly, $\mu^{\circ}_{Ag^+}$, $H^{\circ}_{Ag^+}$, and $S^{\circ}_{Ag^+}$ turn out to be 18.43 kcal, 25.23 kcal, and 17.37 cal deg^{-1} at 25 °C. These numbers simply tell by how much these quantities exceed the corresponding values for hydrogen ions. Hence, if we devise a cell in which the change in state

$$\tfrac{1}{2}H_2(g) + Ag^+ \longrightarrow H^+ + Ag(c)$$

takes place, then $E^{\circ} = 0.799$ V, $\Delta\tilde{\mu}^{\circ} = -EF = -18.43$ kcal, and $\Delta\tilde{H}^{\circ} = -25.23$ kcal. To calculate $\Delta\tilde{S}^{\circ}$, we cannot simply take the negative of $S^{\circ}_{Ag^+}$, for the entropies of $H_2(g)$ and $Ag(c)$ are not zero, even by convention. However, we can evaluate $\Delta\tilde{S}^{\circ}$ from $(\Delta\tilde{H}^{\circ} - \Delta\tilde{\mu}^{\circ})/T$ or by recourse to values of the entropies given in Appendix D. Thus we find that $\Delta \tilde{S}^\circ = -22.80$ cal deg^{-1}. We shall now turn to the question how, by combining a pair of electrodes, we could construct a cell in which a reaction like that illustrated above might take place.

Cells Without Liquid Junction Potentials

Although the cell that we considered on page 439 appears to consist of two hydrogen electrodes placed opposite each other, the electromotive force of the cell is not simply the difference between the electromotive forces of the two electrodes. This apparent discrepancy can be attributed to the transference of ions across the liquid junction separating the two solutions, which also means that the chloride ions play an important role in the operation of the cell. Now it is possible to eliminate the effect of the junction almost completely by joining the solutions to each other with a concentrated salt bridge. When this is done, the cell is said to be free of a liquid junction potential, and it is written as

$$\text{Pt,H}_2(\text{g}) \mid \text{H}^+ \parallel \text{H}^+ \mid \text{H}_2(\text{g}),\text{Pt}$$
$$\quad p \qquad m_1 \quad m_2 \qquad p$$

The double vertical line indicates absence of a liquid junction potential, and the ions conjugate to the hydrogen ions, say chloride, are no longer of consequence except as they affect the activities of the hydrogen ions. When one faraday of electricity passes through such a cell, the net change in state, we can readily see, is

$$\text{H}^+ \longrightarrow \text{H}^+$$
$$m_2 \qquad m_1$$

The electromotive force will therefore be given by the expression

$$E = \frac{RT}{F} \ln \frac{a_2}{a_1} \tag{20.66}$$

in which a_2 and a_1 are the activities of the hydrogen ions in the two solutions. We can obtain equation 20.66 simply by taking the difference between the electrode potentials of the two hydrogen electrodes involved. At this point you may ask now we can determine the single-ion activities that appear in equation 20.66. The answer is that we cannot, at least not exactly. If we assume, however, that the activity coefficients for ions of equal and opposite charge are the same in a particular solution (this is a prediction of the Debye-Hückel theory), we can adopt for reference purposes a dilute solution that will give us a basis for consistent measurement of hydrogen-ion activities.

If a hydrogen electrode is placed opposite a silver electrode without liquid junction, the resulting cell will be

$$\text{Pt,H}_2(g) \mid H^+ \parallel Ag^+ \mid Ag(c)$$

and its electromotive force will be given by the expression

$$E = E° + \frac{RT}{F} \ln \frac{a_{Ag^+}}{a_{H^+}} \tag{20.67}$$

in which, of course, $E°$ is the difference between the standard values for the two electrodes, in this case simply $E°_{Ag^+/Ag}$.

The foregoing discussion can be generalized to cover any combination of electrodes. If the electrode reactions are written in such a way that electrons appear on the left-hand side of the chemical equations, then the standard electromotive force of a whole cell will be the standard electrode potential of the right-hand electrode minus that for the left-hand electrode. If the standard electromotive force of the whole cell is positive, then positive electricity would tend to flow from left to right *within* the cell if the substances were all in their standard states. On the other hand, if the electrode with the greater standard electrode potential is on the left-hand side, then the standard electromotive force of the whole cell would be negative, indicating that positive electricity would tend to flow from right to left within the cell under standard conditions. Table 20-1 gives some typical standard electrode potentials.

In manipulating the numbers for computational purposes, one must remember that electromotive forces are intensive properties; the corresponding Gibbs free-energy changes, on the other hand, are extensive and hence require that we consider the number of electrons (or faradays) involved in the reaction. Consider the following example:

$$\text{Pb}(c) \mid Pb^{++} \parallel Fe^{++}, Fe^{+++} \mid Pt$$

If two faradays of electricity pass through the cell, the net change in state is

$$\text{Pb}(c) + 2Fe^{+++} \longrightarrow Pb^{++} + 2Fe^{++}$$

Hence

$$\Delta\bar{\mu} = \Delta\bar{\mu}° + RT \ln \frac{a_{Pb^{++}} a_{Fe^{++}}^2}{a_{Fe^{+++}}^2} \tag{20.68}$$

and since $\Delta\bar{\mu} = -2EF$,

$$E = E° - \frac{RT}{F} \ln \frac{a_{Pb^{++}}^{1/2} a_{Fe^{++}}}{a_{Fe^{+++}}} \tag{20.69}$$

in which expression $E° = E°_{Fe^{++},Fe^{+++}/Pt} - E°_{Pb^{++}/Pb} = 0.770 - (-0.126) = 0.896$ V. The standard reaction potential, $\Delta\bar{\mu}°$, will equal $-2EF = 2(-0.896) \times 23.06 = -41.33$ kcal. This value of $\Delta\bar{\mu}°$ will, of course, equal $\mu°_{Pb^{++}} + 2\mu°_{Fe^{++}} - 2\mu°_{Fe^{+++}}$.

TABLE 20-1

Standard Electrode Potentials at 25 °C

(Values calculated for aqueous systems from data in Appendix D. All substances are presumed to be in their standard states, using molality as the reference function for the ions and assuming other materials to be pure solids, liquids, or gases.)

ELECTRODE REACTION	$E°$ (VOLTS)	ELECTRODE REACTION	$E°$ (VOLTS)
$Li^+ + e^- \rightleftharpoons Li$	−3.045	$AgI + e^- \rightleftharpoons Ag + I^-$	−0.151
$Ca(OH)_2 + 2e^- \rightleftharpoons Ca + 2OH^-$	−3.026	$Sn^{++} + 2e^- \rightleftharpoons Sn$	−0.141
$K^+ + e^- \rightleftharpoons K$	−2.926	$Pb^{++} + 2e^- \rightleftharpoons Pb$	−0.126
$Rb^+ + e^- \rightleftharpoons Rb$	−2.925	$Fe^{+++} + 3e^- \rightleftharpoons Fe$	−0.016
$Cs^+ + e^- \rightleftharpoons Cs$	−2.923	$2H^+ + 2e^- \rightleftharpoons H_2$	0
$Ba^{++} + 2e^- \rightleftharpoons Ba$	−2.906	$AgBr + e^- \rightleftharpoons Ag + Br^-$	0.073
$Sr^{++} + 2e^- \rightleftharpoons Sr$	−2.899	$HgO + H_2O + 2e^- \rightleftharpoons$	
$Ca^{++} + 2e^- \rightleftharpoons Ca$	−2.868	$Hg + 2OH^-$	0.098
$Na^+ + e^- \rightleftharpoons Na$	−2.714	$Sn^{4+} + 2e^- \rightleftharpoons Sn^{++}$	0.154
$Mg^{++} + 2e^- \rightleftharpoons Mg$	−2.357	$Cu^{++} + e^- \rightleftharpoons Cu^+$	0.161
$Al^{+++} + 3e^- \rightleftharpoons Al$	−1.677	$AgCl + e^- \rightleftharpoons Ag + Cl^-$	0.222
$Mn(OH)_2 + 2e^- \rightleftharpoons$		$Cu^{++} + 2e^- \rightleftharpoons Cu$	0.340
$Mn + 2OH$	−1.557	$Fe(CN)_6^= + e^- \rightleftharpoons Fe(CN)_6^{4-}$	0.356
$ZnS + 2e^- \rightleftharpoons Zn + S^=$	−1.488	$Cu^+ + e^- \rightleftharpoons Cu$	0.518
$Zn(OH)_2 + 2e^- \rightleftharpoons Zn + 2OH^-$	−1.247	$I_2 + 2e^- \rightleftharpoons 2I^-$	0.535
$Mn^{++} + 2e^- \rightleftharpoons Mn$	−1.182	$Cu^{++} + Cl^- + e^- \rightleftharpoons CuCl$	0.561
$FeS + 2e^- \rightleftharpoons Fe + S^=$	−0.965	$2H^+ + C_2H_2 + 2e^- \rightleftharpoons C_2H_4$	0.731
$H_2O + SO_4^= + 2e^- \rightleftharpoons$		$Fe^{+++} + e^- \rightleftharpoons Fe^{++}$	0.770
$SO_3^= + 2OH^-$	−0.936	$Hg_2^{++} + 2e^- \rightleftharpoons 2Hg$	0.796
$Fe(OH)_2 + 2e^- \rightleftharpoons Fe + 2OH^-$	−0.891	$Ag^+ + e^- \rightleftharpoons Ag$	0.799
$2H_2O + 2e^- \rightleftharpoons H_2 + 2OH^-$	−0.828	$2Hg^{++} + 2e^- \rightleftharpoons Hg_2^{++}$	0.908
$Zn^{++} + 2e^- \rightleftharpoons Zn$	−0.762	$4H^+ + PbO_2 + 2e^- \rightleftharpoons$	
$HgS + 2e^- \rightleftharpoons Hg + S^=$	−0.707	$Pb^{++} + 2H_2O$	1.000
$S + 2e^- \rightleftharpoons S^=$	−0.444	$Br_2 + 2e^- \rightleftharpoons 2Br^-$	1.078
$Hg_2Cl_2 + 2e^- \rightleftharpoons 2Hg + 2Cl^-$	−0.412	$MnO_2 + H_2O + 2e^- \rightleftharpoons$	
$Fe^{++} + 2e^- \rightleftharpoons Fe$	−0.409	$Mn(OH)_2 + 2OH^-$	1.178
$Cd^{++} + 2e^- \rightleftharpoons Cd$	−0.402	$4H^+ + O_2 + 2e^- \rightleftharpoons 2H_2O$	1.229
$PbSO_4 + 2e^- \rightleftharpoons Pb + SO_4^=$	−0.355	$Cl_2 + 2e^- \rightleftharpoons 2Cl^-$	1.360
$Co^{++} + 2e^- \rightleftharpoons Co$	−0.282	$Co^{+++} + e^- \rightleftharpoons Co^{++}$	1.951
$Ni^{++} + 2e^- \rightleftharpoons Ni$	−0.236	$2H^+ + O_3 + 2e^- \rightleftharpoons O_2 + H_2O$	2.075
$AgCN + e^- \rightleftharpoons Ag + CN^-$	−0.160	$F_2 + 2e^- \rightleftharpoons 2F^-$	2.890

EXERCISE:

Consider the following electrolytic cell:

$$Zn(c)\,|\,Zn^{++}SO_4^{=}(m_1)\,|\,Hg_2SO_4(c)\,|\,Hg(liq)\,|\,Hg_2SO_4(c)\,|\,Cu^{++}SO_4^{=}(m_1)\,|\,Cu(c)$$

(A) Write the over-all cell reaction, and set up the expression for the electromotive force of the cell.

(B) From standard electrode potential data at 25 °C, calculate the value for the equilibrium constant

$$\frac{a_{Zn^{++}}a_{Cu}}{a_{Cu^{++}}a_{Zn}} = \frac{a_{Zn^{++}}}{a_{Cu^{++}}} = K$$

EXERCISE:

From electrode potential data given in Table 20-1 calculate the standard Gibbs energy change in calories for the reaction

$$Cl_2(g) + H_2O(liq) \rightleftarrows H^{+}(aq) + Cl^{-}(aq) + HClO(aq)$$

Problems

1. The molal freezing-point lowerings of aqueous solutions of potassium chloride are summarized in the following table:

m	$\Delta T_f/m$
0.001	3.66
0.005	3.648
0.01	3.610
0.02	3.566
0.05	3.503
0.10	3.451
0.20	3.394
0.50	3.314

Calculate the mean-activity coefficient, $\gamma_{\pm}$, of potassium chloride for $m = 0.10$ and 0.50 by use of equation 20.19. You may assume that $\alpha = 0.001$ per degree for water.

2. The molal freezing-point lowerings of aqueous solutions of lead nitrate are summarized in the following table:

m	$\Delta T_f/m$
0.001	5.368
0.005	5.090
0.01	4.898
0.02	4.657
0.05	4.276
0.10	3.955
0.20	3.560
0.50	2.940

Calculate the mean-activity coefficient, $\gamma_\pm$, of lead nitrate for $m = 0.10$ and 0.50 by use of equation 20.19. As indicated in Problem 1, $\alpha = 0.001$ per degree for water.

3. If Π is the osmotic pressure of a solution of an electrolyte, show that

$$\ln \gamma_\pm = -X - \int_0^m Xd \ln m$$

if

$$X = 1 - \frac{\Pi V_A 1,000}{vm\, M_A\, RT}$$

(Assume that the partial molar volume of the solvent is independent of concentration. Compare with Prob. 14, Chap. Nineteen.)

4. (A) Taking into account the dependence of volume on composition, show that equation 20.38 should, more precisely, be

$$\mu_B(\text{el}) = \frac{G(\text{el})}{n_B}\left(1 + \frac{n_A V_A}{2V}\right)$$

   (B) Show that the partial molar electrical chemical potential of the solvent should be

$$\mu_A(\text{el}) = -\frac{G(\text{el})V_A}{2V}$$

   Observe that, if the contribution of the solute to the volume of solution can be neglected, the following equation holds approximately:

$$\mu_A(\text{el}) = -\frac{G(\text{el})}{2n_A}$$

5. The solubility of a certain salt ($C^{++}A^=$) is equal to 0.02 mole per liter of water. If you assume the validity of the limiting Debye-Hückel Law for ion activities, what is the value of $\Delta\bar{\mu}^0$ for the process

$$CA(c) \rightleftarrows \underbrace{C^{++} + A^=}_{\text{dissolved}}$$

Let the pure solid represent the standard state of the solid, and use the molar concentration as the reference function for ion activity.

6. Neglecting the coefficient of thermal expansion of a dilute solution of an electrolyte, and assuming that the dielectric constant of the solvent is independent of temperature, show that $H(\text{el})$ and $S(\text{el})$ are given by the expressions

$$H(\text{el}) = \tfrac{3}{2}G(\text{el})$$

$$S(\text{el}) = G(\text{el})/2T$$

Since the electrical Gibbs energy is negative, the electrical entropy is also negative. What would this mean about the degree of solute order in a solution of an electrolyte compared with that of an otherwise equivalent solution of uncharged particles? Explain.

Actually the dielectric constant, D, is quite markedly dependent on temperature. Taking this into account show that

$$H(\text{el}) = \frac{3G(\text{el})}{2}\left(1 + \frac{T}{D}\frac{\partial D}{\partial T}\right)$$

and

$$S(\text{el}) = \frac{G(\text{el})}{2T}\left(1 + \frac{3T}{D}\frac{\partial D}{\partial T}\right)$$

The magnitude of $\partial D/\partial T$ is such that $S(\text{el})$ now turns out to be positive. What does this mean with respect to the effect of ionic charge on the degree of order in the solvent?

7. Our derivation of the electrical potential in the neighborhood of an ion tacitly involved the assumption that the ions are point particles. Suppose, however, that a pair of oppositely charged ions cannot come nearer to each other than some distance, a, between their centers (a would equal the sum of the ionic radii). What effect would this have on the theory?

The effect of ionic size can be determined approximately as follows. For $r > a$, assume once more the validity of equation 20.32. For $r < a$, however, write that

$$\psi = \frac{z_+ e}{Dr} + B$$

B, a constant, is the effect of the environment and $z_+ e/Dr$ is the potential attributable to the central ion. You can determine B and the constant C, of equation 20.32, by equating, at the point $r = a$, the two expressions for ψ and the corresponding two expressions for the first partial derivatives of ψ with respect to r. By matching ψ, you will obtain

$$\frac{C_1 e^{-\kappa a}}{a} = \frac{z_+ e}{Da} + B$$

By differentiating the two expressions for ψ and setting $r = a$, you will obtain another condition to be fulfilled. By eliminating C_1 and B, show that, for $r \lessgtr a$,

$$\psi = \frac{z_+e}{Dr} - \frac{z_1 e\kappa}{D(1 + \kappa a)}$$

It will be seen that the effect of finite radius is to diminish the environment potential by the factor $1/(1 + \kappa a)$ or, in other words, to increase the radius of the hypothetical shell of ions from $1/\kappa$ to $1/\kappa + a$. Although the rest of the derivation is somewhat involved, it can be shown further that equation 20.40 would be replaced by

$$\ln \gamma_\pm = -\frac{z_+z_-e^2\kappa}{2DkT(1 + \kappa a)}$$

8. Consider the cell

$$\text{Zn(c)} \mid \text{Zn}^{++}2\text{Cl}^- \ (m = 0.555) \mid \text{AgCl(c)} \mid \text{Ag(c)}$$

(A) What is the cell reaction?

(B) At 0 °C the electromotive force is 1.015 V, and at 25 °C it is 1.005 V. Assuming that $\Delta \tilde{H}$ is constant, calculate in calories the value of $\Delta \tilde{H}$ for the cell reaction.

9. The electromotive force of the cell

$$\text{Zn(c)} \mid \text{Zn}^{++}2\text{Cl}^- \ (m) \mid \text{AgCl(c)} \mid \text{Ag(c)}$$

at 25 °C is given for various values of the molality, m, in the following table:*

m	E
0.002941	1.1983
0.007814	1.16502
0.01236	1.14951
0.02144	1.13101
0.04242	1.10897

(A) Show that for this cell

$$E = E^\circ - \frac{RT}{2F} \ln 4m^3\gamma_\pm^3$$

(B) Hence show that by extrapolating $E + (3RT/2F) \ln m$ to $m = 0$, one should be able to calculate E°. By plotting $E + (3RT/2F) \ln m$ against $\sqrt{m}$ (or by any other means that seem desirable), evaluate E° in volts.

(C) Using the result of part B, evaluate $\gamma_\pm$ for ZnCl_2 for the five molalities given in the table. (Use the molality as the reference function for activity.)

---

* Scatchard and Tefft, *J. Am. Chem. Soc.*, **52**, 2272 (1930).

10. The equilibrium partial pressures of HCl above certain aqueous solutions of molality m at 25 °C are as follows:

m (moles of HCl per kg of H_2O)	p (HCl)
4.0	0.0182 Torr
5.0	0.0530

(A) Assuming that the cation transference number is 0.828, calculate the electromotive force of the following cell:

$$Pt, H_2(g) \mid H^+Cl^- \mid H^+Cl^- \mid H_2(g), Pt$$
$$m = 4.0 \quad m = 5.0$$

(B) Calculate the electromotive force of the following cell:

$$Ag(c) \mid AgCl(c), H^+Cl^- \mid H_2(g), Pt, H_2(g) \mid H^+Cl^-, AgCl(c) \mid Ag(c)$$
$$m = 4.0 \qquad\qquad m = 5.0$$

11. At 1 atm pressure, white tin and gray tin are in equilibrium at 18 °C. The heat of transition of white tin to gray tin is -0.50 kcal mol^{-1}.

(A) Estimate the electromotive force in volts for the following cell at 0 °C and at 25 °C.

$$Sn(c) \text{ (white)} \mid Sn^{++}2Cl^- \text{ (aq)} \mid Sn(c) \text{ (gray)}$$

(B) The densities of white and gray at 18 °C are 7.28 and 5.75 g cm^{-3}. Calculate $(\partial E/\partial p)_T$ in V atm^{-1} for the same cell at 18 °C and 1 atm.

Exact and Inexact Differentials; Line Integrals

I. Theorem: For the differential expression $M(T,V)dT + N(T,V)dV$ to be exact, it is necessary and sufficient that $(\partial M/\partial V) = (\partial N/\partial T)$.

(A) To prove the *necessary* part of the theorem, suppose that $MdT + NdV$ is an exact differential, let us say df, with f a function of T and V. Then

$$M = \frac{\partial f}{\partial T} \text{ and } N = \frac{\partial f}{\partial V} \qquad (A.1)$$

and

$$\frac{\partial M}{\partial V} = \frac{\partial^2 f}{\partial V \partial T} = \frac{\partial^2 f}{\partial T \partial V} = \frac{\partial N}{\partial T} \qquad (A.2)$$

The condition is therefore necessary.

(B) To prove the *sufficient* part of the theorem, suppose that $(\partial M/\partial V) = (\partial N/\partial T)$. Now consider a function, g, defined as $g = \int MdT$, with the integration carried out on the assumption that V is constant. It follows that

$$\frac{\partial g}{\partial T} = M \qquad (A.3)$$

Moreover,

$$\frac{\partial^2 g}{\partial T \partial V} = \frac{\partial^2 g}{\partial V \partial T} = \frac{\partial M}{\partial V} = \frac{\partial N}{\partial T} \qquad (A.4)$$

Hence

$$\frac{\partial}{\partial T}\left(N - \frac{\partial g}{\partial V}\right) = 0 \tag{A.5}$$

Therefore $N - \partial g/\partial V$ must be a function of V only; let us say $\phi(V)$. We can write, then, that

$$N - \frac{\partial g}{\partial V} = \phi(V) \tag{A.6}$$

Now let us define a function, $f(T,V)$, as follows:

$$f = g + \int \phi(V)dV \tag{A.7}$$

Then

$$N = \frac{\partial f}{\partial V} \tag{A.8}$$

But, since g and f differ only by a function of V, we can also assert from the definition of g that

$$M = \frac{\partial f}{\partial T} \tag{A.9}$$

Therefore

$$MdT + NdV = \frac{\partial f}{\partial T}dT + \frac{\partial f}{\partial V}dV = df \tag{A.10}$$

Thus we see that $MdT + NdV$ is an exact differential—namely, df; so the condition originally specified is *sufficient*.

II. (A) Green's Theorem in Two Dimensions: Suppose that $M(T,V)$ and $\partial M/\partial V$ are continuous functions of T and V within and on the bound-

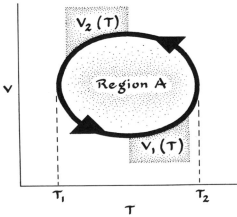

ary of some region, A. (See the accompanying figure.) Now consider the double integral

$$\iint\limits_{A} \frac{\partial M}{\partial V} \, dA$$

This integral can be evaluated by iterated integrals as follows:

$$\iint\limits_{A} \frac{\partial M}{\partial V} \, dA = \int_{T_1}^{T_2} \left\{ \int_{V_1(T)}^{V_2(T)} \frac{\partial M}{\partial V} \, dV \right\} dT$$

$$= \int_{T_1}^{T_2} [M(T,V_2) - M(T,V_1)] dT \qquad (A.11)$$

Examination of the latter integral will disclose that it is precisely equal to the line integral of MdT integrated in a clockwise direction about the boundary of region A. If we use the *counter*-clockwise direction to denote integration in the positive sense,* it is evident that

$$\iint\limits_{A} \frac{\partial M}{\partial V} \, dA = - \oint MdT \qquad (A.12)$$

Let us similarly consider functions $N(T,V)$ and $\partial N/\partial T$, both assumed to be continuous within and on the boundary of the same region, A. It can be shown that

$$\iint\limits_{A} \frac{\partial N}{\partial T} \, dA = \oint NdV \qquad (A.13)$$

From equations A.12 and A.13 it will be seen that

$$\iint\limits_{A} \left(\frac{\partial M}{\partial V} - \frac{\partial N}{\partial T} \right) dA = - \oint MdT + NdV \qquad (A.14)$$

Equation A.14 is known as Green's theorem
(B) From Green's theorem we see that, if

$$\frac{\partial M}{\partial V} = \frac{\partial N}{\partial T} \qquad (A.15)$$

throughout the region of interest, then

$$\oint MdT + NdV = 0 \qquad (A.16)$$

In other words, the line integral about a cycle is equal to zero. This is equivalent to saying that the line integral between any two points in the region is independent of the path. It can be shown, conversely, that, if the line integral

---

* This is the usual convention adopted by mathematicians. In thermodynamics, however, we use the reverse convention when calculating the work attending a cyclical process, so that we evaluate W by integrating pdV in a clockwise sense. We employ the mathematicians' convention in this appendix to avoid confusion should the reader seek further mathematical explanation.

$$\int_C MdT + NdV$$

is independent of the path, then $(\partial M/\partial V) = (\partial N/\partial T)$ and $MdT + NdV$ must be an exact differential—that is, the differential of some function $f(T,V)$.

Some Useful Integrals

$$\int_0^\infty e^{-a^2x^2}\,dx = \frac{\sqrt{\pi}}{2a}$$

$$\int_0^\infty x^2 e^{-a^2x^2}\,dx = \frac{\sqrt{\pi}}{4a^3}$$

$$\int_0^\infty x^4 e^{-a^2x^2}\,dx = \frac{3\sqrt{\pi}}{8a^5}$$

$$\int_0^\infty x^{2n} e^{-a^2x^2}\,dx = \frac{(2n-1)(2n-3)\ldots 5\cdot 3\cdot 1\sqrt{\pi}}{2^{n+1}a^{2n+1}}$$

$$\int_0^\infty x e^{-a^2x^2}\,dx = \frac{1}{2a^2}$$

$$\int_0^\infty x^3 e^{-a^2x^2}\,dx = \frac{1}{2a^4}$$

$$\int_0^\infty x^{2n+1} e^{-a^2x^2}\,dx = \frac{n!}{2a^{2n+2}}$$

$$\int_0^x e^{-x^2}\,dx = \frac{\sqrt{\pi}}{2} - \frac{e^{-x^2}}{2x}\left(1 - \frac{1}{2x^2} + \frac{1\cdot 3}{4x^4} - \frac{1\cdot 3\cdot 5}{8x^6} + \cdots\right)$$

(asymptotic for large x)

Table of Some Useful Constants and Conversion Factors

Avogadro's number	$N_A = 6.022\ 17 \times 10^{23}\ \text{mol}^{-1}$
Electronic charge	$e = 1.602\ 19 \times 10^{-19}\ \text{C}$
	$= 4.803\ 25 \times 10^{-10}\ \text{cm}^{3/2}\ \text{g}^{1/2}\ \text{s}^{-1}\ (\text{esu})$
Planck's constant	$h = 6.626\ 20 \times 10^{-27}\ \text{erg s}$
Speed of light	$c = 2.997\ 925 \times 10^{10}\ \text{cm s}^{-1}$
Boltzmann's constant	$k = 1.380\ 62 \times 10^{-16}\ \text{erg K}^{-1}$
Stefan-Boltzmann constant	$\sigma = 5.669\ 61 \times 10^{-5}\ \text{erg cm}^{-2}\ \text{s}^{-1}\ \text{K}^{-4}$
Acceleration of gravity	$g = 980.665\ \text{cm s}^{-2}$
Standard atmosphere	$1\ \text{atm} = 1\ 013\ 250\ \text{dyn cm}^{-2}$
	$= 1\ 033.227\ \text{g cm}^{-2}$
	$= 760\ \text{Torr (mm Hg)}$
Calorie (defined)	$1\ \text{cal} = 4.184\ 0\ \text{J}$
	$= 0.041\ 293\ \text{liter atm}$
Ice point	$0\ °\text{C} = 273.15\ \text{K}$
Molar volume perfect gas At 0 °C and 1 atm	$V_m = 22\ 413.6\ \text{cm}^3\ (\text{ml})$
Gas constant	$R = 8.314\ 34\ \text{J K}^{-1}\ \text{mol}^{-1}$
	$= 1.987\ 17\ \text{cal K}^{-1}\ \text{mol}^{-1}$
	$= 0.082\ 056\ 1\ \text{liter atm K}^{-1}\ \text{mol}^{-1}$
	$\ln x = 2.302\ 585\ \lg x$
	$R \ln x = 4.575\ 63\ \lg x\ \text{cal K}^{-1}\ \text{mol}^{-1}$

At 0 °C
$$RT = 2\ 271.06\ \text{J mol}^{-1}$$
$$= 542.795\ \text{cal mol}^{-1}$$
$$= 22.413\ 6\ \text{liter atm mol}^{-1}$$

At 25 °C
$$RT = 2\ 478.92\ \text{J mol}^{-1}$$
$$= 592.475\ \text{cal mol}^{-1}$$
$$= 24.465\ 0\ \text{liter atm mol}^{-1}$$
$$RT \ln x = 1\ 364.22\ \lg x\ \text{cal mol}^{-1}$$

Faraday constant
$$F = 96\ 486.7\ \text{C}$$

$$\frac{R}{F} \ln x = 0.000\ 198\ 415\ \lg x\ \text{V K}^{-1}$$

At 25 °C
$$\frac{RT}{F} \ln x = 0.059\ 157\ 5\ \lg x\ \text{V}$$

Table of Thermodynamic Data

The following table contains thermodynamic data for a number of selected elements, ions, and compounds. Most of the data have been taken from Technical Notes published by the U.S. National Bureau of Standards, although other sources of data have also been used. (See Appendix F-IV.) It should be emphasized that the reliability and usefulness of thermodynamic calculations depend in large measure on the internal consistency of the data. Such internal consistency is not always realized in the following table, so persons interested in precise calculations are advised to look into new bulletins as they are released.

The order followed by the elements (O, H, He, Ne, Ar, Kr, Xe, F, Cl, Br, I, S, N, etc.) is that employed in the National Bureau of Standards publications. After oxygen and hydrogen, the order is substantially that of going down the columns of the periodic table, starting with the right hand column and moving toward the left. A compound is listed under that one of its elements which appears last in the specified order. For example, H_2O is listed under hydrogen, and HNO_3 is listed under nitrogen.

The data include standard enthalpies of formation (ΔH_f°) and standard Gibbs free energies (or chemical potentials) of formation (ΔG_f° or $\Delta \mu_f^\circ$) from the elements in their standard states, all for 25 °C. Also included are standard entropies (S°), based on the Third Law, and heat capacities at constant pressure (C_p°), likewise at 25 °C. The values for ions in aqueous solution, such as $Na^+(aq)$, are based on the convention that the properties listed for $H^+(aq)$ are equal to 0.

By adding values for neutral combinations of oppositely charged ions, one obtains logically correct values for salts in aqueous solution. The heats of forma-tion of gaseous ions are based on the convention that the heat of formation of the electron gas is equal to 0. (On the other hand, the entropy and heat capacity of an ideal electron gas are not assumed to be equal to 0.)

The standard state for ions in solution is based on the use of molality (con-centration in moles per kilogram of solvent) as the reference function. Accord-ingly, the standard state is a hypothetical state of unit molality behaving thermally as if it were infinitely dilute. The standard state for gases is similarly chosen to correspond to a gas behaving ideally at a fugacity of 1 atmosphere.

THERMODYNAMIC DATA FOR 298.15 K (25 °C)

Substance and State	ΔH_f° kcal mol^{-1}	$\Delta G_f^\circ(\Delta \mu_f^\circ)$ kcal mol^{-1}	S° cal deg^{-1} mol^{-1}	C_p° cal deg^{-1} mol^{-1}
Oxygen				
O(g)	59.553	55.389	38.467	5.237
O$_2$(g)	0	0	49.003	7.016
O$_3$(g)	34.1	39.0	57.08	9.37
Hydrogen				
H(g)	52.095	48.581	27.391	4.9679
H$^+$(g)	367.161	362.6	26.01	4.968
H$^+$(aq)	0	0	0	0
H$^-$(g)	33.20	31.58	26.015	4.968
H$_2$(g)	0	0	31.208	6.889
^{2}H$_2$(g)	0	0	34.620	6.978
^{1}H^{2}H(g)	0.076	−0.350	34.343	6.978
H$_2^+$(g)	357.23			
OH$^-$(g)	−33.67	−33.25	41.23	6.965
OH$^-$(aq)	−54.970	−37.594	−2.57	−35.5
H$_2$O(liq)	−68.315	−56.687	16.71	17.995
^{2}H$_2$O(liq)	−70.411	−58.195	18.15	20.16
^{1}H^{2}HO(liq)	−69.285	−57.817	18.95	
H$_2$O(g)	−57.796	−54.634	45.104	8.025
^{2}H$_2$O(g)	−59.560	−56.059	47.378	8.19
^{1}H^{2}HO(g)	−58.628	−55.719	47.658	8.08
H$_2$O$_2$(liq)	−44.88	−28.78	26.2	21.3
H$_2$O$_2$(g)	−32.58	−25.24	55.6	10.3
H$_2$O$_2$(aq, undissoc.)	−45.69	−32.05	34.4	
Helium				
He(g)	0	0	30.1244	4.9679

Substance and State	ΔH_f° kcal mol^{-1}	$\Delta G_f^\circ(\Delta\mu_f^\circ)$ kcal mol^{-1}	S° cal deg^{-1} mol^{-1}	C_p° cal deg^{-1} mol^{-1}
Neon				
Ne(g)	0	0	34.9471	4.9679
Argon				
Ar(g)	0	0	36.9822	4.9679
Krypton				
Kr(g)	0	0	39.1905	4.9679
Xenon				
Xe(g)	0	0	40.5290	4.9679
Fluorine				
F(g)	18.88	14.80	37.917	5.436
F$^-$(g)	−64.7			
F$^-$(aq)	−79.50	−66.64	−3.3	−25.5
F$_2$(g)	0	0	48.44	7.48
HF(g)	−64.8	−65.3	41.508	6.963
HF(aq, undissoc.)	−76.50	−70.95	21.2	
HF$_2^-$(aq)	−155.34	−138.18	22.1	
Chlorine				
Cl(g)	29.082	25.262	39.457	5.220
Cl$^-$(g)	−58.8			
Cl$^-$(aq)	−39.952	−31.372	13.5	−32.6
Cl$_2$(g)	0	0	53.288	8.104
HCl(g)	−22.062	−22.777	44.646	6.96
Bromine				
Br(g)	26.741	19.701	41.805	4.968
Br$^-$(g)	−55.9			
Br$^-$(aq)	−29.05	−24.85	19.7	−33.9
Br$_2$(liq)	0	0	36.384	18.090
Br$_2$(g)	7.387	0.751	58.641	8.61
HBr(g)	−8.70	−12.77	47.463	6.965
BrF(g)	−22.43	−26.09	54.70	7.88
BrCl(g)	3.50	−0.23	57.36	8.36
Iodine				
I(g)	25.535	16.798	43.184	4.968
I$^-$(g)	−47.0			
I$^-$(aq)	−13.19	−12.33	26.6	−34.0
I$_2$(c)	0	0	27.757	13.011
I$_2$(g)	14.923	4.627	62.28	8.82
I$_3^-$(aq)	−12.3	−12.3	57.2	

Substance and State	ΔH_f° kcal mol^{-1}	$\Delta G_f^\circ(\Delta\mu_f^\circ)$ kcal mol^{-1}	S° cal deg^{-1} mol^{-1}	C_p° cal deg^{-1} mol^{-1}
HI(g)	6.33	0.41	49.351	6.969
IF(g)	−22.86	−28.32	56.42	7.99
ICl(g)	4.25	−1.30	59.140	8.50
IBr(g)	9.76	0.89	61.822	8.71
Sulfur				
S(c, rhombic)	0	0	7.60	5.41
S(c, monoclinic)	0.08			
S(g)	66.636	56.951	40.084	5.658
S$^=$(aq)	7.9	20.5	−3.5	
S$_8$(g)	24.45	11.87	102.98	37.39
SO$_2$(g)	−70.944	−71.748	59.30	9.53
SO$_2$(aq, undissoc.)	−77.194	−71.871	38.7	
SO$_3$(g)	−94.58	−88.69	61.34	12.11
SO$_3^=$(aq)	−151.9	−116.3	−7.	
SO$_4^=$(aq)	−217.32	−177.97	4.8	−70.
HS$^-$(aq)	−4.2	2.88	15.0	
H$_2$S(g)	−4.93	−8.02	49.16	8.18
H$_2$S(aq)	−9.5	−6.66	29.	
HSO$_3^-$(aq)	−149.67	−126.15	33.4	
HSO$_4^-$(aq)	−212.08	−180.69	31.5	−20.
H$_2$SO$_3$ (aq, undissoc.)	−145.51	−128.56	55.5	
H$_2$SO$_4$(liq)	−194.548	−164.938	37.501	33.20
SF$_6$(g)	−289.	−264.2	69.72	23.25
Nitrogen				
N(g)	112.979	108.886	36.613	4.968
N$_2$(g)	0	0	45.77	6.961
N$_3^-$(aq)	65.76	83.2	25.8	
NO(g)	21.57	20.69	50.347	7.133
NO$_2$(g)	7.93	12.26	57.35	8.89
NO$_3^-$(aq)	−49.56	−26.61	35.0	−20.7
N$_2$O(g)	19.61	24.90	52.52	9.19
N$_2$O$_4$(g)	2.19	23.38	72.70	18.47
N$_2$O$_5$(g)	2.7	27.5	85.0	20.2
NH$_3$(g)	−11.02	−3.94	45.97	8.38
NH$_3$(aq, undissoc.)	−19.19	−6.35	26.6	
NH$_4^+$(aq)	−31.67	−18.97	27.1	19.1
N$_2$H$_4$(g)	22.80	38.07	56.97	11.85
HN$_3$(g)	70.3	78.4	57.09	10.44
HN$_3$(aq)	62.16	76.9	34.9	
HNO$_3$(liq)	−41.61	−19.31	37.19	26.26

Substance and State	ΔH_f° kcal mol^{-1}	$\Delta G_f^\circ(\Delta\mu_f^\circ)$ kcal mol^{-1}	S° cal deg^{-1} mol^{-1}	C_p° cal deg^{-1} mol^{-1}
HNO$_3$(g)	-32.28	-17.87	63.64	12.75
NH$_2$OH(aq)	-23.5			
NH$_2$OH$_2{}^+$(aq)	-32.8			
NH$_4$OH				
(aq, undissoc.)	-87.505	-63.04	43.3	
NH$_4$NO$_3$(c)	-87.37	-43.98	36.11	33.3
N$_2$H$_5$OH(g)	-49.0	-18.9	63.	
NH$_4$F(c)	-110.89	-83.36	17.20	15.60
NH$_4$Cl(c)	-75.15	-48.51	22.6	20.1
NH$_4$Br(c)	-64.73	-41.9	27.	23.
NH$_4$I(c)	-48.14	-26.9	28.	
(NH$_4$)$_2$SO$_4$(c)	-282.23	-215.56	52.6	44.81
Phosphorus				
P(c, white)	0	0	9.82	5.698
P(c, red, triclinic)	-4.2	-2.9	5.45	5.07
P(g)	75.20	66.51	38.978	4.968
P$_4$(g)	14.08	5.85	66.89	16.05
PO$_3{}^-$(aq)	-233.5			
PO$_4{}^\equiv$(aq)	-305.3	-243.5	$-53.$	
P$_2$O$_7{}^{4-}$(aq)	-542.8	-458.7	$-28.$	
P$_4$O$_6$(c)	-392.0			
P$_4$O$_{10}$(c)	-713.2	-644.8	54.70	50.60
PH$_3$(g)	1.3	3.2	50.22	8.87
HPO$_3$(c)	-226.7			
HPO$_3{}^=$(aq)	-231.6			
HPO$_4{}^=$(aq)	-308.83	-260.34	-8.0	
H$_2$PO$_3{}^-$(aq)	-231.7			
H$_2$PO$_4{}^-$(aq)	-309.82	-270.17	21.6	
H$_3$PO$_3$(c)	-230.5			
H$_3$PO$_4$(c)	-305.7	-267.5	26.41	25.35
H$_3$PO$_4$				
(aq, undissoc.)	-307.92	-273.10	37.8	
H$_4$P$_2$O$_7$(c)	-535.6			
H$_4$P$_2$O$_7$				
(aq, undissoc.)	-542.2	-485.7	64.	
PF$_3$(g)	-219.6	-214.5	65.28	14.03
PCl$_3$(g)	-68.6	-64.0	74.49	17.17
PCl$_5$(g)	-89.6	-72.9	87.11	26.96
POCl$_3$(g)	-133.48	-122.60	77.76	20.30
PBr$_3$(g)	-33.3	-38.9	83.17	18.16

Substance and State	ΔH_f° kcal mol^{-1}	$\Delta G_f^\circ(\Delta\mu_f^\circ)$ kcal mol^{-1}	S° cal deg^{-1} mol^{-1}	C_p° cal deg^{-1} mol^{-1}
Arsenic				
As(c, gray)	0	0	8.4	5.89
As(g)	72.3	62.4	41.61	4.968
As$_4$(g)	34.4	22.1	75.	
AsO$_4$$^\equiv$(aq)	−212.27	−155.00	−38.9	
As$_2$O$_5$(c)	−221.05	−187.0	25.2	27.85
As$_4$O$_6$				
(c, octahedral)	−314.04	−275.46	51.2	45.72
As$_4$C$_6$(g)	−289.0	−262.4	91.	
AsH$_3$(g)	15.88	16.47	53.22	9.10
HAsO$_4$$^=$				
(aq, undissoc.)	−216.62	−170.82	−0.4	
H$_2$AsO$_4$$^-$				
(aq, undissoc.)	−217.39	−180.04	28.	
H$_3$AsO$_4$(c)	−216.6			
H$_3$AsO$_4$				
(aq, undissoc.)	−215.7	−183.1	44.	
AsF$_3$(g)	−220.04	−216.46	69.07	15.68
AsCl$_3$(liq)	−72.9	−62.0	51.7	
AsCl$_3$(g)	−62.5	−59.5	78.17	18.10
AsBr$_3$(g)	−31.	−38.	86.94	18.92
Antimony				
Sb(c)	0	0	10.92	6.03
Sb(g)	62.7	53.1	43.06	4.97
SbO$^+$(aq)		−42.33		
SbO$_2$$^-$(aq)		−81.32		
Sb$_2$O$_5$(c)	−232.3	−198.2	29.9	
SbH$_3$(g)	34.681	35.31	55.61	9.81
HSbO$_2$				
(aq, undissoc.)	−116.6	−97.4	11.1	
H$_3$SbO$_4$(aq)	−216.8			
SbCl$_3$(c)	−91.34	−77.37	44.0	25.8
SbCl$_3$(g)	−75.0	−72.0	80.71	18.33
SbCl$_5$(g)	−94.25	−79.91	96.04	28.95
SbOCl(g)	−25.5			
SbBr$_3$(g)	−46.5	−53.5	89.09	19.17
Bismuth				
Bi(c)	0	0	13.56	6.10
Bi(g)	49.5	40.2	44.669	4.968
Bi^{+++}(g)	1196.4			

Substance and State	ΔH_f° kcal mol^{-1}	$\Delta G_f^\circ(\Delta \mu_f^\circ)$ kcal mol^{-1}	S° cal deg^{-1} mol^{-1}	C_p° cal deg^{-1} mol^{-1}
Bi_2O_3(c)	-137.16	-118.0	36.2	27.13
$Bi(OH)_3$(c)	-170.0			
$BiCl_3$(g)	-63.5	-61.2	85.74	19.04
$BiOCl$(c)	-87.7	-77.0	28.8	
Carbon				
C(c, graphite)	0	0	1.372	2.038
C(c, diamond)	0.4533	0.6930	0.568	1.4615
C(g)	171.291	160.442	37.7597	4.9805
CO(g)	-26.416	-32.780	47.219	6.959
CO_2(g)	-94.051	-94.254	51.06	8.87
CO_2(aq, undissoc.)	-98.90	-92.26	28.1	
$CO_3^=$(aq)	-161.84	-126.17	-13.6	
CH_4(g)	-17.88	-12.13	44.492	8.439
$HCOO^-$(aq)	-101.71	-83.9	22.	-21.0
HCO_3^-(aq)	-165.39	-140.26	21.8	
HCHO(g)	$-28.$	$-27.$	52.26	8.46
HCOOH(liq)	-101.51	-86.38	30.82	23.67
HCOOH(g)	-90.48			
HCOOH				
(aq, un-ionized)	-101.68	-89.0	39.	
H_2CO_3(aq)	-167.22	-184.94	44.8	
CH_3OH(liq)	-57.04	-39.76	30.3	19.5
CH_3OH(g)	-47.96	-38.72	57.29	10.49
CH_3OH(aq)	-58.779	-41.92	31.8	
CF_4(g)	$-221.$	$-210.$	62.50	14.60
CCl_4(liq)	-32.37	-15.60	51.72	31.49
CCl_4(g)	-24.6	-14.49	74.03	19.91
$COCl_2$(g)	-52.3	-48.9	67.74	13.78
CH_3Cl(g)	-19.32	-13.72	56.04	9.74
CH_2Cl_2(g)	-22.10	-15.75	64.56	12.18
$CHCl_3$(liq)	-32.14	-17.62	48.2	27.2
$CHCl_3$(g)	-24.65	-16.82	70.65	15.70
CBr_4(c)	4.5	11.4	50.8	34.5
CBr_4(g)	19.	16.	85.55	21.79
CH_3Br(g)	-8.4	-6.2	58.86	10.14
CH_3I(g)	3.1	3.5	60.71	10.54
CH_2I_2(g)	27.0	22.9	74.0	13.79
CS_2(liq)	21.44	15.60	36.17	18.1
CS_2(g)	28.05	16.05	56.82	10.85
COS(g)	-33.96	-40.47	55.32	9.92
CN^-(g)	16.00	10.76	46.81	6.963

Substance and State	ΔH_f° kcal mol^{-1}	$\Delta G_f^{\circ}(\Delta \mu_f^{\circ})$ kcal mol^{-1}	S° cal deg^{-1} mol^{-1}	C_p° cal deg^{-1} mol^{-1}
CN$^-$(aq)	36.0	41.2	22.5	
CNO$^-$(aq)	−34.9	−23.3	25.5	
HCN(g)	32.3	29.8	48.20	8.57
HCN				
(aq, un-ionized)	25.6	28.6	29.8	
CH$_3$NH$_2$(g)	−5.49	7.67	58.15	12.7
HCNO				
(aq, un-ionized)	−36.90	−28.0	34.6	
NH$_4$HCO$_3$(c)	−203.0	−159.2	28.9	
CO(NH$_2$)$_2$(c)	−79.56	−47.04	25.00	22.26
CO(NH$_2$)$_2$(aq)	−75.954			
CNF(g)			53.67	9.99
CNCl(g)	32.97	31.32	56.42	10.75
CNS$^-$(aq)	18.27	22.15	34.5	−9.6
C$_2$O$_4$$^=$(aq)	−197.2	−161.1	10.9	
C$_2$H$_2$(g)	54.19	50.00	48.00	10.50
C$_2$H$_4$(g)	12.49	16.28	52.45	10.41
C$_2$H$_6$(g)	−20.24	−7.86	54.85	12.58
HC$_2$O$_4$$^-$(aq)	−195.6	−166.93	35.7	
CH$_2$CO(g)	−14.6	−14.8	59.16	12.37
(COOH)$_2$(c)	−197.7			28.
CH$_3$COO$^-$(aq)	−116.16	−88.29	20.7	−1.5
C$_2$H$_4$O(g)	−12.58	−3.12	57.94	11.45
CH$_3$CHO(g)	−39.72	−30.81	59.8	13.7
CH$_3$COOH(liq)	−115.8	−93.2	38.2	29.7
CH$_3$COOH(g)	−103.31	−89.4	67.5	15.9
CH$_3$COOH				
(aq, un-ionized)	−116.10	−94.78	42.7	
C$_2$H$_5$OH(liq)	−66.37	−41.80	38.4	26.64
C$_2$H$_5$OH(g)	−56.19	−40.29	67.54	15.64
C$_2$H$_5$OH(aq)	−68.9	−43.44	35.5	
(CH$_3$)$_2$O(g)	−43.99	−26.93	63.64	15.39
C$_2$F$_4$(g)	−155.5	−147.2	71.69	19.23
C$_2$F$_6$(g)	−310.	−290.	79.4	25.5
C$_2$Cl$_4$(g)	−2.9	5.4	81.5	22.69
C$_2$Cl$_6$(g)	−33.9	−13.8	95.3	32.7
C$_2$H$_5$Cl(g)	−26.81	−14.45	65.94	15.01
C$_2$H$_5$Br(g)	−15.42	−6.34	68.50	15.42
C$_2$H$_5$I(g)	−1.84	4.57	73.1	16.0
(CN)$_2$(g)	73.84	71.07	57.79	13.58
CH$_3$CN(g)	20.9	25.0	58.67	12.48

Substance and State	ΔH_f° kcal mol^{-1}	$\Delta G_f^\circ(\Delta\mu_f^\circ)$ kcal mol^{-1}	S° cal deg^{-1} mol^{-1}	C_p° cal deg^{-1} mol^{-1}
$C_3H_6(g)$	4.879	14.990	63.80	15.27
$C_3H_8(g)$	-24.820	-5.614	64.51	17.57
$(C_2H_5)_2O(liq)$	-66.82			
$(C_2H_5)_2O(g)$	-60.26			
$C_6H_6(g)$	19.820	30.989	64.34	19.52
Silicon				
$Si(c)$	0	0	4.50	4.78
$Si(g)$	108.9	98.3	40.12	5.318
$SiO_2(c, quartz)$	-217.72	-204.75	10.00	10.62
$SiH_4(g)$	8.2	13.6	48.88	10.24
$Si_2H_6(g)$	19.2	30.4	65.14	19.31
$Si_3H_8(g)$	28.9			
$H_2SiO_3(c)$	-284.1	-261.1	32.	
H_2SiO_3 (aq, undissoc.)	-282.7	-258.0	26.	
$H_4SiO_4(c)$	-354.0	-318.6	46.	
H_4SiO_4 (aq, undissoc.)	-351.0	-314.7	43.	
$SiF_4(g)$	-385.98	-375.88	67.49	17.60
$SiF_6^=(aq)$	-571.0	-525.7	29.2	
$SiCl_4(g)$	-157.03	-147.47	79.02	21.57
$SiBr_4(g)$	-99.3	-103.2	90.29	23.21
$SiC(c, cubic)$	-15.6	-15.0	3.97	6.42
Germanium				
$Ge(c)$	0	0	7.43	5.580
$Ge(g)$	90.0	80.3	40.103	7.345
$GeH_4(g)$	21.7	27.1	51.87	10.76
Tin				
$Sn(c, white)$	0	0	12.32	6.45
$Sn(c, gray)$	-0.50	0.03	10.55	6.16
$Sn(g)$	72.2	63.9	40.243	5.081
$Sn^{++}(aq)$	-2.1	-6.5	$-4.$	
$Sn^{4+}(aq)$	7.3	0.6	$-28.$	
$SnO(c)$	-68.3	-61.4	13.5	10.59
$SnO_2(c)$	-138.8	-124.2	12.5	12.57
$SnH_4(g)$	38.9	45.0	54.39	11.70
$Sn(OH)_2(c)$	-134.1	-117.5	37.	
$Sn(OH)_4(c)$	-265.3			
$SnCl_2(c)$	-77.7			
$SnCl_4(liq)$	-122.2	-105.2	61.8	39.5
$SnCl_4(g)$	-112.7	-103.3	87.4	23.5

Substance and State	ΔH_f° kcal mol^{-1}	$\Delta G_f^\circ(\Delta\mu_f^\circ)$ kcal mol^{-1}	S° cal deg^{-1} mol^{-1}	C_p° cal deg^{-1} mol^{-1}
Lead				
Pb(c)	0	0	15.49	6.32
Pb(g)	46.6	38.7	41.889	4.968
Pb$^+$(g)	219.116			
Pb^{++}(g)	567.25			
Pb^{++}(aq)	-0.4	-5.83	2.5	
PbO(c, yellow)	-51.94	-44.91	16.42	10.94
PbO(c, red)	-52.34	-45.16	15.9	10.95
PbO$_2$(c)	-66.3	-51.95	16.4	15.45
Pb$_3$O$_4$(c)	-171.7	-143.7	50.5	35.1
HPbO$_2^-$(aq)		-80.90		
Pb(OH)$_2$(c)	-123.3			
PbCl$_2$(c)	-85.90	-75.08	32.5	
PbS(c)	-24.0	-23.6	21.8	11.83
PbSO$_4$(c)	-219.87	-194.36	35.51	24.667
Boron				
B(c)	0	0	1.40	2.65
B(g)	134.5	124.0	36.65	4.971
B$_2$O$_3$(c)	-304.20	-285.30	12.90	15.04
B$_2$O$_3$(g)	-201.67	-198.85	66.85	15.98
B$_2$H$_6$(g)	8.5	20.7	55.45	13.60
H$_3$BO$_3$(c)	-261.55	-231.60	21.23	19.45
H$_3$BO$_3$				
(aq, un-ionized)	-256.29	-231.56	38.8	
BF$_3$(g)	-271.75	-267.77	60.71	12.06
BCl$_3$(g)	-96.50	-92.91	69.31	14.99
Aluminum				
Al(c)	0	0	6.77	5.82
Al(g)	78.0	68.3	39.30	5.11
Al^{+++}(aq)	$-127.$	$-116.$	-76.9	
AlO$_2^-$(aq)	-219.6	-196.8	$-5.$	
Al$_2$O$_3$(c, corundum)	-400.5	-378.2	12.17	18.89
AlCl$_3$(c)	-168.3	-150.3	26.45	21.95
Thallium				
Tl(c)	0	0	15.34	6.29
Tl(g)	43.55	35.24	43.225	4.968
Tl$^+$(g)	185.883			
Tl$^+$(aq)	1.28	-7.74	30.0	
Tl^{+++}(aq)	47.0	51.3	$-46.$	
TlOH(c)	-57.1	-46.8	21.	

Substance and State	ΔH_f° kcal mol^{-1}	$\Delta G_f^\circ(\Delta \mu_f^\circ)$ kcal mol^{-1}	S° cal deg^{-1} mol^{-1}	C_p° cal deg^{-1} mol^{-1}
TlF(c)	−77.6			
TlCl(c)	−48.79	−44.20	26.59	12.17
TlCl$_3$(c)	−75.3			
Zinc				
Zn(c)	0	0	9.95	6.07
Zn(g)	31.245	22.748	38.450	4.968
Zn^{++}(g)	665.09			
Zn^{++}(aq)	−36.78	−35.14	−26.8	11.
ZnO(c)	−83.24	−76.08	10.43	9.62
Zn(OH)$_2$(c)	−153.74	−132.68	19.5	17.3
ZnCl$_2$(c)	−99.20	−88.296	26.64	17.05
ZnS(c, sphalerite)	−49.23	−48.11	13.8	11.0
ZnSO$_4$(c)	−234.9	−209.0	28.6	
Cadmium				
Cd(c)	0	0	12.37	6.21
Cd(g)	26.77	18.51	40.066	4.968
Cd^{++}(g)	627.04			
Cd^{++}(aq)	−18.14	−18.542	−17.5	
CdO(c)	−61.7	−54.6	13.1	10.38
Cd(OH)$_2$(c)	−134.0	−113.2	23.	
Cd(OH)$_2$(aq)	−128.08	−93.73	−22.6	
CdCl$_2$(c)	−93.57	−82.21	27.55	17.85
CdSO$_4$(c)	−223.06	−196.65	29.407	23.80
Mercury				
Hg(liq)	0	0	18.17	6.688
Hg(g)	14.655	7.613	41.79	4.968
Hg$^+$(g)	256.82			
Hg^{++}(g)	690.83			
Hg^{++}(aq)	40.9	39.30	−7.7	
Hg$_2^{++}$(aq)	41.2	36.70	20.2	
HgO(c, red)	−21.71	−13.995	16.80	10.53
HgF(g)	1.0	−2.4	53.98	8.26
HgCl(g)	20.1	15.0	62.09	8.68
HgCl$_2$(c)	−53.6	−42.7	34.9	
Hg$_2$Cl$_2$(c)	−63.39	−50.377	46.0	
HgI(g)	31.64	21.14	67.26	8.99
HgS(c, red)	−13.9	−12.1	19.7	11.57
HgSO$_4$(c)	−169.1			
Hg$_2$SO$_4$(c)	−177.61	−149.589	47.96	31.54

Substance and State	ΔH_f° kcal mol^{-1}	$\Delta G_f^{\circ}(\Delta \mu_f^{\circ})$ kcal mol^{-1}	S° cal deg^{-1} mol^{-1}	C_p° cal deg^{-1} mol^{-1}
Copper				
Cu(c)	0	0	7.923	5.840
Cu(g)	80.86	71.37	39.74	4.968
Cu$^+$(g)	260.513			
Cu$^+$(aq)	17.13	11.95	9.7	
Cu^{++}(g)	729.93			
Cu^{++}(aq)	15.48	15.66	-23.8	
CuO(c)	-37.6	-31.0	10.19	10.11
Cu$_2$O(c)	-40.3	-34.9	22.26	15.21
CuCl(c)	-32.8	-28.65	20.6	11.6
CuCl$_2$(c)	-52.6	-42.0	25.83	13.82
CuS(c)	-12.7	-12.8	15.9	11.43
CuSO$_4$(c)	-184.36	-158.2	26.	23.9
Silver				
Ag(c)	0	0	10.17	6.059
Ag(g)	68.01	58.72	41.321	4.9679
Ag$^+$(g)	243.59			
Ag$^+$(aq)	25.234	18.433	17.37	5.2
Ag$_2$O(c)	-7.42	-2.68	29.0	15.74
AgF(c)	-48.9			
AgCl(c)	-30.370	-26.244	23.0	12.14
AgBr(c)	-23.99	-23.16	25.6	12.52
AgI(c)	-14.78	-15.82	27.6	13.58
Ag$_2$S (c, orthorhombic)	-7.79	-9.72	34.42	18.29
Ag$_2$SO$_4$(c)	-171.10	-147.82	47.9	31.40
AgNO$_3$(c)	-29.73	-8.00	33.68	22.24
AgCN(c)	34.9	37.5	25.62	15.95
Gold				
Au(c)	0	0	11.33	6.075
Au(g)	87.5	78.0	43.115	4.968
Au$^+$(g)	301.73			
Au(OH)$_3$(c)	-101.5	-75.77	45.3	
AuCl(c)	-8.3			
AuCl$_3$(c)	-28.1			
Nickel				
Ni(c)	0	0	7.14	6.23
Ni(g)	102.7	91.9	43.519	5.583
Ni$^+$(g)	280.243			

Substance and State	ΔH_f° kcal mol^{-1}	$\Delta G_f^\circ(\Delta \mu_f^\circ)$ kcal mol^{-1}	S° cal deg^{-1} mol^{-1}	C_p° cal deg^{-1} mol^{-1}
Ni^{++}(g)	700.32			
Ni^{++}(aq)	−12.9	−10.9	−30.8	
NiO(c)	−57.3	−50.6	9.08	10.59
Ni$_2$O$_3$(c)	−117.0			
Ni(OH)$_2$(c)	−126.6	−106.9	21.	
NiCl$_2$(c)	−72.976	−61.918	23.34	17.13
Cobalt				
Co(c)	0	0	7.18	5.93
Co$^+$(g)	284.348			
Co^{++}(g)	679.17			
Co^{++}(aq)	−13.9	−13.0	−27.	
Co^{+++}(aq)	22.	32.	−73.	
CoO(c)	−56.87	−51.20	12.66	13.20
Co$_3$O$_4$(c)	−213.	−185.	24.5	29.5
CoCl$_2$(c)	−74.7	−64.5	26.09	18.76
Iron				
Fe(c)	0	0	6.52	6.00
Fe(g)	99.5	88.6	43.112	6.137
Fe$^+$(g)	283.16			
Fe^{++}(g)	657.8			
Fe^{++}(aq)	−21.3	−18.85	−32.9	
Fe^{+++}(aq)	−11.6	−1.1	−75.5	
FeO(c)	−65.0			
Fe$_2$O$_3$(c)	−197.0	−177.4	20.89	24.82
Fe$_3$O$_4$(c)	−267.3	−242.7	35.0	34.28
Fe(OH)$_2$(c)	−136.0	−116.3	21.	
Fe(OH)$_3$(c)	−196.7	−166.5	25.5	
FeCl$_2$(c)	−81.69	−72.26	28.19	18.32
FeCl$_3$(c)	−95.48	−79.84	34.0	23.10
FeS(c)	−23.9	−24.0	14.41	12.08
FeSO$_4$(c)	−221.9	−196.2	25.7	24.04
Fe$_2$(SO$_4$)$_3$(c)	−617.0			
Fe$_3$C(c)	6.0	4.8	25.0	25.3
Fe(CN)$_6$$^{\equiv}$(aq)	134.3	174.3	64.6	
Fe(CN)$_6$$^{4-}$(aq)	108.9	166.09	22.7	
Platinum				
Pt(c)	0	0	9.95	6.18
Pt(g)	135.1	124.4	45.960	6.102
Pt$^+$(g)	343.3			
Pt^{++}(g)	772.9			

Substance and State	ΔH_f° kcal mol^{-1}	$\Delta G_f^\circ(\Delta\mu_f^\circ)$ kcal mol^{-1}	S° cal deg^{-1} mol^{-1}	C_p° cal deg^{-1} mol^{-1}
$Pt(OH)_2(c)$	-84.1			
$PtCl_2(c)$	-26.5			
$PtCl_4^=(aq)$	-120.3	-88.1	40.	
$PtCl_6^=(aq)$	$-161.$	$-117.$	52.6	
Manganese				
$Mn(c)$	0	0	7.65	6.29
$Mn(g)$	67.1	57.0	41.49	4.97
$Mn^+(g)$	240.0			
$Mn^{++}(g)$	602.1			
$Mn^{++}(aq)$	-52.76	-54.5	-17.6	12.
$MnO(c)$	-92.07	-86.74	14.27	10.86
$MnO_2(c)$	-124.29	-111.18	12.68	12.94
$MnO_4^-(aq)$	-129.4	-106.9	45.7	
$MnO_4^=(aq)$	$-156.$	-119.7	14.	
$Mn(OH)_2(amorph)$	-166.2	-147.0	23.7	
$MnCl_2(c)$	-115.03	-105.29	28.26	17.43
Chromium				
$Cr(c)$	0	0	5.68	5.58
$Cr(g)$	94.8	84.1	41.68	4.97
$Cr^+(g)$	252.29			
$Cr^{++}(g)$	634.2			
$Cr^{++}(aq)$	-34.3			
$CrO_2(c)$	$-143.$			
$CrO_3(c)$	-140.9			
$CrO_4^=(aq)$	-210.60	-173.96	12.00	
$Cr_2O_3(c)$	-272.4	-252.9	19.4	28.38
$Cr_2O_7^=(aq)$	-356.2	-311.0	62.6	
$HCrO_4^-(aq)$	-209.9	-182.8	44.0	
Tungsten				
$W(c)$	0	0	7.80	5.80
$W(g)$	203.0	192.9	41.549	5.093
$WO_2(c)$	-140.94	-127.61	12.08	13.41
$WO_3(c)$	-201.45	-182.62	18.14	17.63
$WO_4^=(aq)$	-257.1			
Titanium				
$Ti(c)$	0	0	7.32	5.98
$Ti(g)$	112.3	101.6	43.066	5.839
$Ti^+(g)$	271.1			
$TiO_2(c, rutile)$	-225.8	-212.6	12.03	13.15

Substance and State	ΔH_f° kcal mol^{-1}	$\Delta G_f^{\circ}(\Delta \mu_f^{\circ})$ kcal mol^{-1}	S° cal deg^{-1} mol^{-1}	C_p° cal deg^{-1} mol^{-1}
$TiF_4(g)$	−371.0			
$TiCl_4(g)$	−182.4	−173.7	84.8	22.8
Beryllium				
$Be(c)$	0	0	2.27	3.93
$Be^{++}(g)$	715.398			
$Be^{++}(aq)$	−91.5	−90.75	−31.0	
$BeO(c)$	−145.7	−138.7	3.38	6.10
$BeO_2^=(aq)$	−189.0	−153.0	−38.	
Magnesium				
$Mg(c)$	0	0	7.81	5.95
$Mg(g)$	35.30	27.04	35.502	4.968
$Mg^{++}(g)$	561.299			
$Mg^{++}(aq)$	−111.58	−108.7	−33.0	
$MgO(c)$	−143.81	−136.10	6.44	8.88
$Mg(OH)_2(c)$	−220.97	−199.23	15.10	18.41
$MgF_2(c)$	−268.5	−255.8	13.68	14.72
$MgCl_2(c)$	−153.28	−141.45	21.42	17.06
$MgBr_2(c)$	−125.3	−120.4	28.0	
$MgI_2(c)$	−87.0	−85.6	31.0	
$MgSO_4(c)$	−307.1	−279.8	21.9	23.06
$MgCO_3(c)$	−261.9	−241.9	15.7	18.05
Calcium				
$Ca(c)$	0	0	9.90	6.05
$Ca(g)$	42.6	34.5	36.992	4.968
$Ca^{++}(g)$	460.29			
$Ca^{++}(aq)$	−129.74	−132.30	−12.7	
$CaO(c)$	−151.79	−144.37	9.50	10.23
$Ca(OH)_2(c)$	−235.68	−214.76	19.93	20.91
$CaF_2(c)$	−291.5	−279.0	16.46	16.02
$CaF_2(g)$	−186.8	−188.9	65.55	12.25
$CaCl_2(c)$	−190.2	−178.8	25.0	17.35
$CaCl_2(g)$	−112.7	−114.54	69.35	14.18
$CaBr_2(c)$	−163.2	−158.6	31.	
$CaBr_2(g)$	−95.2			
$CaI_2(c)$	−127.5	−126.4	34.	
$CaI_2(g)$	−65.			
$CaSO_4$				
(c, anhydrite)	−342.76	−315.93	25.5	23.82
$CaC_2(c)$	−14.3	−15.5	16.72	14.99
$CaCO_3(c, calcite)$	−288.46	−269.80	22.2	19.57

Substance and State	ΔH_f° kcal mol^{-1}	$\Delta G_f^\circ(\Delta\mu_f^\circ)$ kcal mol^{-1}	S° cal deg^{-1} mol^{-1}	C_p° cal deg^{-1} mol^{-1}
Strontium				
Sr(c)	0	0	12.5	6.3
Sr(g)	39.3	31.3	39.32	4.968
Sr^{++}(g)	427.96			
Sr^{++}(aq)	−130.45	−133.71	−7.8	
SrO(c)	−141.5	−134.3	13.0	10.76
Sr(OH)$_2$(c)	−229.2			
SrCl$_2$(c)	−198.1	−186.7	27.45	18.07
SrSO$_4$(c)	−347.3	−320.5	28.	
SrCO$_3$(c)	−291.6	−272.5	23.2	19.46
Barium				
Ba(c)	0	0	15.0	6.71
Ba(g)	43.2	35.	40.663	4.968
Ba^{++}(g)	396.86			
Ba^{++}(aq)	−128.50	−134.02	2.3	
BaO(c)	−132.3	−125.5	16.83	11.42
Ba(OH)$_2$(c)	−225.8			
BaCl$_2$(c)	−205.2	−193.7	29.56	17.96
BaSO$_4$(c)	−352.1	−325.6	31.6	24.32
BaCO$_3$(c)	−290.7	−271.9	26.8	20.40
Lithium				
Li(c)	0	0	6.96	5.89
Li(g)	38.09	30.6	33.14	4.968
Li$^+$(g)	163.90			
Li$^+$(aq)	−66.55	−70.22	2.70	
Li$_2$O(c)	−143.10	−134.35	9.06	12.93
LiOH(c)	−116.48	−104.92	10.23	11.85
LiCl(c)	−97.69			
Sodium				
Na(c)	0	0	12.26	6.73
Na(g)	25.60	18.475	36.714	4.968
Na$^+$(g)	145.589			
Na$^+$(aq)	−57.433	−62.589	13.96	
Na$_2$O(c)	−99.90	−90.61	17.94	16.52
NaOH(c)	−101.766	−90.77	15.40	14.23
Na$_2$O$_2$(c)	−122.66	−107.47	22.66	21.34
NaF(c)	−137.52	−130.28	12.24	11.20
NaCl(c)	−98.279	−91.79	17.24	11.98
NaBr(c)	−86.38	−83.48	20.75	12.28

Substance and State	ΔH_f° kcal mol^{-1}	$\Delta G_f^\circ(\Delta\mu_f^\circ)$ kcal mol^{-1}	S° cal deg^{-1} mol^{-1}	C_p° cal deg^{-1} mol^{-1}
$NaI(c)$	-68.80	-68.01	23.54	12.48
$Na_2SO_3(c)$	-260.6	-239.5	34.9	28.71
$Na_2SO_4(c)$	-331.55	-303.38	35.76	30.55
$NaNO_3(c)$	-111.54	-87.45	27.85	22.24
$Na_2CO_3(c)$	-270.26	-250.50	33.17	26.53
$NaHCO_3(c)$	-226.5	-203.6	24.4	20.94
Potassium				
$K(c)$	0	0	15.46	7.05
$K(g)$	21.33	14.50	38.297	4.968
$K^+(g)$	122.907			
$K^+(aq)$	-60.271	-67.466	24.15	
$K_2O(c)$	-86.80	-76.99	22.50	20.00
$KOH(c)$	-101.51	-90.57	18.86	15.51
$KF(c)$	-135.90	-128.81	15.90	11.71
$KCl(c)$	-104.33	-97.70	19.74	12.26
$KClO_3(c)$	-93.50	-69.29	34.2	23.96
$KClO_4(c)$	-102.80	-71.79	36.10	26.86
$KBr(c)$	-94.12	-90.92	22.93	12.52
$KI(c)$	-78.37	-77.20	25.43	12.61
$K_2SO_4(c)$	-342.66	-314.62	42.0	31.08
$KNO_3(c)$	-117.76	-93.96	31.81	23.01
$KMnO_4(c)$	-194.4	-170.6	41.0	28.10
$K_2CO_3(c)$	-274.90	-254.44	37.17	27.35
Rubidium				
$Rb(c)$	0	0	18.35	7.36
$Rb(g)$	19.33	13.35	40.3	4.968
$Rb^+(g)$	117.136			
$Rb^+(aq)$	-60.018	-67.45	28.79	
$RbCl(c)$	-103.99		22.90	
Cesium				
$Cs(c)$	0	0	20.33	7.70
$Cs(g)$	18.18	11.88	41.942	4.968
$Cs^+(g)$	109.445			
$Cs^+(aq)$	-61.673	-67.41	31.75	
$CsCl(c)$	-105.82		24.18	

Symbols and Notation

I. Non-Greek characters

a	van der Waals constant, activity
A	area, Madelung constant, Helmholtz free energy
b	van der Waals constant
c	speed of light, concentration by volume
(c)	crystalline state
C	heat capacity, number of components, Celsius
D	dissociation energy, dielectric constant
e	electronic charge
E	energy (of atomic or molecular level)
E	electron affinity, strength of electric field, electromotive force
f	fugacity, restoring force
F	degrees of freedom, faraday
g	acceleration of gravity, degeneracy of energy levels
(g)	gaseous state
G	Gibbs free energy
h	height, Planck's constant
H	enthalpy
$\mathcal{H}$	apparent enthalpy

I ionization potential, moment of inertia, ionic strength

k Boltzmann's constant, Hooke's Law constant

K kelvin (degrees), equilibrium constant, rotational quantum number, Henry's Law constant

l length, liquid

(liq)   liquid state

L length

m mass, molality

M molecular weight

n number of moles, translational and vibrational quantum numbers

N number of molecules

p pressure, probability

P number of phases, probability

q partition function (particle)

Q heat, partition function (system)

R gas constant

s solid

S entropy

t Celsius temperature, time

T absolute temperature

u energy density of radiation

U energy (thermodynamic)

v vapor, velocity

V volume, potential energy

$\mathcal{V}$ apparent volume

w weight percent

W work

x mole fraction

z number of faradays

Z generalized extensive property, compression factor

II. Greek characters

α second virial coefficient, coefficient of thermal expansion, degree of dissociation, Lagrangian multiplier

β correction parameter for nonideal gases, Lagrangian multiplier

γ heat capacity ratio (C_p/C_v), surface tension, fugacity coefficient, activity coefficient

$\delta(\ )$ variation of ( )

θ reduced temperature, characteristic temperature, freezing point lowering, boiling point elevation

κ parameter used in Debye-Hückel theory, isothermal compressibility

λ wavelength, absolute activity

μ Joule-Thomson coefficient, dipole moment, chemical potential

ν frequency of light, vibrational frequency, number of ions obtained from dissociation of molecule of electrolyte

ξ extent of reaction

π reduced pressure

Π osmotic pressure, continued product

ρ density

σ symmetry number

Σ sum

ϕ reduced volume, volume fraction

ψ electrical potential

Ω thermodynamic probability

III. Subscripts and superscripts

(A) Quantities derived from intensive variables (using pressure, p, as example)

p_i partial pressure of ith substance

p_l pressure of liquid

p_v pressure of vapor

p_0 vapor pressure of pure liquid

p_{0i} vapor pressure of pure liquid i

p^* pressure in reference state for activity or fugacity

p_{id} ideal pressure

(B) Quantities derived from extensive variables (using enthalpy, H, and chemical potential, μ, as examples)

H_i partial molar enthalpy of ith substance

H^* enthalpy in reference state

H_i^* partial molar enthalpy of ith substance in reference state

H_i° standard enthalpy of ith substance

H_m molar enthalpy

H_{mi} molar enthalpy of pure ith substance

$\Delta \tilde{H}$ enthalpy of reaction

$\Delta \tilde{H}^\circ$ standard enthalpy of reaction

ΔH_f° standard enthalpy of formation

$\mu_i (=G_i)$ chemical potential (partial molar Gibbs free energy) of ith substance

μ° standard chemical potential

μ_i° standard chemical potential of ith substance

μ_i^* chemical potential of ith substance in reference state

μ_{0i} chemical potential of pure ith substance

$\Delta \tilde{\mu}$ reaction potential (negative of affinity)

$\Delta \tilde{\mu}^\circ (= \Delta G^\circ)$ standard reaction potential

(c) Miscellaneous

$\langle v \rangle$	average speed of molecules
$\tilde{v}$	most probable speed of molecules
$\sqrt{\langle v^2 \rangle}$	root-mean-square speed of molecules
$\langle V_m \rangle$	average molar volume
C_p	heat capacity at constant pressure
C_v	heat capacity at constant volume
K_b	molal boiling-point-elevation constant
K_f	molal freezing-point-lowering constant
n_i	number of moles of ith substance
N_i	number of molecules of ith substance
p_x, p_y, p_z	components of molecular momenta
T_0	normal freezing (or boiling) point of pure substance
x_i	mole fraction of ith substance
$\dot{x}, \dot{y}, \dot{z}$	components of molecular speeds

Recommended Reading and Selected References

I. General Thermodynamics

1. E. F. Caldin, *Chemical Thermodynamics*, Oxford University Press, Oxford, 1958.
2. E. A. Guggenheim, *Thermodynamics*, 3rd edition, North-Holland Publishing Company, Amsterdam, Interscience Publishers, New York, 1957.
3. I. M. Klotz, *Chemical Thermodynamics*, revised edition, W. A. Benjamin, New York, 1964.
4. G. N. Lewis and M. Randall (revised by Pitzer and Brewer), *Thermodynamics*, McGraw-Hill, Book Co., New York, 1961.
5. I. Prigogine and R. Defay (translated by D. H. Everett), *Chemical Thermodynamics*, Longmans, Green and Co., London, 1954.
6. F. D. Rossini, *Chemical Thermodynamics*, John Wiley & Sons, New York 1950.
7. M. W. Zemansky, *Heat and Thermodynamics*, 4th edition, McGraw-Hill Book Co., New York, 1957.

II. Statistical Mechanics

1. N. Davidson, *Statistical Mechanics*, McGraw-Hill Co., New York, 1962.
2. R. H. Fowler and E. A. Guggenheim, *Statistical Thermodynamics*, Cambridge University Press, Cambridge, 1939.
3. T. L. Hill, *Introduction to Statistical Thermodynamics*, Addison-Wesley Publishing Co., Reading, Mass., 1960.
4. J. E. Mayer and M. G. Mayer, *Statistical Mechanics*, John Wiley & Sons, New York, 1940.
5. O. K. Rice, *Statistical Mechanics, Thermodynamics, and Kinetics*, W. H. Freeman and Company, San Francisco, 1967.
6. R. C. Tolman, *The Principles of Statistical Mechanics*, Oxford University Press, Oxford, 1938.

III. Special Topics

1. P. J. Flory, *Principles of Polymer Chemistry*, Cornell University Press, Ithaca, New York, 1953.
2. H. S. Harned and R. B. Owen, *The Physical Chemistry of Electrolyte Solutions*, 2nd edition, Reinhold Publishing Corporation, New York, 1950.
3. J. H. Hildebrand and R. L. Scott, *The Solubility of Non-electrolytes*, 3rd edition, Reinhold Publishing Corporation, New York, 1950.
4. M. Karplus and R. N. Porter, *Atoms & Molecules*, W. A. Benjamin, Inc., New York, 1970.
5. H. Margenau and G. M. Murphy, *The Mathematics of Physics and Chemistry*, 2nd edition, D. Van Nostrand Co., New York, 1956.
6. Linus Pauling, *The Nature of the Chemical Bond*, 3rd edition, Cornell University Press, Ithaca, N.Y., 1960.
7. Linus Pauling and E. B. Wilson, Jr., *Introduction to Quantum Mechanics*, McGraw-Hill Book Co., New York, 1935.
8. J. Wilks, *The Third Law of Thermodynamics*, Oxford University Press, Oxford, 1961.

IV. Data and Notation References

1. D. D. Wagman, W. H. Evans, V. B. Parker, I. Halow, S. M. Bailey, R. H. Schumm, and K. L. Churney, "Selected Values of Chemical Thermodynamic Properties," National Bureau of Standards Technical Notes 270-3 (1968), 270-4 (1969), 270-5 (1971) and 270-6 (1971), U.S. Government Printing Office, Washington, D.C. 20402.
2. "Selected Values of Properties of Chemical Compounds," Thermodynamics Research Center Data Project, B. J. Zwolinski, Director, Thermodynamics Research Center, Texas A & M University, College Station, Texas 77843 (Loose-leaf data sheets, extant 1971).
3. JANAF Thermochemical Tables, Second Edition, D. R. Stull and H. Prophet, Project Directors, Office of Standard Reference Data, National Bureau of Standards, U.S. Government Printing Office, Washington, D.C. 20402, 1971.

4. F. D. Rossini, D. D. Wagman, W. H. Evans, S. Levine, and I. Jaffe, "Selected Values of Chemical Thermodynamic Properties," National Bureau of Standards Circular 500 (1952), U.S. Government Printing Office, Washington, D.C.

5. "Selected Values of Properties of Hydrocarbons and Related Compounds," American Petroleum Institute Research Project 44, B. J. Zwolinski, Director, Thermodynamics Research Center, Texas A & M University, College Station, Texas 77843 (Looseleaf data sheets, extant 1971).

6. F. D. Rossini, *Values of the Fundamental Constants for Chemistry*, Pure and Applied Chemistry, Vol. 9 No. 3, p. 453, Butterworths, London, 1964.

7. M. L. McGlashan, *Manual of Symbols and Terminology for Physicochemical Quantities and Units*, Pure and Applied Chemistry, Vol. 21 No. 1, Butterworths, London, 1970.

Index